Probability and Random Processes for Engineers and Scientists

Probability and Random Processes for Engineers and Scientists

A. BRUCE CLARKE
Department of Mathematics
Western Michigan University
RALPH L. DISNEY
Department of Industrial Engineering
The University of Michigan

John Wiley & Sons Inc., New York · London · Sydney · Toronto

10 9 8 7 6 5 4 3
Library of Congress Catalog Card Number: 76–104008

SBN 471 15980 8

Printed in the United States of America

Preface

This book is designed to introduce junior-level students in engineering, the physical and social sciences, and management programs, to the important areas of probability and stochastic processes. We have tried to present the material at a level that our experience has shown to be intelligible to these students. We have used preliminary versions of the manuscript in several classes in probability and stochastic processes. The material in Chapters 7 to 14 has been used in a junior-level course in industrial engineering for about five semesters. The material in Chapters 1 to 14 has been used for four summer conference programs at The University of Michigan. The success with both of these groups encourages us to believe that the level and the presentation are reasonable for both the young student with a current, up-to-date, general undergraduate mathematics background, and for the practicing engineer and scientist whose introduction to the subject is most often through applications.

We believe that the topics covered are just beginning to emerge as useful areas of applications in engineering, the physical and social sciences, and business administration. Unfortunately, introductory text material is lacking at present. Therefore there is a need for an elementary level text to give undergraduate nonmathematics students an appropriate introduction to these subjects. There are many good intermediate and advanced books on these topics, but our experience shows that they are inappropriate as introductions.

With the specified audience in mind, we have attempted to write about probability and stochastic processes in a way that will give the student material that he can use. Consequently, we have tried to be explicit, precise, and problem-oriented. The book does not "come off" as a mathematics text. It is not intended that it should. Our feeling is that the engineer should be acquainted with probability and stochastic processes and be encouraged to explore these fields because of their usefulness to him. Once he has the "at home" feeling with the material, then we hope that he will be motivated to explore more deeply.

Thus, the book leaves much to be desired in terms of "theorem, proof, theorem, proof." We have tried to present the ideas (and, indeed, in some cases, theorems and proofs) in such a manner that the student does not have to "unlearn" later as he goes deeper into the field. We have tried to present the ideas usefully so that the student can gain enough manipulative skill to go beyond manipulation. We have tried to tie together ideas that occasionally appear to be disjointed (for example, elementary queueing theory appears first in Chapter 4 to motivate the introduction of random sequences and Markov processes, and then is discussed in more detail in Chapters 11 to 13 as a special case of Markov processes, not as a separate entity). In this way we have tried to write an applied mathematics text in probability and stochastic processes. In doing this we have been driven by the thought that these attempts must tread a fine line lest they become poor mathematics poorly applied. The reader must judge whether we have stepped off the line.

It has been our experience that most students are primarily determinists. Little else could be expected from the current background they receive. As a result, the student must be helped into the world of probability, random phenomena, and stochastic processes. Thus we think that an understanding of the material in this book can be expedited by a knowledgeable teacher. There are places in the book where the teacher must expand on the material, and places he can skip or "leave as an exercise." Where these places are depends strongly on the background and curricula of the students. (Chapter 1 might not be required in a course in which students have a good background in set theory.) Consequently, we believe that material in this book must be taught as well as learned. With this in mind, we have at times been brutally brief.

The material represents the topics that we believe can be covered in detail in about two academic semesters. Latitude in content is provided by the inclusion of some chapters and sections on special topics that could be omitted without loss in the development of the main concepts. On the other hand, some chapters (such as Chapters 5 and 6) could properly form introductions to a course in statistics rather than to work in stochastic processes. Thus the user is left the freedom to choose the areas best suited to his needs and future uses.

We have found that Chapters 1 to 6 form a one-semester package of material in basic probability theory suitable for students with minimal mathematical prerequisites (calculus). In all likelihood, students emerging from the new mathematics of precollege education will find Chapter 1 repetitive. Hence, for these students, one could dispense with parts of Chapter 1 and parts of Chapter 3 so as to cover the material in Chapter 7. In general, we have not yet found that all our students are well enough prepared so that we can omit Chapter 1 completely or the very important concept of a function and a random variable in Chapter 3.

We have found that Chapters 7 to 14 form a one-semester package of material that follows the above material reasonably well. The maturity of the student is an important point to be considered in deciding how fast to move here. If the student has some background in matrix methods, he can move through Chapter 9 more rapidly. If the student has a good background in elementary differential equations, he can move through Chapter 11 more rapidly.

Our original attempt was to write a book suitable for a one-semester course that would introduce the beginning student to random processes. This can be done with this book at some sacrifice of other material. Chapters 1 to 4, and 7 and 8, plus certain topics in Chapter 6, form such a course. In the process one sacrifices certain "classical" probability material on moments and special distributions (Chapter 5 and 6). We have tried to introduce some of this material elsewhere in the book and this sacrifice is not calamitous for a study of discrete state random processes.

At an elementary level of presentation, it is nearly impossible to acknowledge all of the sources used in a book. It will be apparent to the scholar in the field that we owe a debt to many others. To those, known and unknown to us, whose work has made it possible even to consider trying to acquaint the beginning student in probability theory with the theory of random processes and its application, we are, of course, indebted.

To Raymond Carroll, of The University of Michigan Engineering Summer Conference Program, and his staff, we are indebted for their help and wise counsel which allowed us to launch the course giving rise to this book. Especially, we must thank Donald Danford and his office staff for the trials and tribulation they suffered while we wrote, revised, rewrote, and rerevised the original notes. Not always did we meet their time schedule but always they met ours with an outward calm and efficiency.

We are indebted to Miguel Taube-Netto for preparing answers to the exercises and, in the process, for finding errors that would otherwise have escaped our attention. Errors still in the book, of course, are our responsibility.

Mrs. Anne Rubin, Mrs. Constance Applegate and Miss Julianne Black are to be thanked for their help in typing, proofreading and, in general, holding our hands so that we did not cut our throats in difficult periods of preparation.

To our wives we are thankful for encouragement and also for their proofreading, typing, counsel, and the many small things that allowed us even to consider this venture. To our children we are thankful for staying out of the way and for not complaining too bitterly when they could not go skiing.

A. B. CLARKE
R. L. DISNEY

Kalamazoo, Michigan
Ann Arbor, Michigan

Possible Course Structures

One semester course:

Introduction to Random Processes:
Chapters 1–4
Chapter 6, Sections 1–3, 5, 7
(omitting moment properties)
Chapter 7
Chapter 8

One-semester course:

Probability (Introduction to Statistics):
Chapters 1–6

Two-semester course:

Probability and Random Processes:
Semester 1: Chapters 1–6
Semester 2: Chapters 7–14

Contents

Probability and Random Processes for Engineers and Scientists

1 • Sample Spaces

1. Random Experiments and Sample Spaces

Probability theory is rooted in the real life situation where a person performs an experiment the outcome of which may not be certain. Such an experiment is called a *random experiment*. Mathematically, the concept of a random experiment is considered a primitive notion that is not otherwise defined.

Examples of random experiments are plentiful in real-life situations. The strength of a manufactured part is uncertain prior to its manufacture. The process of manufacturing the part can be construed as the performance of a random experiment. Games of chance are classic examples of random experiments; the outcome of the game is not known with certainty prior to playing the game. Firing a rocket is an example of performing a random experiment, as is conducting a sales campaign.

In each of the examples it is clear that the outcome of the performed experiment is not certain *a priori*. The exact strength of the manufactured part is not certain *before* the part is manufactured. The face appearing on the cast die is not certain prior to casting the die. The increased sales are not *a priori* known. Success, failure, or partial success of a rocket firing is not certain prior to firing.

Thus the study of probability can be initiated with the primitive idea of a random experiment. Associated with the experiment are the possible results, or outcomes, of the experiment; for example, associated with the die tossing experiment are the possible outcomes: an ace turns up, a two turns up, etc. Associated with the manufacture of a part are the possible strengths of the completed part. The important concept here is that associated with any random experiment there is a set, S, of all possible outcomes. This set S is called the *sample space*.

We must point out that the sample space is not determined completely by the experiment. It is partially determined by the purpose for which the experiment is carried out. If a coin is tossed twice, for some purposes it might be sufficient to consider only three possible outcomes: two heads, two tails, a combination of one head with one tail. These three outcomes would then constitute the sample space S. However, one might be playing a game in which the exact *order* of appearance of the sides of the coin is important. In this case the sample space S must be considered as containing *four* possible outcomes: head-head, tail-tail, head-tail, tail-head. Other possible sample spaces would be obtained if we took into account such things as the exact locations of the fall of the coins, the number of times they spin in the air, etc. Frequently we use a larger sample space than is strictly necessary because it is easier to use. For instance, in the preceding illustration the sample space with four possible outcomes is easier to discuss, since for an "honest" coin those four outcomes are all "equally likely," although this is not true in the sample space with three outcomes.

EXAMPLE 1.1. A lot consists of five manufactured articles labeled 1, 2, 3, 4, 5. An experiment consists of selecting three of these articles for testing. If order is disregarded, by direct enumeration we find that the sample space consists of the following 10 possible outcomes:

123, 124, 125, 134, 135, 145, 234, 235, 245, 345

If order is taken into account, there will be 60 possible outcomes. For instance, the sample consisting of articles 1, 2, and 3 can be selected in the following six orders:

123, 132, 213, 312, 231, 321

Which one of these possible sample spaces should be used depends on the method of testing and evaluating. For example, if the articles are tested sequentially under the rather loose rule that the lot is rejected only if at least two of the articles tested are defective, then, clearly the second sample space should be used. In this case, notice that if the first two articles selected and tested were found to be either both good or both defective, it would be unnecessary even to select a third article in order to reach a decision. △

EXAMPLE 1.2. In anthropological studies one "draws a person at random" and measures his height and weight. To use the resulting information to design an elevator a designer might be concerned only with the weight load on the elevator. Thus, the sample space would consist of the set of all possible weights of the people drawn at random. On the other hand, the same elevator designer might be concerned only with the head room design of the elevator. If this is true, the sample space could be taken to be the set of all possible heights of the people. In most real-life design problems the designer at some time would be concerned with both the set of all possible heights and the set of all possible weights of users of the elevator. More specifically, he would usually be interested in the outcomes that simultaneously gave the height and weight of the possible users. Thus, the sample space could be the set of all possible *pairs* (height, weight). △

We can categorize three types of sample spaces for most applications.

(a) In the experiment of tossing a single coin there may be only two possible outcomes. More generally, the sample space may contain only a finite number of outcomes. We say that the sample space associated with the experiment is a *finite sample space.*

(b) If the outcomes of a sales campaign are given in terms of the *number* of units sold, we can take as an idealization of the sample space the set of all nonnegative integers. Certainly, no real sales campaign will result in an arbitrarily large number of units sold, but the set of nonnegative integers could serve as a mathematical idealization, or *model*, of the campaign. Sets of outcomes in these cases can be put into a one-to-one correspondence with the counting numbers. Hence, such a sample space is said to be *countably infinite.*

(c) The random experiment of manufacturing a part and measuring its strength conceivably could have an entire *interval* of real numbers as possible values. Since an interval of real numbers cannot be enumerated in a sequence (even an infinite sequence), such a sample space is said to be *noncountable* or *nondenumerable*. Certainly no real experiment conducted using real measuring instruments can ever yield such a sample space, since there is a limit to the fineness to which any instrument can measure. Nonetheless, such a sample space frequently can be taken as an idealization of, an approximation to, or a model of a real world situation, which is easier to discuss than a more exact model.

In many cases it is not necessary to distinguish between finite and countably infinite sample spaces. Therefore, if a sample space is either finite or countably infinite, we say that it is a *countable sample space*. In some usages we speak of a countable sample space as a *discrete sample space* and of a noncountable sample space as a *continuous sample space*.

2. Set Notation

Since we have identified a sample space S as the *set* of all possible outcomes of a random experiment, it is convenient to review and to consolidate some *set* ideas that will be useful.

If s is a possible outcome of the experiment, then s is one of the members of S. We indicate the statement "s is a member of S" or "s belongs to S" or "s is an element of S" by the notation

$$s \in S$$

Elements of S are termed *sample points.*

If s is not a possible outcome of the experiment, then, s is not an element of S, and we write

$$s \notin S$$

Several notations are used to specify a set in terms of its elements.

If the set S is finite, it may be convenient to merely list the elements of S (Example 1.1). Thus, for the experiment of tossing a single coin, there are two possible outcomes, *heads* or *tails*. We could then define the sample space as the set S whose elements are: (1) *heads*; (2) *tails*. Notationally, we would write

$$S = \{\text{heads, tails}\}$$

The notation $S = \{\quad\}$ is read as "S is the set whose elements are those specified inside the braces."

In a similar fashion

$$S = \{1, 2, 3, 4, 5, 6\}$$

would denote the set S whose elements are the counting numbers 1, 2, 3, 4, 5, 6. This set could serve as the sample space for a die tossing experiment.

For infinite sets, the listing of the elements is impossible; for large finite sets, the listing is inconvenient. Thus, we seek another notation for these sets. We first note that any set is uniquely specified once its elements are known. For a given set, we could state a rule, an inclusion condition, that could be applied to any object to determine whether the object is or is not to be included in the set. Thus, a set is uniquely defined in terms of a category of possible objects and an inclusion condition.

Suppose that the experiment consists of manufacturing a part and of measuring its strength, and that each manufactured part must have a strength between a and b, with all such numbers being possible strengths. The sample space S then consists of elements that are real numbers which lie in the interval from a to b. Notationally, then, S is defined by

$$S = \{x : a < x < b\}$$

This notation is read as "S is a set whose elements x consist of all numbers in the interval from a to b." To make the notation more explicit, we could say that

"S is the set (the braces indicate this) of elements x, so that (the colon is used to denote these words) x satisfies some statement (the statement is explicit and forms the inclusion condition)." In this case the sample points are real numbers. This type of interval is termed an *open* interval. For further discussion, see exercise 18 at the end of this chapter.

As a further example of set notation, we might be interested in a set of points in a plane that lie on the same given line. Then, S can be considered as a set of points whose coordinates are given by ordered pairs (x, y). The inclusion condition for S is that a point belongs to S if it satisfies an equation of the form $ax + by = c$ for given a, b, c. Symbolically, then,

$$S = \{(x, y) : ax + by = c\}$$

says, "S is the set whose elements are those ordered pairs (x, y) that satisfy the condition $ax + by = c$."

Notice that we might consider the pair (x, y) above to be a vector emanating from the origin with terminus at (x, y). Then, if we consider an element $\mathbf{x} = (x, y)$ as such a vector, the set S can be taken as a set of vectors. The particular set of vectors of interest in the example would be denoted by

$$S = \{\mathbf{x} : \mathbf{a}\mathbf{x}^T = c;\ \mathbf{a} = (a, b)\}$$

[Here $\mathbf{x}^T$ is the column vector $\begin{pmatrix} x \\ y \end{pmatrix}$ called the *transpose* of the vector $\mathbf{x} = (x, y)$.]

In the elevator design (Example 1.2), if the designer is interested in both the height and the weight of the users, a pair of numbers (height, weight) taken in that order would be a point (a sample element) in the sample space consisting of all such pairs. Hence, S can be taken to be a set of vectors. A vector is one of the ordered pairs (height $= x$, weight $= y$).

3. Combinations of Sets

In many applications we are given more than one set, for instance, A_1, $A_2, \ldots,$ and we wish to combine them to arrive at new sets that are of importance. For example, in most applications we have the sample space S, but we are interested in only some of the elements of S. In other words, we are concerned with some elements of S that share a common property as distinguished from other elements of S that do not have this property. Thus, if S is the sample space for the strength of a manufactured part and if

$$S = \{x : a < x < b\}$$

we may be particularly concerned with the outcomes for which $a < x < c$ where c is some number between a and b. For example, c might be the least strength that a part could have to serve a given function. The new set, for example, A, would be given by

$$A = \{x : a < x < c\}$$

and could be considered in this example as the "set of defects."

Other kinds of combinations are also of interest. Suppose that we conduct the experiment of measuring the resistance x (in ohms) of an electrical part and suppose that

$$S = \{x : a < x < b\}$$

Suppose further that

$$A_1 = \{x : a < x < c\}$$

represents the strengths of the parts whose resistance is "too low" for use in a television set. Also suppose that

$$A_2 = \{x : d < x < b\}$$

represents the strengths of the parts whose resistance is "too great" for use in a television set. Then the set

$$A_3 = \{x : x \in A_1 \text{ or } x \in A_2\}$$

represents the set of defective strengths, that is, the resistances that are either too low or too great for use in a television set. Certainly, the manufacturer of television sets is interested in A_3, which is a set formed by "combining" sets A_1 and A_2.

Hence, clearly, many questions of interest require that one be able to take given sets and operate on them to produce new sets. We discuss some of these operations in this section.

(a) **Equal Sets.** Two sets A and B are *equal sets*, $A = B$, if they contain the same sample points. Thus, the sets $A = \{1, 2\}$ and $B = \{2, 1\}$ are equal sets. Notice that $C = \{1, 2, 1, 2, 1\}$ is a set that is also equal to A, since it contains the same elements. However, $D = \{1, 2, 3, 1, 2, 1, 2\}$ is not equal to A, B, or C.

(b) **Subsets.** A set A is said to be a *subset* of B, $A \subset B$, if every sample point, or element, of A is also an element of B. If, in addition, every element of B is also an element of A, then by the above definition $A = B$.

(c) Unions. Suppose A and B are two sets. The *union* of A and B, symbolically $A \cup B$, is the set whose elements belong to either A or B (or both). That is,

$$A \cup B = \{s : s \in A \quad \text{or} \quad s \in B\}$$

Notice that s may belong to both A and B, in which case it is still a member of $A \cup B$.

If A and B are characterized by verbal phrases, then $A \cup B$ is characterized by placing these phrases together with the word "or" between them. For example, when tossing a coin twice, if the sets A and B are defined by

A: "the first coin is a head" = {head-head, head-tail}
B: "the coins match" = {head-head, tail-tail}

then the event $A \cup B$ is defined by

$A \cup B$: "the first coin is a head or the coins match"
= {head-head, head-tail, tail-tail}

Observe that the word "or" is always used in the mathematical or non-exclusive sense of "one or the other or both." The set A_3 in the introduction to this section is another example of a set that is the union of two sets.

(d) Intersections. Suppose that A and B are two sets. The *intersection* of A and B, symbolically $A \cap B$, is the set whose elements belong to both A and B. That is,

$$A \cap B = \{s : s \in A \quad \text{and} \quad s \in B\}$$

If A and B are characterized by verbal phrases, then $A \cap B$ is characterized by placing these phrases together with the word "and" between them. In the preceding illustration we have

$A \cap B$: "the first coin is a head and the coins match"
= {head-head}

Suppose that in testing parts for use in a satellite it is required that a part be "strong enough" and yet "light enough." Thus, suppose that any part whose yield strength exceeds a is strong enough and that any part whose weight is less than b is light enough. Then let A be the set of all parts whose strength exceeds a, and let B be the set of parts whose weight is less than b. Then $A \cap B$ would represent the set of parts that are acceptable for use in the satellite.

Some authors use the symbolism AB as equivalent to $A \cap B$.

(e) Complements. Suppose that A is a subset of S. The *complement* of A (symbolically $\bar{A}$) is the set whose elements belong to S but not to A. Since we are only concerned with the points that do belong to S (the sample space), the set $\bar{A}$ is just those points that are not members of A. Symbolically

$$\bar{A} = \{s : s \in S \quad \text{and} \quad s \notin A\}$$

Verbally, the phrase characterizing A is changed into the phrase characterizing $\bar{A}$ by the insertion of the word "not." In the coin-tossing illustration

$$\bar{A}\text{: "the first coin is not a head"} = \{\text{tail-head, tail-tail}\}$$

Again, in testing electronic parts, if S is the set of all parts whose resistance x is between a and b, if a part is "good" when its resistance exceeds c (that is, $c < x < b$), and if A represents the set of "good" parts in S, then $\bar{A}$ represents the set of defective parts.

Although the definitions are given for two sets, we immediately determine that they extend to any finite number of sets. However, it is convenient to change notation somewhat. Thus we define

$$\begin{aligned}\bigcup_{i=1}^{n} A_i &= A_1 \cup A_2 \cup A_3 \cup \cdots \cup A_n \\ &= \{s : s \in A_1 \quad \text{or} \quad s \in A_2 \quad \text{or} \quad s \in A_3 \quad \text{or} \quad \cdots \quad s \in A_n\}\end{aligned}$$

$$\begin{aligned}\bigcap_{i=1}^{n} A_i &= A_1 \cap A_2 \cap A_3 \cap \cdots \cap A_n \\ &= \{s : s \in A_1 \quad \text{and} \quad s \in A_2 \quad \text{and} \quad \ldots \quad s \in A_n\}\end{aligned}$$

Obviously, these definitions can be extended to the union and intersection of an infinite number of sets.

Two particular sets require further discussion. The sample space S itself is a set. By any reasonable definition of the term *subset*, we must conclude that S is a subset of itself because, otherwise, we would be called on to produce an element of the left side of the inclusion $S \subset S$, which is not an element of the right side, and this is clearly impossible. Thus, $S \subset S$. Let us also postulate the existence of a set $\varnothing$, which contains no elements whatsoever (the *empty set*). Necessarily, $\varnothing$ must be a subset of *every* set, since for the formula $\varnothing \subset A$ to be false we would have to produce an element of $\varnothing$ that is not an element of A, and clearly this is impossible because $\varnothing$ has no elements. Hence, in particular, $\varnothing \subset S$.

We notice that the following obvious identities always hold true.

$$\overline{S} = \varnothing$$
$$\overline{\varnothing} = S$$
$$S \cup A = S$$
$$S \cap A = A$$
$$A \cup \overline{A} = S$$
$$A \cap \overline{A} = \varnothing$$

The statement that two sets A and B have no elements in common can be symbolized by the formula $A \cap B = \varnothing$. In this case, we say that A and B are *mutually exclusive* (or *disjoint*) sets.

Notice also that for sets of numbers $\varnothing \neq \{0\}$. The right side is a set containing one element 0. The left side is a set containing no elements. Thus they cannot be the same set.

4. Venn Diagrams

A convenient way to ascertain some set relations is to draw a Venn diagram of the sets of interest. Thus, for a given sample space S and two subsets A and B, we have the Venn diagram shown in Figure 1.1.

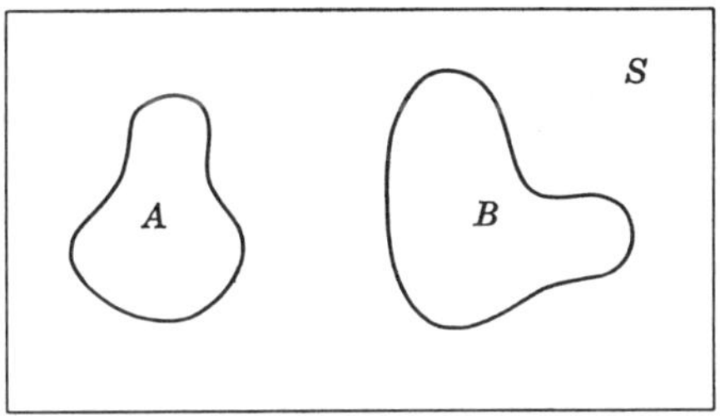

FIGURE 1.1 Venn diagram for sample space S and two sets A and B.

In this figure the points constituting the sample space are symbolically denoted by the ones within the rectangle. The subsets A and B are represented by certain regions within S.

The set $A \cup B$ is represented by the set of points lying in either A or B. $A \cup B$ is represented by the shaded region in Figure 1.2.

More generally, if A and B are arbitrary subsets of S that can overlap, they

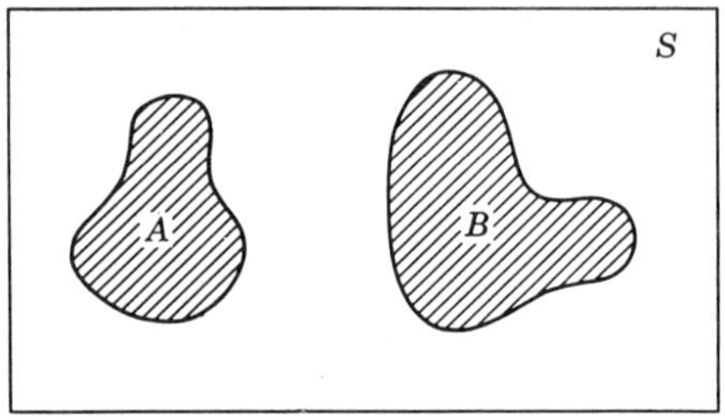

FIGURE 1.2 Venn diagram for sample space S and the set $A \cup B$ when $A \cap B = \varnothing$.

might be represented by a Venn diagram like Figure 1.3. The region representing $A \cap B$ is indicated on this diagram, and $A \cup B$ is represented by the shaded region.

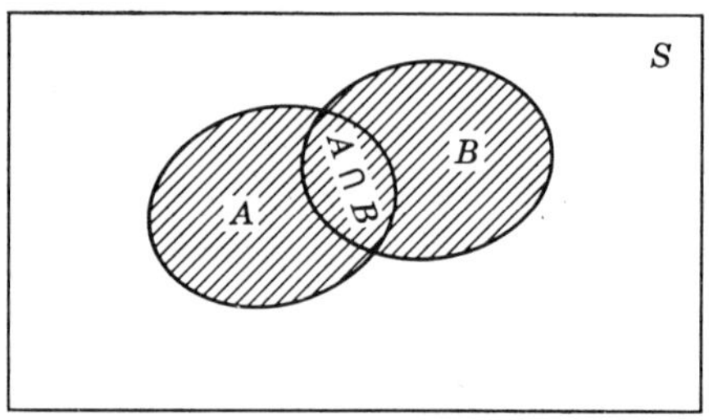

FIGURE 1.3 Venn diagram for sample space S and the set $A \cup B$ when $A \cap B \neq \varnothing$.

For any set A, the complement $\bar{A}$ consists of all points in S that do not lie in A and, hence, $\bar{A}$ is represented by the unshaded region in Figure 1.4.

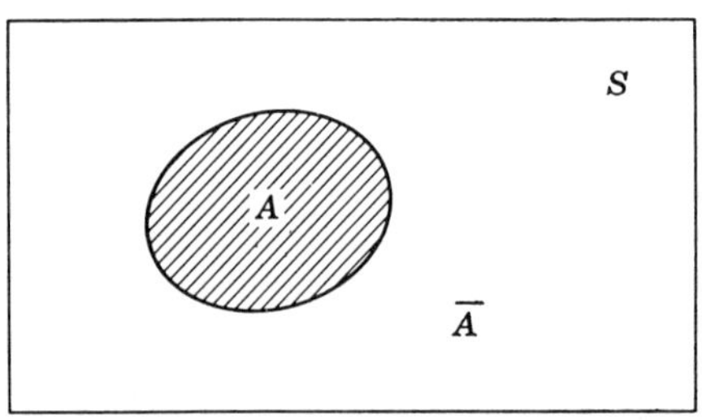

FIGURE 1.4 Venn diagram for sample space S and sets A and $\bar{A}$.

By reference to Figures 1.3 and 1.4, the following identities are observed to hold true for any subsets A and B of S:

$$
\begin{aligned}
A \cap S &= A \\
A \cup S &= S \\
\overline{(\overline{A})} &= A \\
\overline{(A \cup B)} &= \overline{A} \cap \overline{B} \\
\overline{(A \cap B)} &= \overline{A} \cup \overline{B}
\end{aligned}
$$

If we wish to discuss combinations of three sets, A, B, C, a Venn diagram like Figure 1.5 is useful.

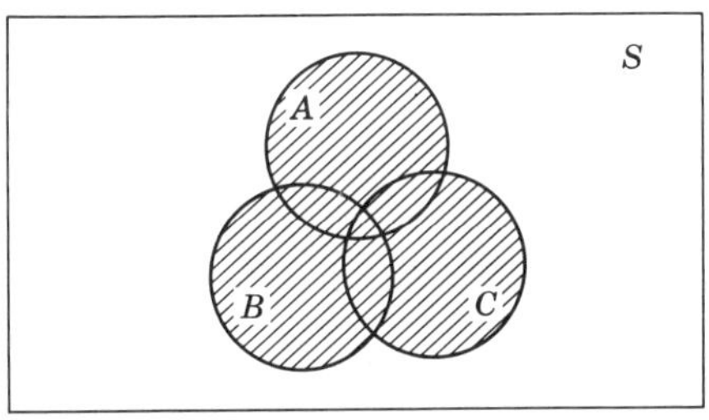

FIGURE 1.5 Venn diagram for sample space S and the sets A, B, C, and $A \cup B \cup C$

By shading in appropriate regions in a diagram such as Figure 1.5, the reader is urged to justify these identities:

$$
\begin{aligned}
&A \cap B \cap C \subset A \cap B \subset A \\
&A \subset A \cup B \subset A \cup B \cup C \\
&A \cap (B \cup C) = (A \cap B) \cup (A \cap C) \\
&A \cup (B \cap C) = (A \cup B) \cap (A \cup C)
\end{aligned}
$$

5. Events

We have seen that a subset of a sample space S can be determined either by a direct listing of its elements or by a statement or formula giving the condition that elements must satisfy in order to belong to the subset. In probability terminology the term *event* is used to describe any subset that is of interest in the discussion or, equivalently, any statement of conditions that defines this subset is also termed an *event*.

For example, in the die-tossing experiment having the sample space $S = \{1, 2, 3, 4, 5, 6\}$, the subset $A = \{2, 4, 6\}$ is defined by the statement "the toss yields an even number." We use the term *event* interchangeably to describe the subset or the statement. Thus, we say that A is an event, and that the statement "the toss yields an even number" defines the same event.

Vaguely speaking, an event can be defined as a statement whose truth or falsity is determined *after* the experiment. (Not until after the die is tossed does one know whether an even number comes up.) The corresponding subset of S is simply the set of all experimental outcomes for which the statement is true.

No error will be introduced if the student interprets the term *event* simply to mean *subset of the sample space*. All subsets of S considered in this book are events.

Suppose that we consider an event to be defined in the die-tossing experiment by the following statement B: "the toss yields the number 2." Thus the event B is a subset of S consisting of just one element:

$$B = \{2\}$$

We must exercise some notational care at this point. Notice, for example, that $2 \in S$ and $2 \in B$ but 2 is not B. B is a subset of S, while 2 is an *element* of S. Hence, $B \neq 2$. Clearly, B is not an element of S, since the elements of S are numbers not sets of numbers.

In the coin-tossing example the sample space is $S = \{\text{head, tail}\}$. Events in this space are: S, $\varnothing$, $A_1 = \{\text{head}\}$, $A_2 = \{\text{tail}\}$.

EXAMPLE 1.3. Three items are selected from a lot and each is tested and classified as either *good* (G) or *bad* (B). The sample space S can be treated as consisting of 8 sample points:

$$S = \{(G, G, G), (G, G, B), (G, B, G), (B, G, G), (G, B, B), (B, G, B), (B, B, G), (B, B, B)\} \qquad \triangle$$

Frequently the sample space is much more complicated than in any of the preceding examples.

EXAMPLE 1.4. The sequence of times between breakdowns of a certain piece of machinery in a factory is observed for a one-month period.

Here, the sample space S would consist of elements (sample points) each of which is itself a finite sequence $(T_1, T_2, \ldots)$ of positive numbers, the times between breakdowns. The event B: "breakdowns occur at least 24 hours apart," would be represented by the subset of S consisting of those sequences $(T_1, T_2, \ldots)$ in S with each $T_i \geq 24$. $\triangle$

EXERCISES

1. Describe a possible sample space for each of the following experiments:
   - **(a)** A coin is tossed 4 times and observed to be either a head or a tail each time.
   - **(b)** A large lot of machine screws is known to contain a number of defectives. Three screws are chosen at random and tested.
   - **(c)** A box of 10 screws is known to contain one defective and nine good screws. Three screws are chosen at random from this box and tested.

2. An electronic system contains 3 components, any one of which may fail during operation. A random experiment now consists of a field test of the system. Define 3 events A_1, A_2, A_3 as follows:

$$A_i\text{: "Component } i \text{ does not fail in the test,"} \quad i = 1, 2, 3$$

   - **(a)** Express each of the following events in words:

$$A_1 \cup A_2,\ A_1 \cup A_2 \cup A_3,\ A_1 \cap A_2,\ A_1 \cap A_2 \cap A_3,$$
$$\bar{A}_1,\ \bar{A}_1 \cap A_3,\ \bar{A}_1 \cup \bar{A}_2,\ \bar{A}_1 \cap \bar{A}_2,\ \bar{A}_3,\ \overline{A_1 \cap A_2},\ \overline{A_1 \cup A_2}$$

   - **(b)** The test is termed "successful" if no more than 1 component fails. It is termed "highly successful" if no components fail. Denote these two events by B and C. Express B and C in terms of A_1, A_2, A_3.

3. Illustrate each of the following events by shading the appropriate region on a Venn diagram with general A and B.
   - **(a)** $A \cap B$
   - **(b)** $(A \cap \bar{B}) \cup (\bar{A} \cap B)$
   - **(c)** $\overline{(A \cup B)}$
   - **(d)** $\bar{A} \cap \bar{B}$
   - **(e)** "exactly 0 of the events A and B occur"
   - **(f)** "exactly 1 of the events A and B occurs"
   - **(g)** "A occurs but not B"

4. Let the sample space S consist of all real numbers, and define the following events (subsets):

$$A = \{x : x > 0\}$$
$$B = \{x : -1 < x < 2\}$$
$$C = \{x : x < 0\}$$

   - **(a)** Describe and sketch on a number line (real axis) the following subsets of S:

$$A \cup C,\ A \cap C,\ \bar{A},\ B \cap C,\ A \cap \bar{B},\ A \cup B$$

(b) Express the following subsets of S in terms of A, B, and C:

$$D = \{x : 0 < x < 2\}$$
$$E = \{x : x = 0\}$$
$$F = \{x : x \leq -1\}$$

5. A standard deck of 52 cards is dealt to four players: South, North, East, West. Define the events:

$$S_i = \text{"South has at least } i \text{ spades,"} \quad i = 0, 1, 2, \ldots, 13,$$

with similar definitions for N_i, E_i, W_i
Which of the following are true statements and which are false? Justify your answers.

(a) $W_6 \cap N_8 = \varnothing$
(b) $S_0 = S$
(c) $S_3 \cap N_3 \cap E_3 \cap W_3 = \varnothing$
(d) $S_4 \cup N_4 \cup E_4 \cup W_4 = S$
(e) $\overline{S_{13}} = N_1 \cup E_1 \cup W_1$

6. Let $S = \{10, 20, 30, 40, 50, 60, 70, 80, 90, 100\}$. Write out the elements of the following subsets of S:

(a) $A = \{a : a \text{ is exactly divisible by } 8\}$
(b) $B = \{b : b \text{ is exactly divisible by } 4\}$
(c) $C = \{c : c/10 \text{ is a prime number}\}$
(d) $D = \{d : d/10 \text{ is an odd number}\}$
(e) $E = \{e : (3/2)e \in S\}$

7. Using the sets defined in exercise 6, determine

(a) $\bar{B}$
(b) $A \cup B$
(c) $A \cap B$
(d) $A \cap C$
(e) $(A \cap C) \cup E$
(f) $(D \cup E) \cap (A \cup C)$
(g) $D \cup B$
(h) $C \cap E$
(i) $\overline{(A \cup B)}$
(j) $\bar{A} \cap \bar{B}$

8. From the results of exercises 6 and 7, show that

(a) $\bar{B} = D$
(b) $A \cup B \subset B$
(c) $\overline{A \cap C} = S$
(d) $A \subset B \subset A \cup B$
(e) $A \cap B \subset A$
(f) $\overline{D \cup E} = \overline{\{80{,}100\}}$
(g) $\overline{A \cup B} = \bar{A} \cap \bar{B}$
(h) $A \subset \bar{C}$
(i) $B \not\subset A$

9. Check the universal validity of the following set relations using a Venn diagram. To do this, sketch the various events in most general position on a rectangle representing S. Then shade in the regions defined by the left and right sides of each identity, and show that they are identical regions.

(a) $\overline{(E_1 \cup E_2)} = \bar{E}_1 \cap \bar{E}_2$
(b) $\overline{(E_1 \cap E_2)} = \bar{E}_1 \cup \bar{E}_2$
(c) $E_1 \cup (E_2 \cup E_3) = (E_1 \cup E_2) \cup E_3$
(d) $E_1 \cap (E_2 \cap E_3) = (E_1 \cap E_2) \cap E_3$
(e) $E_1 \cap (E_2 \cup E_3) = (E_1 \cap E_2) \cup (E_1 \cap E_3)$
(f) $E_1 \cup (E_2 \cap E_3) = (E_1 \cup E_2) \cap (E_1 \cup E_3)$
(g) $E_1 \cup E_2 = E_2 \cup E_1$
(h) $E_1 \cap E_2 = E_2 \cap E_1$
(i) $\overline{E_1 \cap (E_2 \cup E_3)} = \overline{(E_1 \cap E_2)} \cap \overline{(E_1 \cap E_3)}$

Note. Relations (**a**) and (**b**) are called *DeMorgan's Rule*. Relations (**c**) and (**d**) are the *associative laws* of sets, just as $a + (b + c) = (a + b) + c$ is the associative law of addition of numbers. Relations (**e**) and (**f**) are the *distributive laws* of sets analogous to $a(b + c) = ab + ac$ for numbers. Relations (**g**) and (**h**) are the *commutative laws* of sets analogous to $a + b = b + a$ for numbers.

10. Using a Venn diagram, illustrate the following set relation. Assume that A_1, A_2, A_3, A_4, A_5 are mutually exclusive events (sets), and that

$$E \subset (A_1 \cup A_2 \cup A_3 \cup A_4 \cup A_5)$$

Then show that

$$(E \cap A_1) \cup (E \cap A_2) \cup (E \cap A_3) \cup (E \cap A_4) \cup (E \cap A_5) = E$$

11. Consider the series electrical circuit shown below

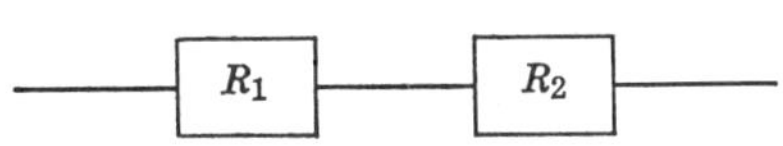

Suppose that components R_1 and R_2 can or cannot be operative. Let the sample space S consist of the four possible combinations of operative and inoperative components, $S = \{(R_1 \text{ operative}, R_2 \text{ operative}); \ldots; (R_1 \text{ inoperative}, R_2 \text{ inoperative})\}$. Let E_1 be the event "the entire circuit is operative," and E_2 be the event "at least one of R_1, R_2 is operative." Are E_1, E_2 mutually exclusive? Interpret $\bar{E}_1$ and $\bar{E}_2$. What can we say about the relation between $\bar{E}_1$ and E_2?

12. Show that, if A_1 and A_2 are any two events, then $A_1 \subset A_1 \cup A_2$ and $A_1 \cap A_2 \subset A_2$. Is this result valid if A_1 and A_2 are mutually exclusive?

13. An assembly of electronic equipment consists of 3 components arranged in a series-parallel circuit as follows:

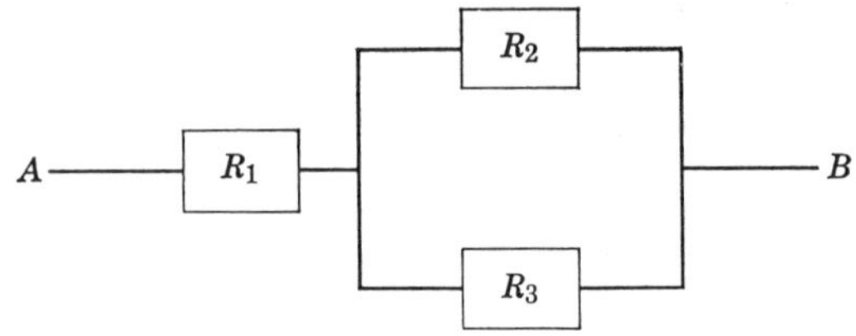

Each component either is operative or fails under load. However, the entire assembly fails only if the path from A to B is broken. Let the sample space S consist of the eight possible combinations of operative or inoperative components. Let E_1 be the event "the assembly is operative;" let E_2 be the event "R_2 has failed but the assembly is operative;" let E_3 be the event "R_3 has failed but the assembly is operative."

(**a**) List the elements of S, E_1, E_2, and E_3.

(**b**) Are the events E_1 and E_2 mutually exclusive?
Are the events E_1 and E_3 mutually exclusive?
Are the events E_2 and E_3 mutually exclusive?

14. Consider the following series-parallel assembly consisting of five components:

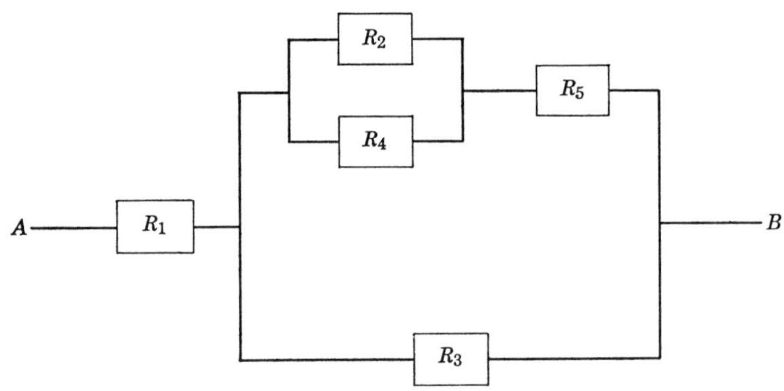

Answer the same questions as those in exercise 13. (In this case S will contain 32 points.)

15. You flip a penny until the first head appears. Describe the sample space of this experiment.

16. You observe a power chain conveyor until the first empty carrier appears. Describe the sample space for this experiment. (*Note.* A power chain conveyor is

a materials-handling device. Basically, it consists of a powered chain that has carrying devices (called carriers) equally spaced along its length. The carriers carry material to be worked on at manufacturing centers along the path of the conveyor. Some may be empty, others loaded.)

17. The set $A - B$, called the *difference of A and B*, is the set of points that are elements of A but not B. Show that if A and B are mutually exclusive, then $A - B = A$. In general show that $A - B = A \cap \bar{B}$. Hence, in **(f)** of exercise 8 show that $D \cup E = S - \{80, 100\}$.

18. If a and b are numbers, $a \leq b$, then the set of all real numbers between a and b is termed an *interval*. Different kinds of intervals are defined depending on whether the end points are included in the set or not. The following notation and terminology is standard.

$[a, b] = \{x : a \leq x \leq b\}$, the *closed* interval from a to b

$(a, b) = \{x : a < x < b\}$, the *open* interval from a to b

$[a, b) = \{x : a \leq x < b\}$, the *left-closed* interval from a to b

$(a, b] = \{x : a < x \leq b\}$, the *right-closed* interval from a to b

Define

$$S = [0, 100],\ A = [10, 100],\ B = [5, 10),\ C = [0, 50],$$
$$D = [0, 25),\ E = (90, 100]$$

and describe the following sets using interval notation wherever possible. Sketch each on a number line (real axis).

(a) $A \cup B$
(b) $A \cap C$
(c) $C \cap D$
(d) $A \cup E$
(e) $A \cap B$
(f) $\bar{A} \cap B$
(g) $A \cap \bar{B}$
(h) $\overline{A \cap B}$
(i) $[(C \cap \bar{D}) \cap D] \cap E$
(j) $B \cup E$

19. **(a)** Assuming that $\varnothing$ is an interval, show that the intersection of two intervals is always an interval.

(b) Classify the types of intervals we obtain by taking the intersection of open, closed, left-closed, and right-closed intervals in any combination.

(c) Is the union of two intervals always an interval? If true, prove it; if false, give a counterexample.

(d) Is the union of the *complements* of two intervals the *complement* of an interval? If true, prove it; if false, give a counterexample.

2 • Combinatorial Probability

1. Introduction

In the previous chapter we learned that the physical behavior of random experiments can be modeled rather naturally using concepts such as sets, subsets, and classes of sets. In addition, we would like to assign numbers to events that, in some sense, tell us what the chances are of a particular event occurring. To this end, we must assign some measure for events termed the *probability of the event* A and symbolized by $\Pr[A]$.

Since much of the earliest development of the theory of probability arose in games of chance, it seemed, at least historically, quite natural to interpret the *probability of the event* A in the following sense.

Consider a sequence of repetitions of the same experiment under identical conditions. Let f_n denote the number of occurrences of the event A in the first n repetitions of the experiment. The ratio f_n/n then gives the *proportion* of "successes" among the first n trials. For example, if the experiment consists of tossing a coin and if the event A corresponds to "heads," then f_n/n gives the fraction of heads in the first n tosses. Intuitively, we feel that, as n increases, the ratio f_n/n should stabilize and approach some fixed number that would measure the likelihood of the event A occurring. Thus we would like to assign probabilities to events in our model in such a way that

$$\Pr[A] = \lim_{n\to\infty} \frac{f_n}{n} \tag{2.1}$$

Of course, the existence of the limit in (2.1) could never be checked in any real situation, because no *infinite* sequence of real experiments is possible. Nonetheless, it is intuitively so attractive that we would hope to have a mathematical model in which this formula holds true.

In repeated tosses of an "honest" coin we expect the proportion of heads to approach $\frac{1}{2}$, so that by (2.1) we assign the value $\frac{1}{2}$ to Pr["heads"]. In fact, we could define an "honest" coin as one in which $\lim_{n\to\infty}(f_n/n) = \frac{1}{2}$. Figure 2.1

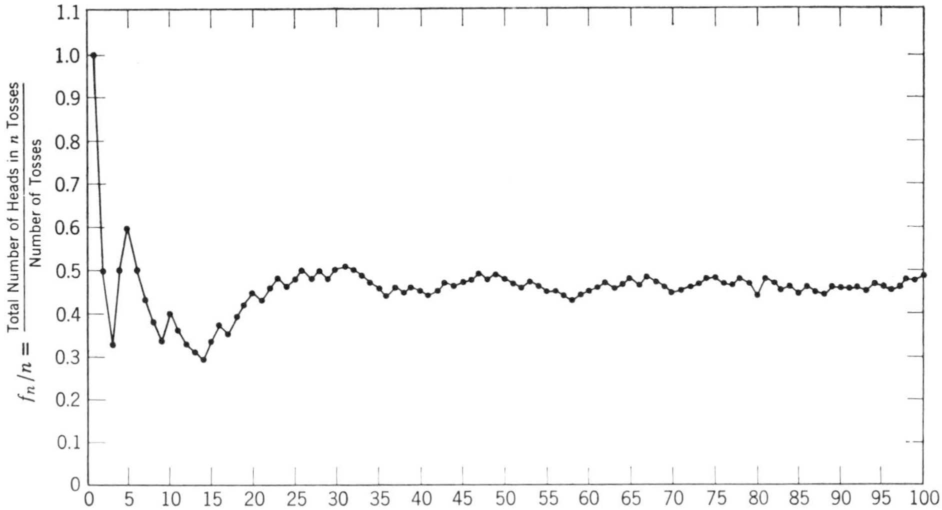

FIGURE 2.1 Proportion of Heads in 100 tosses of a quarter.

gives one possible graph of f_n/n for n up to 100 tosses of a supposedly honest quarter. Although this graph neither proves the existence of the limit of f_n/n nor guarantees a numerical value of $\frac{1}{2}$ for this limit if it exists, it does illustrate the behavior that we expect from a relative frequency sequence.

Clearly, conceptually at least, we could model repetitive experiments such as coin tossings, card playing, and games of chance in the sense described above. Furthermore, we could model probabilities for more complicated random experiments, for example, the sales campaign mentioned in Chapter 1, in this sense. By taking some liberty with existing measuring instruments, we could use the foregoing approach to probability in order to model probabilities for the manufacturing operation that measures the strength of a part. Hence, for a considerable amount of engineering and scientific application, this *relative frequency* sense of probability serves to give an adequate model of the probability measure. That is, for many engineering problems, we can imagine repeating an experiment many times and can imagine measuring the number of "successes." Therefore, any mathematical construction of a probability measure must allow us to interpret the probability in this relative frequency sense.

Historically, probability models were based on (2.1). But this approach is not enough. Consider the rocket-firing experiment of Chapter 1. Surely no one is going to fire a sophisticated rocket system many times to determine the probability of the event "success" since, if for no other reason, a large number of trials could be prohibitively expensive. The experiment of observing the behavior of the national economy for a given calendar year is inherently impossible of repetition. In many areas a large number of repeated experiments is just not possible. Nonetheless, we would hope to be able to assign a probability measure to events in these cases.

Although a relative frequency approach to probability is satisfactory for many engineering applications and games of chance, such an approach is inadequate for many uses. The mathematical construction of a probability measure must be independent of any envisioned application but must be usable for many applications. To this end the modern theory of probability starts with the construction of a set of *axioms of probability* that, we will observe, do not refer to how the probabilities are actually obtained. How we determine the actual probability of a given event depends strongly on the random experiment that is conducted; and, hence, the determination of a real probability is an extra-mathematical problem. The theory of probability properly starts with the assumption that probabilities can be assigned that satisfy the following *axioms of probability*. Whether, in a particular case, these probabilities are determined by repetition of a random experiment, or are based on personal judgement, on communion with nature, or on a visitation with the local soothsayer, is irrelevant to the study of probability theory.

2. Axioms of Probability

If probabilities are to be assigned to events in a way consistent with (2.1), certain conditions obviously are necessary.

(a) Since $0 \leq f_n/n \leq 1$ for every n, then we should require that $0 \leq \Pr[A] \leq 1$ for every every A.

(b) If A is the *certain* event, $A = S$, then $f_n = n$ and $f_n/n = 1$. Thus events certain to occur should have probability 1.

(c) Suppose that A and B are mutually exclusive events; that is, both cannot happen at the same time. Let $f_n(A)$, $f_n(B)$, and $f_n(A \cup B)$ be the number of occurrences of the events A, B, and $A \cup B$, respectively, in the first n trials. Clearly, $f_n(A \cup B) = f_n(A) + f_n(B)$ and, thus, $f_n(A \cup B)/n = f_n(A)/n + f_n(B)/n$. Hence we would expect $\Pr[A \cup B]$ to be the sum of $\Pr[A]$ and $\Pr[B]$.

In view of these properties, we postulate that each event, A, is assigned a number, $\Pr[A]$, in such a way that the following *axioms* are satisfied.

(i) $0 \leq \Pr[A] \leq 1$.

(ii) $\Pr[S] = 1$, where S represents the *certain* event; that is, $S =$ the entire sample space.

(iii) If $A_1, A_2, A_3, \ldots,$ is a sequence (finite or infinite) of *mutually exclusive* events ($A_i \cap A_j = \varnothing$ for $i \neq j$), and if

$$B = A_1 \cup A_2 \cup A_3 \cdots$$

is the event "at least one of the A_i's occur," then

$$\begin{aligned}\Pr[B] &= \Pr[A_1] + \Pr[A_2] + \Pr[A_3] + \cdots \\ &= \sum_i \Pr[A_i]\end{aligned}$$

Certainly these axioms are consistent with our intuitive ideas of how probabilities behave. Furthermore, we can show that, if these axioms are satisfied, formula (2.1) holds true, with appropriate definitions of the meaning of the terms "experiments repeated under identical conditions" and "converges."

To avoid certain mathematical difficulties, we must place some restrictions on the types of subsets of the sample space that we term "events" and for which we define probabilities. These restrictions do not affect any part of our discussion. For our purposes every subset of a sample space constructed here can be considered as an event having a probability.

A random experiment with its associated sample space, events, and probability assignment is termed a *probability system.*

3. Elementary Properties of Probability

A number of properties of probability follow directly from axioms (i), (ii), and (iii).

(a) *For any event A*, $\Pr[\bar{A}] = 1 - \Pr[A]$. △

PROOF. A and $\bar{A}$ clearly are mutually exclusive, and $S = A \cup \bar{A}$. Hence, by axioms ii and iii,

$$1 = \Pr[S] = \Pr[A] + \Pr[\bar{A}]$$

from which the assertion follows. △

(b) *If* $\varnothing$ *is the impossible event, then* $\Pr[\varnothing] = 0$. △

PROOF. Since $\varnothing \cup \varnothing = \varnothing$ and $\varnothing = \varnothing \cap \varnothing$, it follows from axiom (iii) that $\Pr[\varnothing] = \Pr[\varnothing] + \Pr[\varnothing] = 2\Pr[\varnothing]$. Hence, $\Pr[\varnothing] = 0$. △

(c) *If A_1 and A_2 are any events, not necessarily mutually exclusive, then*

$$\Pr[A_1 \cup A_2] = \Pr[A_1] + \Pr[A_2] - \Pr[A_1 \cap A_2] \quad \triangle$$

PROOF. Clearly (see Figure 2.2),

$$A_1 \cup A_2 = A_1 \cup (\bar{A}_1 \cap A_2) \qquad \text{and} \qquad A_2 = (A_1 \cap A_2) \cup (\bar{A}_1 \cap A_2)$$

where in each equation the unions on the right side are of mutually exclusive events.

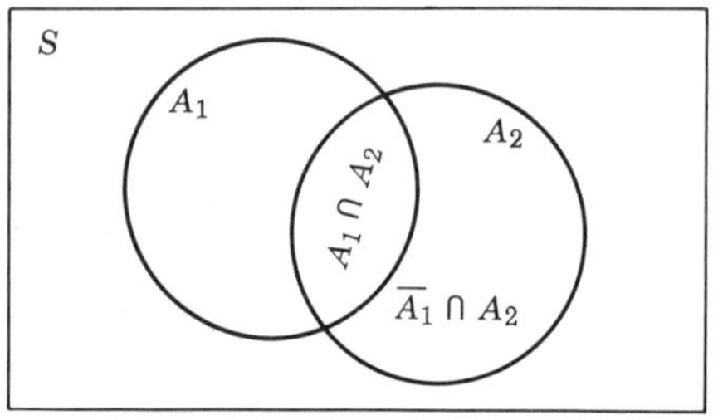

FIGURE 2.2 Venn diagram for the set relation $A_1 \cup A_2 = A_1 \cup (\bar{A}_1 \cap A_2)$.

By axiom (iii)

$$\Pr[A_1 \cup A_2] = \Pr[A_1] + \Pr[\bar{A}_1 \cap A_2]$$

$$\Pr[A_2] = \Pr[A_1 \cap A_2] + \Pr[\bar{A}_1 \cap A_2]$$

The second equation gives $\Pr[\bar{A}_1 \cap A_2] = \Pr[A_2] - \Pr[A_1 \cap A_2]$ which, when substituted into the first equation, yields the given assertion. △

The formula in (c) can be generalized as follows:

(d) *If $A_1, A_2, \ldots, A_k$ are any events, then*

$$\Pr\left[\bigcup_{i=1}^{k} A_i\right] = \Pr[A_1 \cup A_2 \cup \cdots \cup A_k] = \sum_i \Pr[A_i] - \sum_{i<j} \Pr[A_i \cap A_j] + \cdots$$
$$\cdots + (-1)^{k+1}\Pr[A_1 \cap A_2 \cap \cdots \cap A_k]$$

where the successive sums are over all possible events, pairs of events, triples of events, etc. △

PROOF. The proof follows from (c) by induction. The details are omitted here; however, a student familiar with induction arguments should have no difficulty in adding them. △

4. Discrete Sample Spaces; Equally Likely Sample Points

A probability system is particularly simple when the sample space consists of only a finite number N of possible experimental outcomes, sample points, or *elementary events*. To determine the probability of all events in this case, we only need to know the probabilities of the elementary events $p_1, p_2, \ldots, p_N$, $\sum p_i = 1$. Since any event A consists of a certain collection of these sample points, $\Pr[A]$ can be found, using axiom iii, by adding up the probabilities of the separate sample points that make up A.

EXAMPLE 2.1. Three subjects in a psychological learning experiment are each given a sequence of digits to remember for a fixed time. The probabilities that these subjects will remember correctly are 0.4, 0.6, and 0.8, respectively. Now consider the experiment of giving the sequence of digits and of determining how many of the subjects remember them correctly after the fixed time. Thus the sample space for this experiment contains four sample points: 0 subjects remember correctly; 1 remembers correctly; 2 remember correctly; and 3 remember correctly. The probabilities that can be assigned to these elementary events are: .048, .296, .464, and .192, respectively (see Exercise 1). The event "at least one subject remembers incorrectly" $= \{0 \text{ correct}, 1 \text{ correct}, 2 \text{ correct}\}$ thus will have probability $.048 + .296 + .464 = .808$. The event "no more than one remembers incorrectly" $= \{2 \text{ correct}, 3 \text{ correct}\}$ will have probability $.464 + .192 = .656$. △

The situation is even simpler when the N individual sample points all have equal probabilities (*equally likely sample points*),

$$p_1 = p_2 = \cdots = p_N = \frac{1}{N}$$

Calculation of probabilities in this case reduces simply to counting. If the event A consists of k sample points, then

$$\Pr[A] = \frac{1}{N} + \frac{1}{N} + \cdots + \frac{1}{N} = \frac{k}{N}$$

EXAMPLE 2.2. A group of four radios consists of two good radios, labeled g_1 and g_2, and two defective radios, labeled d_1 and d_2. If three radios are selected at random from this group, what is the probability of the event A: "two of the three selected are defective?"

By direct enumeration we observe that there are four ways of selecting three radios (if order is disregarded):

$$g_1 g_2 d_1,\ g_1 g_2 d_2,\ g_1 d_1 d_2,\ g_2 d_1 d_2$$

The phrase "selected at random" can be interpreted here to mean that each of these four sample points are equally likely and, therefore, each has probability $\frac{1}{4}$. Since the event A can occur in 2 ways, $A = g_1d_1d_2, g_2d_1d_2$, we have

$$\Pr[A] = \frac{2}{4} = \frac{1}{2} \qquad \triangle$$

Systems with equally likely sample points occur frequently when we discuss various games of chance. Because of their simple structure, the discussion of these systems occupies much space in many elementary works on probability. Unfortunately, most applications of interest in engineering, business, or science do not fall into this category.

The discussion of finite sample spaces immediately extends to random experiments that have an infinite sequence of possible sample points (*countable sample spaces*). Again, only the probabilities of the elementary events must be given; the probabilities of other events can be obtained by addition. Of course, the equally likely case cannot occur here, since the sum of infinitely many equal numbers cannot add up to one.

EXAMPLE 2.3. Consider the experiment of tossing an honest coin repeatedly and of counting the number of tosses required until the first "head" occurs. Clearly, the terms of the sequence 1, 2, 3, ... are all possibilities for the number of tosses required. Suppose that the corresponding probabilities are $p_1, p_2, p_3, \ldots$. That is,

$$p_i = \Pr[\text{first "head" occurs on the } i\text{th toss}], \qquad \text{for } i = 1, 2, 3, \ldots$$

Since there are 2^i equally likely ways of tossing an honest coin i times, only one of which consists of $i - 1$ "tails" followed by a "head," we observe that

$$p_i = 1/2^i, \qquad i = 1, 2, 3, \ldots$$

As a check on this formula, notice that

$$\sum_{i=1}^{\infty} p_i = \sum_{i=1}^{\infty} 1/2^i = \frac{1/2}{1 - 1/2} = 1$$

by using the formula for the sum of a geometric series.

If A is the event "the first head occurs on an even-numbered toss," then

$$\begin{aligned} \Pr[A] &= p_2 + p_4 + p_6 + \cdots \\ &= \sum_{k=1}^{\infty} 1/2^{2k} = \frac{1/4}{1 - 1/4} = 1/3 \end{aligned} \qquad \triangle$$

5. Counting: Permutations and Combinations

Many mathematical and physical problems are concerned with counting. Frequently we need to count the number of ways in which various objects can be chosen. Many shortcuts and formulas have been developed to simplify the problem of counting. Most of them are based on this *fundamental counting principle*:

When two distinct acts are to be performed in succession, if the first act can be performed in m distinct ways, and if, then, the second act can be performed in n distinct ways, there are a total of mn distinct ways in which the two acts can be performed.

This principle generalizes immediately to a sequence of more than two acts.

EXAMPLE 2.4. How many even two-digit numbers can be constructed using the digits 3, 4, 5, 6, and 7 (repetitions such as 44 being permitted)?

Let the successive acts here be the choosing of the two digits making up the required number, with the units digit chosen first. For the number to be even, the last digit must be 4 or 6, hence, there are 2 possibilities here. After the last digit has been chosen, the remaining digit can be any one of the 5 digits. Thus, there are $2 \cdot 5 = 10$ possible even two-digit numbers that can be constructed in this way. $\triangle$

By a *permutation* we mean a set of objects arranged in some order. Thus, for two permutations to be the same, not only must they contain the same objects but these objects must be arranged in the same order. For example, the permutation (A, B, C) of three letters contains the same objects as the permutation (B, C, A), but they are different permutations because the orders differ.

Given a total of n objects, we frequently want to know the total number of possible permutations of k of these n objects, $k \leq n$, with no repetitions being permitted in any given permutation. This number is denoted by

$$P(n, k) = \textit{the number of permutations of } n \textit{ distinct objects taken } k \textit{ at a time}$$

For example, given the 3 objects, A, B, C, the number of permutations of 2 of these objects can be found by direct enumeration: (AB), (BA), (AC), (CA), (BC), (CB). Thus,

$$P(3, 2) = 6$$

A general formula for $P(n, k)$ can be derived by using the fundamental counting principle. The first element of the permutation can be chosen in n ways; the second can then be any one of the remaining objects and, hence, can be chosen in $n-1$ ways, the third in $n-2$ ways, and so forth. By the fundamental principle the total number of such permutations is obtained by multiplying k such factors together. The last factor is $n-(k-1)=n-k+1$. Hence

$$P(n, k) = \underbrace{n(n-1)(n-2)\cdots(n-k+1)}_{k \text{ factors}} \tag{2.2}$$

For example,

$$P(3, 2) = 3 \cdot 2 = 6$$

$$P(12, 4) = 12 \cdot 11 \cdot 10 \cdot 9 = 11{,}880$$

If formula (2.2) is multiplied and divided by $(n-k)!$ a more compact form results:

$$\begin{aligned} P(n, k) &= \frac{n(n-1)\cdots(n-k+1)\cdot(n-k)(n-k-1)\cdots 2 \cdot 1}{(n-k)!} \\ P(n, k) &= \frac{n!}{(n-k)!} \end{aligned} \tag{2.3}$$

For computation purposes, however, (2.2) is preferable.

Notice that, when $n=k$, the number of possible complete permutations of k distinct objects is given by

$$P(k, k) = k! \tag{2.4}$$

(Observe also that, for formula (2.3) to be valid when $n=k$, we must define $0! = 1$. We always follow this convention.)

Sometimes we wish to count the number of permutations of n objects taking k at a time when the same object is allowed to be repeated any number of times in a given permutation (*permutations with repetitions allowed*). In this case the fundamental counting principle yields

$$\underbrace{n \cdot n \cdot \cdots \cdot n}_{k \text{ factors}} = n^k \tag{2.5}$$

for the number of such permutations.

EXAMPLE 2.5. In a coding problem, we wish to find the number of different 3-letter groups that can be formed from the regular alphabet. If the same letter cannot be used twice in the same group, this number is

$$P(26, 3) = 26 \cdot 25 \cdot 24 = 15{,}600$$

by using (2.2).
If repetitions are permitted, the number of groups is

$$26^3 = 17{,}576$$

by using (2.5). △

EXAMPLE 2.6. If a foreman wants to make up a work crew of 3 men for 3 special jobs and has 10 men in his department to choose from, how many different assignments can he choose? In this case, the first job can be given to any one of the 10, the second to any one of the remaining 9, and the third to any one of the remaining 8. Thus there are 720 possible crew-assignments. In terms of permutations there are

$$P(10, 3) = 10 \cdot 9 \cdot 8 = 720$$

possibilities, using formula 2.2. △

EXAMPLE 2.7. In the preceding example, suppose that the 10 men have 2 foremen, and each work crew consists of 3 men and 1 foreman. How many crew-foreman teams and job assignments can be made? As above, there are 720 possible crew-assignments. There are now 2 possible foremen for each crew. Hence, using the fundamental counting principle, there are $2 \cdot 720 = 1{,}440$ possible crew-foreman team assignments to choose from. △

Notice that, in both of the preceding examples, seemingly identical crews are considered different if the men work at different jobs.

By a *combination* we mean any set of objects, order being disregarded. Thus the combination $\{A, B, C\}$ is identical with the combination $\{B, C, A\}$, since each contains the same set of elements.

As in the case of permutations, a formula exists for the number of possible combinations (subsets) of k objects that can be chosen from a given set of n objects, $k \leq n$. Let us (temporarily) denote this number by

$$C(n, k) = \textit{the number of combinations of } n \textit{ distinct objects taken } k \textit{ at a time}$$

For example, by direct enumeration, we determine that there are only 3 possible combinations of the 3 letters A, B, and C taken 2 at a time, namely, $\{A, B\}$, $\{A, C\}$, $\{B, C\}$. Thus $C(3, 2) = 3$.

Any *permutation* of n objects taken k at a time may be constructed by first choosing the *combination* or set of objects involved, and by then arranging these elements in some *order*. Thus, by the fundamental counting principle, the number of possible permutations must equal the number of possible combinations [namely, $C(n, k)$] *multiplied* by the number of ways in which any one combination of k objects can be rearranged to give different permutations [which, by formula (2.4), is $P(k, k) = k!$]. It follows that

$$P(n, k) = C(n, k)\, k!$$

Thus

$$\begin{aligned} C(n, k) &= \frac{P(n, k)}{k!} \\ &= \frac{n(n-1)\cdots(n-k+1)}{k!} \\ &= \frac{n!}{k!(n-k)!} \end{aligned} \tag{2.6}$$

For example

$$C(3, 2) = \frac{3 \cdot 2}{2!} = 3$$

$$C(12, 4) = \frac{12 \cdot 11 \cdot 10 \cdot 9}{4!} = 495$$

Notice that, from formula (2.6), it follows that $C(n, k)$ is the same as the *binomial coefficient* $\binom{n}{k}$ appearing in the *binomial theorem* of elementary algebra.

$$(a+b)^n = a^n + \binom{n}{1} a^{n-1} b + \cdots + \binom{n}{k} a^{n-k} b^k + \cdots + \binom{n}{n} b^k \tag{2.7}$$

$$(a+b)^n = \sum_{k=0}^{n} \binom{n}{k} a^{n-k} b^k$$

where

$$\binom{n}{k} = \frac{n(n-1)\cdots(n-k+1)}{k!}$$

To avoid using two symbols for the same expression, henceforth the symbol $\binom{n}{k}$ will be used in place of $C(n, k)$ to represent the number of combinations of n objects taken k at a time. Other notations in use are ${}_nC_k$, ${}^nC_k$, C_k^n.

EXAMPLE 2.8. A sample of 5 items is chosen from a lot of 50 items produced by a factory. How many different samples could be chosen?

If the items are selected as a batch rather than in a special order, the number of different samples will simply be the number of combinations of 50 objects taken 5 at a time. This number is

$$\binom{50}{5} = \frac{50 \cdot 49 \cdot 48 \cdot 47 \cdot 46}{5 \cdot 4 \cdot 3 \cdot 2 \cdot 1} = 2{,}118{,}760 \qquad \triangle$$

EXAMPLE 2.9. If we return to Example 2.6, we observe that there are

$$\binom{10}{3} = \frac{720}{3!} = 120$$

distinguishably different *crews* of size 3 that can be chosen from the 10 men available if their different jobs are disregarded. If the foreman is included in the crew (Example 2.7), then there are

$$\binom{10}{3} \cdot \binom{2}{1} = 240$$

distinguishably different crew-foreman teams that can be chosen. △

6. Stirling's Formula

We have learned that several important formulas involve the evaluation of factorials. The estimation of $n!$ for large n is facilitated by using the following approximation, known as Stirling's formula:

$$n! \sim \sqrt{2\pi n}\, n^n e^{-n} \tag{2.8}$$

The symbol $\sim$ here means that this is an *asymptotic* formula. That is, the *ratio* of the two sides in (2.8) approaches 1 as $n \to \infty$,

$$\lim_{n \to \infty} \frac{n!}{\sqrt{2\pi n}\, n^n e^{-n}} = 1$$

Hence, if (2.8) is used to give an approximation to $n!$ for large n, the *relative* error will be small. The *absolute* error, the *difference* between the two sides of (2.8), does not become small. In fact, it grows as $n \to \infty$. This phenomenon is illustrated by Table 2.1.

TABLE 2.1

n	$n!$	$\sqrt{2\pi n}\, n^n\, e^{-n}$	Relative Error	Asolute Error
2	2	1.919	0.04	0.08
5	120	118.0	0.02	2.0
10	3.628×10^6	3.599×10^6	0.008	2.9×10^4
			$\downarrow$	$\downarrow$
			0	∞

7. Further Examples

EXAMPLE 2.10. If two "honest" dice are tossed, what is the probability of the event "a total of seven comes up?"

The equally likely sample points here would consist of all pairs of integers from 1 to 6, with repetitions allowed. Hence, by formula (2.5), $N = 6^2 = 36$. The required event comprises the 6 points: (1, 6), (2, 5), (3, 4), (4, 3), (5, 2), (6, 1). Thus $k = 6$, and the required probability is $6/36 = 1/6$. △

EXAMPLE 2.11. An "honest" coin is tossed 6 times. What is the probability of getting exactly 3 heads?

Again, using (2.5), we find that $N = 2^6$. The number of "*successes*" is equal to the number of ways in which 3 of the 6 tosses can be chosen to be heads. Thus $k = \binom{6}{3}$, and the required probability is

$$\binom{6}{3}\Big/2^6 = \frac{6 \cdot 5 \cdot 4}{1 \cdot 2 \cdot 3 \cdot 2^6} = \frac{5}{16}$$

More generally, we see that the probability of x heads on 6 throws is $\binom{6}{x}\Big/2^6$. △

EXAMPLE 2.12. A bridge hand of 13 cards is dealt from a deck of 52 cards. What is the probability that all 13 cards are spades?

The equally likely sample points can be taken to be all possible combinations of the 52 cards taken 13 at a time, $N = \binom{52}{13}$. Only one of these combinations contains all 13 spades. Therefore the required probability is

$$1\Big/\binom{52}{13} \approx \frac{1}{(6.35)10^{11}} = (1.57)10^{-12}$$

An alternate approach is to take as sample points all possible *permutations* of 52 cards taken 13 at a time, $N = P(52, 13)$. The number of possible successful permutations is $P(13, 13) = 13!$. Thus, the required probability is $13!/P(52, 13) = 1/(P(52, 13)/13!) = 1\Big/\binom{52}{13}$, as before. △

EXAMPLE 2.13. A poker hand of 5 cards is dealt from a deck of 52. What is the probability of having exactly one pair (that is, two cards of the same denomination)?

Take combinations as sample points, $N = \binom{52}{5}$. A hand containing exactly one pair can be constructed by the following series of steps:

(a) choose the denomination of the pair, 13 ways

(b) choose the pair from this denomination, $\binom{4}{2}$ ways

(c) choose the denominations of the other 3 cards, $\binom{12}{3}$ ways

(d) choose the other 3 cards, 4^3 ways

By using the counting principle of Section 1, the total number of possible "successes" is

$$(13)\binom{4}{2}\binom{12}{3}(4^3)$$

Thus the required probability is

$$13\binom{4}{2}\binom{12}{3}4^3\Big/\binom{52}{5} = \frac{(13)(4!)(12!)(4^3)(5!)(47!)}{(2!)(2!)(3!)(9!)(52!)} = 0.423$$

In finding the number of possible "successes" in this kind of problem it is important to determine a list of steps, such as (a), (b), (c), and (d) above, which, if performed in this order, will allow us to construct any given successful case in exactly one way, so that the number of possibilities for each step can be counted. By the counting principle of Section 5, these numbers then can be multiplied to obtain the final answer.

The list of steps given here for constructing a 5-card hand with one pair is by no means the only sequence of steps by which such a hand can be constructed. The student should try to construct an alternate sequence of steps, and to check that the same answer results. △

EXAMPLE 2.14. In many industrial quality control procedures there is a large lot of parts, of which some may be defective. To keep inspection costs low, we sample from the lot instead of looking at every part. From the sample, we try to infer the number of defective parts in the lot. Suppose that a lot has 12 parts and that we take a sample of 3 of them. If 6 of the parts in the lot are defective, what is the probability that all 3 of the sampled parts are good?

Clearly, there are $\binom{12}{3}$ possible samples. Since there are 6 good parts in the lot, there will be a total of $\binom{6}{3}$ possible samples having all items good. Hence the required probability is

$$\binom{6}{3}\Big/\binom{12}{3} = \frac{(6 \cdot 5 \cdot 4)/3!}{(12 \cdot 11 \cdot 10)/3!} = \frac{1}{11} = 0.09$$

If there are *more* than 6 defective parts in the lot, the probability of all sampled parts being good will be even *lower* than 0.09. Hence, if the sampled parts are all actually observed to be good, we could use these probabilities to justify a certain level of confidence that the actual number of defectives in the lot does not exceed 5. △

EXAMPLE 2.15. Consider a randomly chosen group of n people. What is the probability that two or more have the same birthday?

Let us neglect leap years and assume that all 365 days of the year are equally likely for birthdays. The experiment consists of successively choosing n out of the 365 days of the years for birthdays (repetitions being permitted). By formula (2.5), the number of equally likely sample points that make up the sample space is 365^n. If A is the event "two or more have the same birthday," then the complementary event $\bar{A}$ consists of all permutations that have no repetitions, and this contains $P(365, n)$ permutations. Thus

$$\begin{aligned}
\Pr[A] &= 1 - \Pr[\bar{A}] \\
&= 1 - \frac{P(365, n)}{365^n} \\
&= 1 - \frac{(365)(364)(363)\cdots(365 - n + 1)}{365^n} \\
&= 1 - \left(1 - \frac{1}{365}\right)\left(1 - \frac{2}{365}\right)\cdots\left(1 - \frac{n-1}{365}\right)
\end{aligned}$$

The computation of $\Pr[A]$ for moderately small values of n is simplified by taking logarithms, $\ln = \log_e$:

$$\begin{aligned}\ln\{1-\Pr[A]\} &= \ln\left\{\left(1-\frac{1}{365}\right)\left(1-\frac{2}{365}\right)\cdots\left(1-\frac{n-1}{365}\right)\right\}\\ &= \ln\left(1-\frac{1}{365}\right)+\ln\left(1-\frac{2}{365}\right)+\cdots+\ln\left(1-\frac{n-1}{365}\right)\\ &\approx -\frac{1}{365}-\frac{2}{365}-\cdots-\frac{n-1}{365}\\ &= -\frac{1}{365}[1+2+\cdots+(n-1)] = -\frac{1}{365}\cdot\frac{n(n-1)}{2}\\ &= -\frac{n(n-1)}{730}\end{aligned}$$

by using the approximation $\ln(1+x) = x - x^2/2 + \cdots \approx x$, valid for small values of x, and the formula for the sum of an arithmetic progression. In this way $\Pr[A]$ can be calculated for various values of n. In particular, we can show that $\Pr[A] > \frac{1}{2}$ in case n equals 23 or more—a surprisingly small number. Thus a group of 23 or more people has a better than even chance of containing two individuals with the same birthday. △

EXAMPLE 2.16. Two identical decks of n cards each are shuffled and matched against each other, card by card. What is the probability that at least one match occurs?

After shuffling, let us number the cards in the first deck $1, 2, \ldots, n$ consecutively, as they appear, and assign the corresponding numbers to the cards in the second deck. Thus a match will occur if some card in the second deck lies in its natural position, that is, if card number k falls in the kth position in the second deck, for some value of $k = 1, 2, \ldots, n$.

Let A be the event "at least one match occurs." Then

$$A = A_1 \cup A_2 \cup \cdots \cup A_n$$

where A_i is the event "card number i occurs in the ith position in the second deck."

By (d) of Section 3,

$$\Pr[A] = \sum_i \Pr[A_i] - \sum_{i<j} \Pr[A_i \cap A_j] \pm \cdots + (-1)^{n+1}\Pr[A_1 \cap A_2 \cap \cdots \cap A_n]$$

But the number of possible shufflings of the second deck is $n!$, while the number of arrangements in which card number i occupies the ith place is $(n-1)!$. Thus

$$\Pr[A_i] = \frac{(n-1)!}{n!}, \text{ and the first sum} = n\,\frac{(n-1)!}{n!}$$

Similarly,

$$\Pr[A_i \cap A_j] = \frac{(n-2)!}{n!}, \text{ and there are } \binom{n}{2} \text{ distinct pairs } i, j \text{ with } i < j.$$

Continuing in this way, we obtain

$$\begin{aligned}
\Pr[A] &= \sum_i \frac{(n-1)!}{n!} - \sum_{i<j} \frac{(n-2)!}{n!} \pm \cdots + (-1)^{n+1}\frac{1}{n!} \\
&= n\,\frac{(n-1)!}{n!} - \binom{n}{2}\frac{(n-2)!}{n!} \pm \cdots + (-1)^{n+1}\frac{1}{n!} \\
&= 1 - \frac{1}{2!} + \frac{1}{3!} \mp \cdots + (-1)^{n+1}\frac{1}{n!} \\
&= 1 - \left[1 - 1 + \frac{1}{2!} - \frac{1}{3!} \pm \cdots + \frac{(-1)^n}{n!}\right] \\
&\approx 1 - e^{-1} = 0.632\ldots
\end{aligned}$$

since $e^{-1} = 1 - 1 + 1/(2!) - 1/(3!) \pm \cdots$. This series converges very rapidly. Hence the above approximation will be very good, provided that n exceeds 6 or 7. In particular, to a close approximation, the probability of at least one match does not depend on the number of cards in the decks. △

8. Conditional Probability

In many situations we are concerned with two (or more) events whose occurrences are in some way connected. Roughly speaking, we wish to consider a situation where knowledge of the occurrence of one event makes the occurrence of another event more (or less) likely than if this information were not known. For example, consider the experiment of choosing a word at random in an English dictionary. Let A be the event "the letter u occurs in the word," and let B be the event "the letter q occurs in the word." Certainly, event A has a probability but, if we know that B occurs, we are much surer that A must occur too, since q rarely occurs in a word without being followed by u. To consider another example, if buses are known to pass a certain corner at approximately 10-minute intervals, the probability of a bus passing in the next minute would be strongly influenced by the knowledge of whether a bus has passed during the past five minutes.

Thus, if A and B are events, we wish to define a quantity termed the *conditional probability of the event A given that the event B occurs*, or the *conditional probability of A given B*, in symbols:

$$\Pr[A \mid B]$$

Various interpretations of the situation described here are possible. We might consider the experiment to be only partially completed, sufficiently so for the occurrence of B to be known, but not far enough for the occurrence of A to be determined. The above example of the bus passing a corner is of this type. Another interpretation is simply to ignore all experiments in which B fails, and to only consider the ones for which B succeeds. Still another interpretation is to assume that the occurrence of A cannot be directly observed even after the experiment, although B can be directly observed. This would be the case if the experiment consisted of manufacturing an item, B referred to a certain test measurement on the item, and A referred to the existence of some flaw that could only be shown with 100% accuracy by destructive testing.

Since, knowing that B occurs, A can occur only in conjunction with B, it seems reasonable to define the conditional probability as in some way proportional to $\Pr[A \cap B]$. Furthermore, by any reasonable definition, we should have $\Pr[B \mid B] = 1$ (the conditional probability of B occurring given that B occurs should be 1). Therefore, it seems reasonable to define $\Pr[A \mid B]$ to be the relative "size" of the event $A \cap B$ relative to the "size" of the event B, with "size" measured by "probability." Thus we make the following *definition*:

$$\Pr[A \mid B] = \frac{\Pr[A \cap B]}{\Pr[B]} \tag{2.9}$$

This definition only makes sense in case $\Pr[B] > 0$. If $\Pr[B] = 0$, we leave $\Pr[A \mid B]$ undefined.

Formula 2.9 can also be written in the form:

$$\Pr[A \cap B] = \Pr[A \mid B] \Pr[B] \tag{2.10}$$

EXAMPLE 2.17. A company owns two factories that produce similar items. Factory 1 produces 1000 items, 100 of which are defective. Factory 2 produces 4000 items, 200 of which are defective. An item is chosen at random from the production of the company and found to be defective. What is the probability that it came from factory 1?

Let B be the event "the item chosen is defective," and let A be the event "the item chosen came from factory 1." Thus we wish to find $\Pr[A \mid B]$.

We can reason directly as follows. There are a total of 300 defective items, 100 of them from factory 1; hence, the probability that a defective comes from factory 1 is $100/300 = \frac{1}{3}$.

If we apply definition (2.9), we find that $\Pr[A \cap B] = \Pr[\text{item chosen is defective and from factory 1}] = 100/5000$, and that $\Pr[B] = \Pr[\text{item chosen is defective}] = 300/5000$; therefore, by (2.9)

$$\Pr[A \mid B] = \frac{100/5000}{300/5000} = \frac{100}{300} = \frac{1}{3}$$

in agreement with our direct reasoning. △

This type of example assures us that (2.9) is a *reasonable* definition of conditional probability, in accord with what our intuition tells us this phrase should mean.

EXAMPLE 2.18. A box contains two pennies—one honest and one 2-headed. A penny is chosen at random, tossed, and observed to come up heads. What is the probability that the other side is also a head?

We wish to compute $\Pr[A \mid B]$ where B is the event "the coin comes up heads" and A is the event "the coin is 2-headed." We easily determine that $\Pr[B] = \frac{3}{4}$, and $\Pr[A \cap B] = \frac{2}{4}$. (Of the four sides that can come up, three are heads, and two are heads with the 2-headed penny.) Thus, by (2.9)

$$\Pr[A \mid B] = \frac{2/4}{3/4} = \frac{2}{3}$$

A thoughtless person might reason that, since each coin is known to have at least one head, the toss gives no information and, therefore, that the penny has a fifty-fifty chance of being either of the given coins and that the probability of a head on the other side is $\frac{1}{2}$. The student should convince himself of the fallaciousness of this reasoning. △

Formulas 2.9 and 2.10 can be generalized as follows. If A_i, $i = 1, 2, 3, \ldots, n$, are n events, then the conditional probability of A_n, given that $A_1, A_2, \ldots, A_{n-1}$ occur, is given by

$$\Pr[A_n \mid A_1, A_2, A_3, \ldots A_{n-1}] = \frac{\Pr[A_1 \cap A_2 \cap \cdots \cap A_{n-1} \cap A_n]}{\Pr[A_1 \cap A_2 \cap \cdots \cap A_{n-1}]} \tag{2.11}$$

It follows directly that

$$\Pr[A_1 \cap A_2 \cap A_3 \cap \cdots \cap A_n] = \Pr[A_1]\Pr[A_2 \mid A_1]\Pr[A_3 \mid A_1 \cap A_2] \cdots \Pr[A_n \mid A_1 \cap A_2 \cap \cdots \cap A_{n-1}] \quad (2.12)$$

This concept of conditional probability leads us to another formula that is very useful for our subsequent work.

Suppose that $\{B_1, B_2, B_3, \ldots\}$ is a sequence of *mutually exclusive* and *exhaustive* events; that is, no two of the B_i's can occur at the same time, and one of them must occur. Let A be any event. If A occurs, then it must occur in conjunction with one of the B_i's. This statement implies the formula (see Figure 2.3):

$$A = [A \cap B_1] \cup [A \cap B_2] \cup [A \cap B_3] \cup \cdots$$

By axiom (iii) of Section 2,

$$\Pr[A] = \Pr[A \cap B_1] + \Pr[A \cap B_2] + \Pr[A \cap B_3] + \cdots$$

and by formula (2.10),

$$\Pr[A] = \Pr[A \mid B_1]\Pr[B_1] + \Pr[A \mid B_2]\Pr[B_2] + \cdots$$

$$\Pr[A] = \sum_i \Pr[A \mid B_i]\Pr[B_i] \quad (2.13)$$

This formula will be used frequently later.

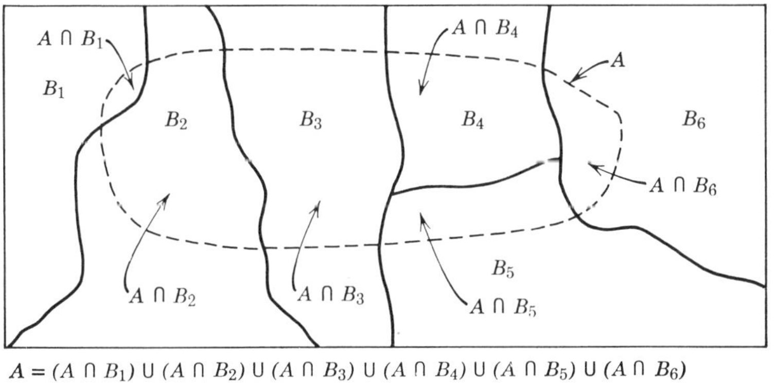

FIGURE 2.3 $A = (A \cap B_1) \cup (A \cap B_2) \cdots \cup (A \cap B_6)$ for $B_1 \cup B_2 \cdots \cup B_6 = S$ and $B_i \cap B_j = \varnothing$, $i \neq j$.

EXAMPLE 2.19. A company producing transistor radios has 3 assembly plants producing 15%, 35%, and 50%, respectively, of its output. Suppose the probabilities that a radio produced by these plants is defective are .01, .05, and .02, respectively. If a radio is chosen at random from the output of the company, what is the probability that it is defective?

By formula (2.13),

$$\begin{aligned}\Pr[\text{defective}] &= \sum_{i=1}^{3} \Pr[\text{defective} \mid \text{from plant } i] \Pr[\text{from plant } i] \\ &= (.01)(.15) + (.05)(.35) + (.02)(.50) \\ &= .029 \quad \triangle\end{aligned}$$

Let us go back to the general situation and suppose that after the experiment we observe that the event A actually occurs, but that it is not known directly which of the mutually exclusive and exhaustive events B_1, B_2, B_3, ... holds true. We then might ask for the conditional probability, $\Pr[B_j \mid A]$, that one of these events B_j occurs, given that A occurs. By using (2.13) and (2.10), we find that

$$\begin{aligned}\Pr[B_j \mid A] &= \frac{\Pr[B_j \cap A]}{\Pr[A]} \\ &= \frac{\Pr[A \mid B_j] \Pr[B_j]}{\sum_i \Pr[A \mid B_i] \Pr[B_i]} \qquad (2.14)\end{aligned}$$

Formula 2.14 is termed *Bayes' formula* and is useful in many applications.

Referring to the preceding example, let us assume that a radio is chosen at random and is found to be defective. What, then, is the probability that it came from plant 2?

By (2.14)

$$\begin{aligned}&\Pr[\text{from plant } 2 \mid \text{defective}] \\ &= \frac{\Pr[\text{defective} \mid \text{from plant } 2] \cdot \Pr[\text{from plant } 2]}{\Pr[\text{defective}]} \\ &= \frac{(.05)(.35)}{(.029)} \\ &= .603\end{aligned}$$

Thus the knowledge that the radio is defective increases the chance that it came from plant 2 from 35% to more than 60%.

9. Independent Events

In the preceding section we observed situations in which the occurrence or nonoccurrence of some event B changed the probabilities of an event A. Certainly important cases exist where a knowledge of the occurrence of an event B does *not* change the probability of an event A, and this fact itself might be important in a discussion of the two events. If a coin is tossed twice, we would reasonably suppose that the probability of the event "a head on the second toss" would not be influenced if we knew that the event "a head on the first toss" had occurred. In life-testing of radio tubes it seems reasonable to suppose that the lifetime of a tube in no way depends on the lifetime of any other tube.

This suggests that we *define* two events, A and B, to be *independent* (or *statistically independent*) if

$$\Pr[A \mid B] = \Pr[A] \tag{2.15}$$

By using (2.10), we observe that this is equivalent to the formula:

$$\Pr[A \cap B] = \Pr[A]\Pr[B] \tag{2.16}$$

Formula 2.16 usually provides a more symmetrical and useful characterization of independence than (2.15). For instance, we need not assume that $\Pr[B] > 0$.

EXAMPLE 2.20. An electronic assembly consists of a transistor and a condensor. The transistor is selected from a lot of 100, 10 of which are defective, and the condensor is selected from a lot of 300, 15 of which are defective. Let A be the event "the transistor part of the assembly is defective," ($\Pr[A] = 10/100 = 0.10$), and let B be the event "the condensor part of the assembly is defective," ($\Pr[B] = 15/300 = 0.05$). Since the choices are from different lots, it seems reasonable to expect A and B to be independent. This can be checked. There are a total of $100 \cdot 300$ possible assemblies, of which $10 \cdot 15$ have both parts defective. Thus

$$\Pr[A \cap B] = \frac{10 \cdot 15}{100 \cdot 300} = 0.005 = (.10)(.05) = \Pr[A]\Pr[B]$$

and, thus, A and B satisfy (2.16) and *are* independent. $\triangle$

By using the concept of independence, we are able to construct many complicated probabilities from simpler ones.

The preceding definition extends to sets of more than two events by generalizing formula (2.16). For k events $A_1, A_2, \ldots, A_k$ to be *independent*, we require not only that

$$\Pr[A_1 \cap A_2 \cap \cdots \cap A_k] = \Pr[A_1]\Pr[A_2]\cdots\Pr[A_k] \tag{2.17}$$

but also that a similar multiplicative formula holds true for every subcollection of the k events.

Notice that, by axiom iii of Section 2, if events are *mutually exclusive*, the probability of their *union* is the *sum* of their probabilities, and by (2.17), if events are *independent*, the probability of their *intersection* is the *product* of their probabilities. One should naturally consider *addition* of probabilities when faced with a set of mutually exclusive events, and *multiplication* of probabilities when faced with a set of independent events.

If events $A_1, A_2, \ldots, A_k$ are such that every pair are independent, then they are said to be *pairwise independent*. It does not follow in this case that the entire set is independent.

EXAMPLE 2.21. Suppose that two coins are tossed. Let A be the event "the first coin is a head," let B be the event "the second coin is a head," and let C be the event "the coins match" (both heads or both tails). Then, $\Pr[B \mid A] = \frac{1}{2} = \Pr[B]$. Hence, A and B are independent events. Also, $\Pr[C \mid A] = \frac{1}{2} = \Pr[C \mid B] = \Pr[C]$. Hence, A and C, as well as B and C, are independent events. The events A, B, and C are *pairwise independent*. But observe that $\Pr[A \cap B \cap C] = \frac{1}{4}$, while

$$\Pr[A]\Pr[B]\Pr[C] = \left(\frac{1}{2}\right)\left(\frac{1}{2}\right)\left(\frac{1}{2}\right) = \frac{1}{8} \neq \Pr[A \cap B \cap C]$$

Thus the events A, B and C are not *independent*, although they are *pairwise independent*. △

One obvious and useful result of independence is the following.

THEOREM. *If events A and B are independent, then so are events A and $\bar{B}$, events $\bar{A}$ and B, and events $\bar{A}$ and $\bar{B}$.* △

PROOF. $A \cap B$ and $A \cap \bar{B}$ are mutually exclusive events whose union is A. Hence

$$\begin{aligned}\Pr[A] &= \Pr[A \cap B] + \Pr[A \cap \bar{B}] \\ &= \Pr[A]\Pr[B] + \Pr[A \cap \bar{B}]\end{aligned}$$

since A and B are independent. It follows that

$$\Pr[A \cap \bar{B}] = \Pr[A](1 - \Pr[B]) = \Pr[A]\Pr[\bar{B}]$$

and the independence of A and $\bar{B}$ follows from (2.16).

The independence of $\bar{A}$ and $\bar{B}$ follows immediately if the events A and B are replaced by $\bar{B}$ and A, while the independence of $\bar{A}$ and B follows by interchanging the roles of A and B. △

This theorem generalizes in obvious fashion to sets of more than two events.

10. Further Examples

EXAMPLE 2.22. Suppose that $(100p)\%$ of the output of a manufacturing process is defective. What is the probability that three randomly chosen parts will have no more than one defective among them?

The probability of no defectives among the three is observed to be

$$(1 - p)^3$$

provided we assume that occurrences of defectives do not depend on defectives appearing in previously chosen parts (that is, the events "the ith chosen part is defective" are independent), since

$$\begin{aligned}\Pr[\text{no defectives}] &= \Pr[\text{part 1 is good}]\Pr[\text{part 2 is good}]\Pr[\text{part 3 is good}] \\ &= (1 - p)^3\end{aligned}$$

Similarly the probability that exactly one is defective is $3p(1 - p)^2$.*

* This obvious fact could be derived formally by using the notations and theory of this section as follows.

Let $\bar{A}_i$ be the event "part i is good" and, thus, A_i is the event "part i is defective." Then the probability of exactly one defect is

$$\Pr[A_1 \cap \bar{A}_2 \cap \bar{A}_3) \cup (\bar{A}_1 \cap A_2 \cap \bar{A}_3) \cup (\bar{A}_1 \cap \bar{A}_2 \cap A_3)]$$

Clearly the events $[A_1 \cap \bar{A}_2 \cap \bar{A}_3]$, $[\bar{A}_1 \cap A_2 \cap \bar{A}_3]$, $[\bar{A}_1 \cap \bar{A}_2 \cap A_3]$ are mutually exclusive. Also the events

$$A_1, A_2, A_3$$

are mutually independent. Hence the probability of exactly one defect is

$$p(1 - p)^2 + (1 - p)p(1 - p) + (1 - p)^2 p = 3p(1 - p)^2$$

using the theorem at the end of Section 9.

(There are three ways of choosing one of the three parts to be the defective one.)

It follows that the probability of no more than one defective is

$$(1-p)^3 + 3p(1-p)^2 \qquad \triangle$$

EXAMPLE 2.23. In Example 2.22 the mutually exclusive cases that make up the required event could easily be enumerated directly. In many situations this is not the case, and we must use counting formulas like the ones given in Section 5. In Example 2.22 suppose that n parts are randomly chosen and that we want the probability of exactly k defectives among these n. Clearly, there are many ways in which exactly k defectives can occur among the n objects, but for *each* of them the probability is given by

$$p^k(1-p)^{n-k} \tag{2.18}$$

assuming that the events "part i defective" are all mutually independent. There are exactly

$$\binom{n}{k} = \frac{n!}{k!(n-k)!}$$

ways in which precisely k of the n parts can be chosen to be the defective ones. Hence the probability of exactly k defectives among the n parts is given by

$$\binom{n}{k} p^k(1-p)^{n-k} \tag{2.19} \triangle$$

EXAMPLE 2.24. In the preceding two examples, the events "the ith part chosen is defective" were mutually independent. However, this is not always the case in practice. Suppose that we have a bin containing N parts, K of which are defective. If n parts are selected at random, and if the first part chosen is defective, this affects the proportion of defectives left in the bin and, hence, affects the conditional probability that the second part chosen will be defective also. The events "the ith part is defective" are dependent, not independent.

The probability of exactly k defectives among n parts chosen can be calculated using the methods of Section 4 (equally likely sample points).

There are a total of $\binom{N}{n}$ equally likely ways of choosing a set of n parts out of the bin, disregarding order.

A set of n parts containing exactly k defectives could be constructed by first choosing k of the K defective parts, $\binom{K}{k}$ ways, and then $n-k$ of the $N-K$ nondefective parts, $\binom{N-K}{n-k}$ ways. Thus, by the counting principle of Section 5, the total number of such sets is $\binom{K}{k}\binom{N-K}{n-k}$.

The probability of k defectives is, therefore, the ratio of the number of sample points containing exactly k defectives to thetotal number of sample points:

$$\frac{\binom{K}{k}\binom{N-K}{n-k}}{\binom{N}{n}} \tag{2.20}$$ △

Formulas (2.19) and (2.20) occur often in quality control procedures. We say that formula (2.19) gives "the probability of exactly k defects in a sample of size n when sampling is done *with replacement* or when sampling from a 'large' population." We say that formula (2.20) gives "the probability of exactly k defects in a sample of size n when sampling is done *without replacement* from a 'finite' population." Here the important distinction is that the events "the ith part chosen is defective" are independent if the sampling procedure does not deplete the population as the sample is drawn (sampling from an infinite population or sampling with replacement) as compared to the case in which the sampling procedure does deplete the population (sampling without replacement from a finite population).

Formula (2.19) is termed the *binomial formula*; formula (2.20) is termed the *hypergeometric formula*.

EXERCISES

1. In Example 2.1 determine the probabilities of the four sample points under the assumption that the correctness of the memories of the three subjects are independent of each other.
2. Four digits are transmitted as a word. In a "noisy" transmission circuit each digit can be transmitted incorrectly with probability 0.2, independent of the transmission of any other digit. Show that each of the events "the ith and jth digits are transmitted correctly, and the kth and lth digits are transmitted

incorrectly" are equally probable with probability .0256, where i, j, k, l is any permutation of 1, 2, 3, 4.

3. Prove that if A, B, and C are events, then

$$\Pr[A \cup B \cup C] \leq \Pr[A] + \Pr[B] + \Pr[C]$$

by using the axioms and the properties developed in Sections 2 and 3.

4. Prove the following special case of property (d), Section 3,

$$\begin{aligned}\Pr[A_1 \cup A_2 \cup A_3] = &\Pr[A_1] + \Pr[A_2] + \Pr[A_3] \\ &- \Pr[A_1 \cap A_2] - \Pr[A_1 \cap A_3] - \Pr[A_2 \cap A_3] \\ &+ \Pr[A_1 \cap A_2 \cap A_3]\end{aligned}$$

by using property (c).

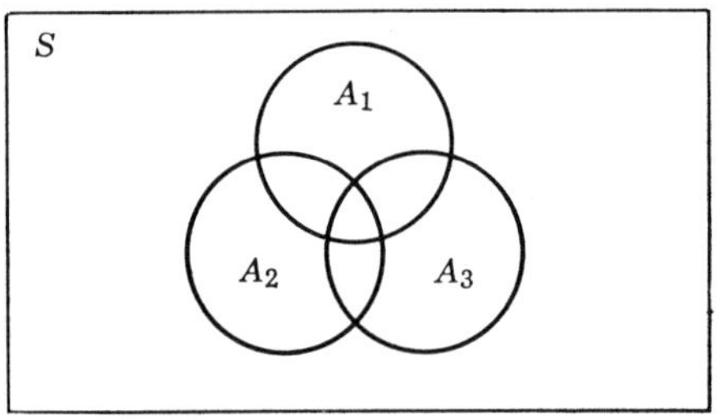

5. Show that for two events, A and B,

$$\Pr[(A \cap \bar{B}) \cup (\bar{A} \cap B)] = \Pr[A] + \Pr[B] - 2\Pr[A \cap B]$$

[The event $(A \cap \bar{B}) \cup (\bar{A} \cap B)$ is the event "exactly one of A and B occurs."]

6. A number x is formed by choosing 6 digits in order, each being either a 0 or a 1, equally likely.
   (a) Describe the sample space and count the number of sample points it contains.
   (b) Compute the probability of each of the following events.
   A: "the number $x \geq 100{,}000$"
   B: "the number x is even"
   C: "no pair of successive digits in x are identical"

7. A number x is formed by arranging the 6 digits: 0, 0, 1, 1, 2, 2 in a random order. Answer the questions in Exercise 6 about this number.

8. An honest coin is tossed until the *second* head appears.
   (a) Describe the same space S and compute the probabilities $\{p_i\}$ of the various sample points.

(b) Compute the probability that an even number of tosses is required. *Hint.* If

$$f(x) = \sum_{k=1}^{\infty} x^{2k-1} = \frac{x}{1-x^2}$$

then

$$f'(x) = \sum_{k=1}^{\infty} (2k-1)x^{2k-2} = \frac{d}{dx}\left(\frac{x}{1-x^2}\right)$$

9. A man, his wife, and four friends enter a row of six seats in a theatre. How many seating arrangements are possible that will allow the wife to sit next to her husband?

10. A man, his wife, and four friends have dinner together at a round table that seats six. How many essentially different seating arrangements are possible that will allow the wife to sit next to her husband?

11. The numbers $\binom{n}{k}$ are termed the binomial coefficients since they occur as the coefficients in the expansion of $(a+b)^n$. By using this interpretation of $\binom{n}{k}$, show that

$$\sum_{k=0}^{n} \binom{n}{k} = 2^n$$

12. Show that

$$\sum_{k=0}^{n} (-1)^k \binom{n}{k} = 0$$

13. Show that

$$\sum_{r=0}^{n} \binom{m}{k-r}\binom{n}{r} = \binom{m+n}{k}$$

where $n \leq k \leq m$.

14. By using Stirlings formula (2.8), show that the number of 13 card hands from a deck of 52 cards is approximately 6.35×10^{11}.

15. An "honest" coin is tossed $2n$ times.

(a) Show that the probability of exactly n heads is $\binom{2n}{n} \Big/ 2^{2n}$ (see Example 2.11).

(b) By using Stirling's formula (2.8), show that the probability in (a) is approximately $1/\sqrt{\pi n}$.

(c) Use (b) with $n = 3$ to approximate the answer in Example 2.11, and compare with the exact value.

Notice that the probability in (a), although this is the most likely case, approaches zero as n increases.

16. If 5 cards are dealt from a deck of 52 cards, compute the probability of getting
    (a) 5 cards of successive denominations, such as 5, 6, 7, 8, 9, etc. (a straight);
    (b) two pairs of different denominations and one odd card.

17. Four items are to be inspected one at a time. Exactly k of the four items are defective and $4-k$ are not. Compute the number of distinguishable arrangements you could have for $k=0, 1, 2, 3, 4$, by direct enumeration.

18. n items are to be inspected, one at a time, of which exactly k are defective and $n-k$ are not. Compute the number of distinguishable arrangements you could have. Check the formula you obtain here with the results of exercise 17.

19. A pocket is known to contain two regular pennies and a two-headed penny. A coin is chosen at random and tossed. If it comes up heads, what is the probability that its other side is also heads?

20. Four urns each contain 3 balls.

    urn 1 contains 3 white balls
    urn 2 contains 2 white and 1 black balls
    urn 3 contains 1 white and 2 black balls
    urn 4 contains 3 black balls

    An urn is chosen at random and a ball chosen from it. If the ball turns out to be white, compute the conditional probability that urn 2 was the urn chosen.

21. In a noisy communication system transmitting binary digits (0's and 1's), the sender may send a digit (for example, 1) and the receiver may receive the other digit (for example, 0). Let A be the event "a 1 is sent." Let B be the event "a 1 is received." Hence $\bar{A}$ is the event "a 0 is sent," and $\bar{B}$ is the event "a 0 is received." Assume that

$$\Pr[\bar{B} \mid A] = .01$$
$$\Pr[B \mid \bar{A}] = .01$$

    Suppose further that $\Pr[A] = .5$. Give a physical interpretation of the above conditional probabilities. Determine the probabilities: $\Pr[A \mid B]$, $\Pr[\bar{A} \mid B]$, $\Pr[A \mid \bar{B}]$, $\Pr[\bar{A} \mid \bar{B}]$. These conditional probabilities are the ones that the receiver might want to know to help him interpret a message he receives.

22. In the design of automobile dashboards, warning lights have replaced gauges. These lights are intended to warn the driver of approaching motor troubles. Hence we would like to have each light turn on when trouble is imminent, with a high probability, and to have the light not turn on when there is no trouble. Suppose T is the event "engine trouble," and O is the event "light turns on." Let $\Pr[O \mid T] = p$ and $\Pr[\bar{O} \mid \bar{T}] = q$. Suppose that $\Pr[T] = .01$. By using Bayes' rule determine the following in terms of p and q.
    (a) $\Pr[T \mid O]$
    (b) $\Pr[T \mid \bar{O}]$

(c) $\Pr[T \cap O]$

(d) $\Pr[T \cap \bar{O}]$

How large should p and q be if the designer wants $\Pr[T \mid O] \geq .99$ and $\Pr[T \mid \bar{O}] \leq .02$?

23. One of three possible hypotheses H_1, H_2, H_3 can "explain" the occurrence of a road accident. The probability of the accident if H_1 is correct is .75. The probability of the accident if H_2 is correct is .80. The probability of the accident if H_3 is correct is .85. H_1 has probability .4 of occurring, H_2 has probability of .5 of occurring; and H_3 has probability of .1 of occurring. Determine the "most probable" hypothesis to explain an observed accident.

24. Suppose that a supplier of material sends raw material that is 1% defective. If we sample 10 parts from a new batch of material and find two defective parts, could we reasonably doubt that this batch is more than 1% defective? Discuss.

25. Suppose that two events, A and B, are mutually exclusive and $\Pr[B] > 0$. Under what conditions will A and B be independent?

26. Suppose that $\Pr[A] > 0$ and $\Pr[B] > 0$, and that A and B are independent. Can A and B be mutually exclusive?

27. Show that every event A is independent of the impossible event $\varnothing$ and of the certain event S.

28. Each of 4 coins has probability p of turning up a head on any toss and probability q of turning up a tail. An experiment consists of tossing the 4 coins. Enumerate the elements of the sample space. Let event E_i be defined as "exactly i heads appear"; $i = 0, 1, 2, 3, 4$. Enumerate each of the events E_i in terms of the sample points. What probabilities would we assign to the events E_i if the coins are considered to be independently tossed?

29. A company makes shafts and bushings. The manufacturing and inspection procedures yield shafts of diameters 0.95, 1.00, 1.05 inches, and bushings of diameters 1.00, 1.05, 1.10 inches.
Define these events.

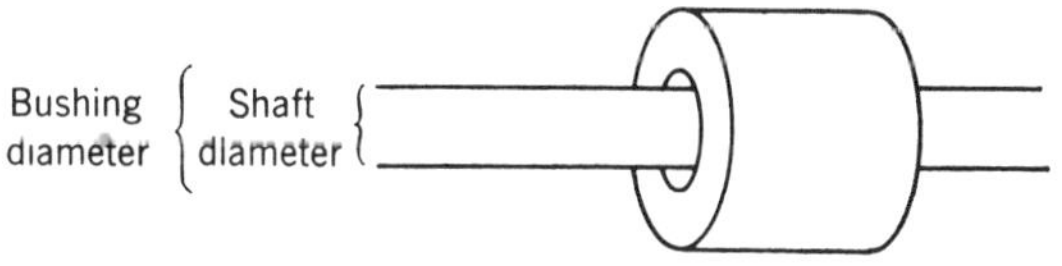

$A_1 =$ "shaft has diameter 0.95"
$A_2 =$ "shaft has diameter 1.00"
$A_3 =$ "shaft has diameter 1.05"
$B_1 =$ "bushing has diameter 1.00"
$B_2 =$ "bushing has diameter 1.05"
$B_3 =$ "bushing has diameter 1.10"

and assume that the quality control procedures were such that

$$\Pr[A_1] = .10, \Pr[A_2] = .70, \Pr[A_3] = .20$$
$$\Pr[B_1] = .25, \Pr[B_2] = .60, \Pr[B_3] = .15$$

A shaft and a bushing are chosen at random and fitted together. If the bushing diameter exceeds the shaft diameter by 0.05 inches, it is said to be a "fit"; if it exceeds by 0.10 inches, it is said to be a "loose fit"; otherwise, it is "no fit." Compute the probabilities of the three events "fit," "loose fit," and "no fit." Assume that the A-events are all independent of the B-events (the shaft and bushing are chosen independently).

30. An electronic component consists of three parts. Each part has probability .99 of performing satisfactorily. The component fails if 2 or more of the parts do not perform satisfactorily. Assuming that the satisfactory performance of one part in no way depends on the performance of the other parts, determine the probability that the component does not perform satisfactorily.

31. A given telephone call is equally likely to occur at any one of the N times $t_1 \leq t_2 \leq \cdots \leq t_N$. Let E_1 be the event "the call occurs after t_j." Let E_2 be the event "the call occurs at or before t_j." Determine each of the following in terms of j, $j = 1, 1, 2, \ldots, N$.
    (a) $\Pr[E_1]$
    (b) $\Pr[E_2]$
    (c) $\Pr[E_1 \cap E_2]$
    (d) $\Pr[E_1 \mid E_2]$

32. Manufacturing Station I produces 10 nondefectives and 3 defectives. Manufacturing Station II produces 3 nondefectives and 5 defectives. Two products from Station I's production bin are randomly chosen and sent to Station II's production bin. One product is then drawn from Station II's bin. Find the probabilities of
    (a) transferring 2 nondefectives
    (b) transferring 1 of each kind
    (c) drawing a defective from Station II's bin

33. A power chain conveyer is a materials-handling device consisting of a power-driven chain with carriers equally spaced along the chain. As the chain moves, the carriers pass by work stations and bring work to the work stations. Suppose the probability that a given carrier is loaded is p, independent of all other carriers. Determine the probability that i full carriers will arrive before the first empty carrier.

34. Using the discussion in exercise 33, determine the probability that k out of the next n carriers will be loaded.

35. An airport shuttle bus makes 10 stops on its rounds. Five people get on the bus at the first stop. Assume that each individual is equally likely to get off at any

one of the remaining stops, and that they act independently. What is the probability that each of the five will get off the bus at a different stop? Answer the same question, assuming that 7 men get on the bus at the first stop.

36. **(a)** An assembly of electronic equipment consists of 3 components arranged as follows.

Each has probability 0.1 of failing under load. However, the entire assembly fails only if the path from A to B is broken. Let A_i be the event "component i fails," and let C be the event "the assembly fails."

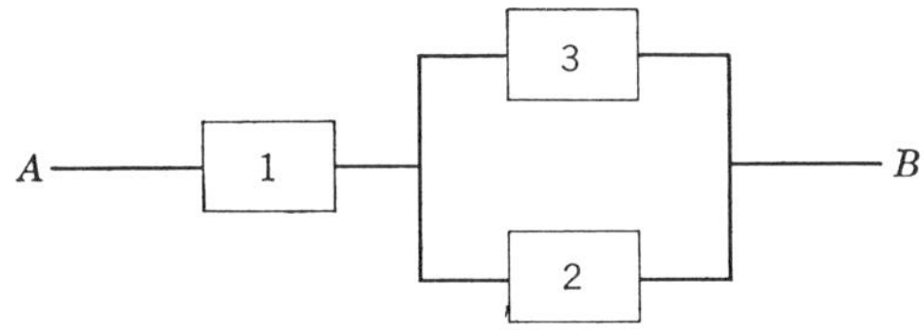

(i) Write C symbolically in terms of A_1, A_2, A_3.

(ii) Compute the probability that the assembly fails. (Assume that component failures are independent.)

(b) If the assembly consists of 5 components arranged as follows each having probability 0.1 of failing, answer the questions given in part (a).

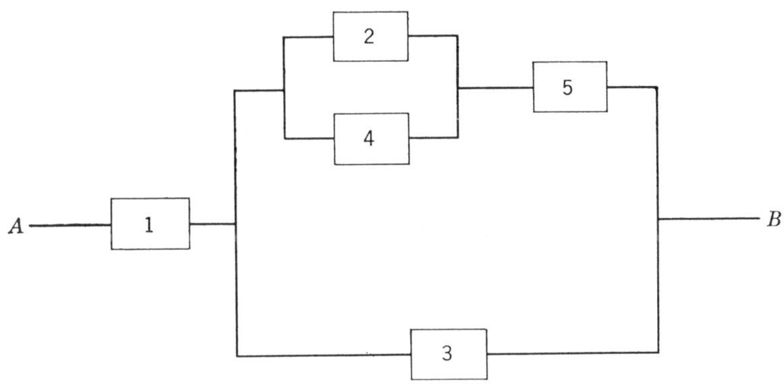

(Assume failures of the individual components are all independent.) (See Exercises 13 and 14 in Chapter 1.)

3 • Random Variables

1. Introduction

At this point in our construction of probability theory we have a complete probability structure. We have discussed a random experiment, a set of outcomes defining the sample space, certain subsets of the outcomes defining events, and an assignment of numbers to each event, which satisfy the axioms of probability. One troublesome point remains that we must attack. Thus far we have treated the sample space as the set of all possible outcomes of an experiment. Consequently, we have dealt with sample spaces such as

$$S = \{1, 2, 3, 4, 5, 6\}$$
$$S = \{\text{head; tails}\}$$

That is, we have observed that some experiments yield sample spaces whose elements are numbers, but that other, quite reasonable, experiments do not naturally yield numerically valued elements. Other examples yielding such nonnumerical sample spaces are:

1. The rocket-firing experiment, where outcomes might be termed "success" or "failure."
2. The quality check experiment whose outcomes might be "good" or "defective."

The important point, of course, is that nothing in our construction requires the sample space to be numerically valued; on the other hand, there is nothing that forbids numerically valued outcomes. For mathematical purposes, this is inconvenient. It is desirable to have numbers associated with the outcomes. Hence, we need one more construction to extend the basic ideas of probability. The construction we now start leads to what is called a *random variable*.

2. The Definition of a Random Variable

Suppose that to each experimental outcome or sample point s we apply some rule of measurement or counting (X) that yields a *number*; for example, choose a person at random from a group and *measure* his height, or choose a sample from a production lot and *count* the number of defective items it contains. Such a *rule* for assigning numbers to sample points is termed a *random variable.*

The terminology used here is traditional, and clearly a random variable is not a variable at all in the usual sense. A more closely related concept is the concept of a *function* in elementary mathematics. Recall that an equation such as

$$y = y(x) = x^2 + 3$$

is said to define y as a function of x, meaning that it specifies a *rule* that for every *number* x assigns a *number* for y. The situation is exactly comparable when we say that the measurement of height is a *rule* that assigns a *number* to each *person.* Thus we have a more precise *definition.*

A RANDOM VARIABLE X *is a function that to each sample point in the sample space S assigns a number* (usually a *real* number).

The use of a single letter, such as X, for this function, in place of $X(s)$ or $X(\,.\,)$, exactly corresponds to the use of a single letter, such as y or f, in place of $y(x)$ or $f(x)$, for a function in calculus. The use of a single letter is logically superior and, in any case, is traditional. If it is necessary to specify the value of the random variable for a *particular* sample point s, we will use the notation $X(s)$. Thus, if X is the height function or random variable, the statement "Joe is 70 inches tall" might be written as

$$\text{“}X(\text{Joe}) = 70\text{”}$$

The sample space S is termed the *domain* of the random variable or function, and the collection of all numbers that are *values* for X is termed the *range* of the random variable. Thus the range is a certain subset of the set of all real numbers.

Notice, in particular, that two or more *different* sample points might give the *same* value for X (two different people might have the same height), but two different numbers in the range cannot be assigned to the same sample point (one person cannot have two different heights) (see Figure 3.1).

For a penny-tossing experiment with only two sample points, we might define the function X as (see Figure 3.2)

$$X(\text{head}) = 0$$

$$X(\text{tail}) = 1$$

Observe that, usually, no unique rule or measurement must be applied. In general, many rules can be assigned to the same sample points. The choice of a

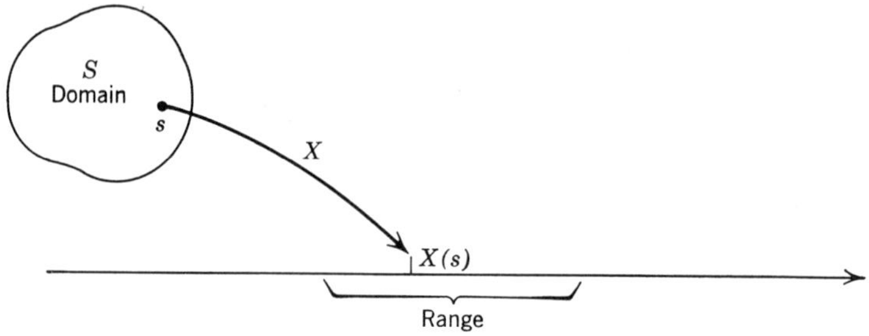

FIGURE 3.1 Random variable X as a function.

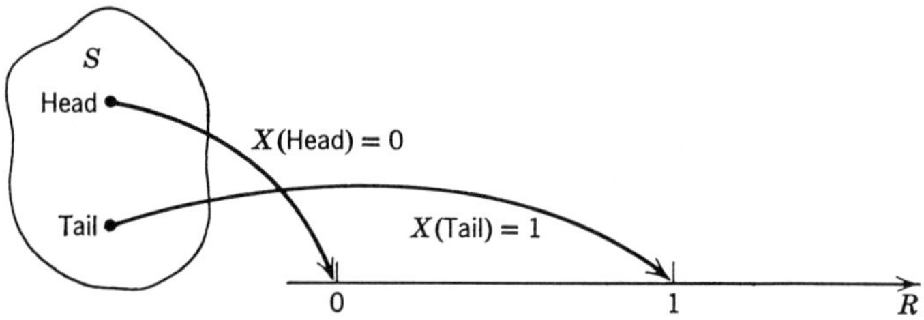

FIGURE 3.2 One random variable associated with coin tossing.

random variable for use in a particular study is not fixed within the framework of probability, and the study itself dictates which one of the possible random variables is most appropriate. There is nothing to keep us from defining X as in Figure 3.2, if this is a useful function. On the other hand, we could just as well choose other functions, say, Y, Z, or U, with

$$Y(\text{heads}) = 1$$
$$Y(\text{tails}) = 0$$

or

$$Z(\text{heads}) = 7 \times 10^6$$
$$Z(\text{tails}) = 4$$

or even

$$U(\text{heads}) = 0$$
$$U(\text{tails}) = 0$$

The student should be aware that, in practice, the motivation for the choice of a random variable is often one of convenience alone, and is dictated by the kind of questions we want to answer.

Henceforth, we shall begin with a given random variable, and shall work with numbers in its range. We shall not often make explicit the sample space from which this random variable arises. This point should not be confusing; and we must be aware that some well-defined sample space lies behind each random variable.

3. Events Defined by Random Variables

In Chapter 2 *probabilities* were defined as numbers assigned to certain subsets of the sample space (events). More formally, *probability* is a set-function —a rule that assigns a number between 0 and 1 to a set of points in S. Its domain is the *set of events* of a random experiment, and its range is contained in the interval [0, 1]. We have learned that a random variable X is also a function whose range is a set of numbers, but whose domain is the set of sample points s making up the sample space S, not subsets of S.

If X is a random variable and x is a fixed real number, we can define the event A_x to be the subset of S consisting of all sample points s to which the random variable X assigns the number x,

$$A_x = \{s : X(s) = x\}$$

As an event, A_x, will have a probability (see Figure 3.3)

$$p = \Pr[A_x]$$

Thus we see that for every number x we can use the random variable X to construct an event or subset A_x of the sample space containing all sample

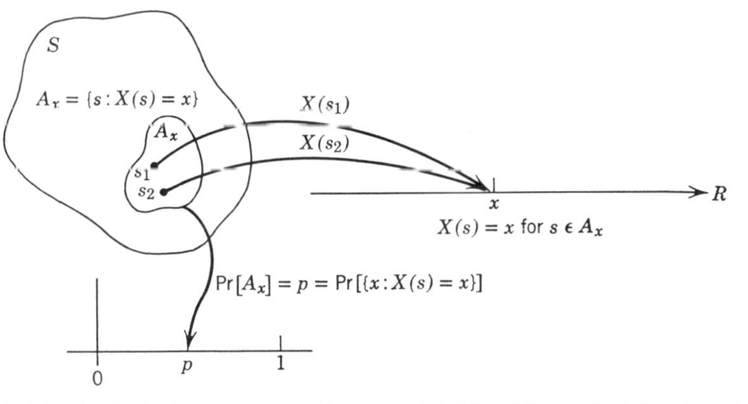

FIGURE 3.3 Relation between a random variable X and its probability function Pr.

points whose X-value is x. This event, A_x, then has a probability $p = \Pr[A_x]$. Generally, we might interpret p as the probability that the random variable X takes on the value x.

More specifically, suppose that S consists of 5 sample points $S = \{s_1, s_2, s_3, s_4, s_5\}$, with individual probabilities $p_1 = \Pr[\{s_1\}]$, $p_2 = \Pr[\{s_2\}]$, $p_3 = \Pr[\{s_3\}]$, $p_4 = \Pr[s_4\}]$, $p_5 = \Pr[\{s_5\}]$, $\sum p_i = 1$. Define the random variable X by $X(s_1) = X(s_3) = 0$, $X(s_2) = X(s_4) = 1$, $X(s_5) = 2$. Thus, $A_0 = \{s : X(s) = 0\} = \{s_1, s_3\}$, and $p = \Pr[A_0] = \Pr[\{s_1, s_3\}] = p_1 + p_3$, etc. (see Figure 3.4). Similarly, $A_1 = \{s_2, s_4\}$, $\Pr[A_1] = p_2 + p_4$, and $A_2 = \{s_5\}$, $\Pr[A_2] = p_5$.

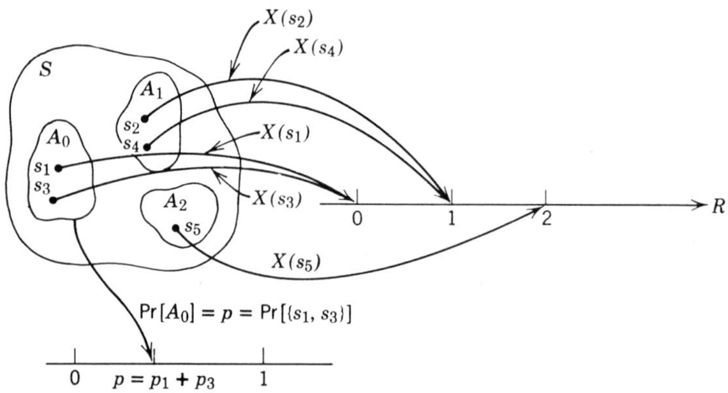

FIGURE 3.4 Relation between a random variable X and its probability function Pr—a specific example.

For simplicity we introduce the notation $[X = x]$ for the set A_x. Thus

$$[X = x] = \{s : X(s) = x\} \tag{3.1}$$

This is a better notation than A_x since it displays the random variable X under consideration as well as the value x. The probability of this set is then

$$\Pr[X = x] = \Pr[\{s : X(s) = x\}] \tag{3.2}$$

This probability is termed *the probability that the random variable X equals x.*

In the preceding example

$$\Pr[X = 0] = p_1 + p_3$$
$$\Pr[X = 1] = p_2 + p_4$$
$$\Pr[X = 2] = p_5$$

Thus, by means of the random variable, we have assigned probabilities to the numbers 0, 1, 2 on the real line, although the original probabilities were only defined for subsets of the abstract set S. We say that the random variable has *induced* probabilities on the real line.

The events $[X = x]$ were defined by (3.1). This convenient notation obviously can be extended to define other types of events in terms of a random variable. For fixed numbers x, a, b, define

$$\begin{aligned} [X \le x] &= \{s : X(s) \le x\} \\ [X > x] &= \{s : X(s) > x\} \\ [a < X < b] &= \{s : a < X(s) < b\}, \text{ etc.} \end{aligned} \tag{3.3}$$

In each case we have defined an event or subset of the abstract sample space S. These events have probabilities that are denoted by

$$\begin{aligned} \Pr[X \le x] &= \text{“the probability that } X \text{ takes a value} \le x\text{”} \\ \Pr[X > x] &= \text{“the probability that } X \text{ takes a value} > x\text{”} \\ &= 1 - \Pr[X \le x] \\ \Pr[a < X < b] &= \text{“the probability that } X \text{ takes a value} \\ &\qquad \text{strictly between } a \text{ and } b\text{”, etc.} \end{aligned}$$

EXAMPLE 3.1. An honest coin is tossed three times. The sample space then consists of eight equally likely sample points: $S = \{HHH, \ldots, TTT\}$. Let X be the random variable that counts the number of heads in each sample point. Thus, X has range $\{0, 1, 2, 3\}$. Table 3.1 lists the value of X

TABLE 3.1

s	$X(s)$		
HHH	3	$[X = 0] = \{TTT\}$,	$\Pr[X = 0] = \frac{1}{8}$
HHT	2	$[X = 1] = \{HTT, THT, TTH\}$	$\Pr[X = 1] = \frac{3}{8}$
HTH	2	$[X = 2] = \{HHT, HTH, THH\}$	$\Pr[X = 2] = \frac{3}{8}$
HTT	1	$[X = 3] = \{HHH\}$	$\Pr[X = 3] = \frac{1}{8}$
THH	2		
THT	1	$[X = x] = \varnothing$ for $x \ne 0, 1, 2, 3$	
TTH	1		
TTT	0		

for each sample point, and for the various events $[X = x]$. Notice that

$$[X \leq 1] = \{TTT, HTT, THT, TTH\} = [X = 0] \cup [X = 1]$$

$$\Pr[X \leq 1] = \Pr[X = 0] + \Pr[X = 1] = \frac{1}{8} + \frac{3}{8} = \frac{1}{2}$$

$$[X \leq 1] \cup [X > 1] = S$$

$$\Pr[X \leq 1] + \Pr[X > 1] = 1$$

$$\Pr[X > 1] = 1 - \Pr[X \leq 1] = \frac{1}{2}$$

In each of the above unions the events are mutually exclusive.

$$[0 < X < 3] = [1 \leq X \leq 2] = \overline{[X = 0] \cup [X = 3]}$$

Hence

$$\Pr[0 < X < 3] = 1 - \Pr([X = 0] \cup [X = 3]) = 1 - \left(\frac{1}{8} + \frac{1}{8}\right) = \frac{3}{4}$$

If we tabulate $[X \leq x]$ and $\Pr[X \leq x]$ for $x = -1, 0, 1, 2, 3, 4$, we observe that each event is included in the following one and, hence, that the probabilities increase (or at least do not decrease) with increasing x (Table 3.2):

$$[X \leq -1] \subset [X \leq 0] \subset [X \leq 1] \subset [X \leq 2] \subset [X \leq 3] \subset [X \leq 4]$$

$$\Pr[X \leq -1] \leq \Pr[X \leq 0] \leq \Pr[X \leq 1] \leq$$
$$\Pr[X \leq 2] \leq \Pr[X \leq 3] \leq \Pr[X \leq 4] \qquad \triangle$$

TABLE 3.2

x	$[X \leq x]$	$\Pr[X \leq x]$
-1	$\varnothing$	0
0	$\{TTT\}$	$\frac{1}{8}$
1	$\{TTT, TTH, THT, HTT\}$	$\frac{4}{8}$
2	$\{TTT, TTH, THT, HTT, HHT, HTH, THH\}$	$\frac{7}{8}$
3	$\{TTT, TTH, THT, HTT, HHT, HTH, THH, HHH\} = S$	1
4	S	1

4. Distribution Functions

If X is a random variable and x is a number, we have defined the event $[X \leq x] = \{s : X(s) \leq x\}$, and its probability $\Pr[X \leq x]$. Clearly, for a given random variable, this event and its probability depend on the value used for x. Thus, $\Pr[X \leq x]$ is a number whose value depends on x. In other words, it is a *function* of x. This function is termed the *distribution function* (or *cumulative distribution function*) of the random variable, and is denoted by $F(x)$ or $F_X(x)$.

$$F(x) = \Pr[X \leq x], \; -\infty < x < \infty \tag{3.4}$$

The distribution function is an ordinary function of the type familiar to every student. Its domain is the set of all real numbers, and its range is a set of numbers between 0 and 1 (since every value of $F(x)$ is a probability). This is in contrast to the random variable X, which is a function whose domain is the sample space, or the probability function Pr, whose domain is a set of events.

Most of the important information about a probability system that refers to measurements given by the random variable X is determined by the behavior of $F(x)$. This function is used extensively in the material that follows. Sometimes we want to compute certain distribution functions, exactly or approximately. Sometimes the distribution function is given, and from it we want to deduce properties of the system, or properties of extensions of the system. When this happens we find that the concepts of sample space, event space, and probability measure, which are of paramount importance in building the theory of probability, often fade into the background and become almost unseen, and functions such as the distribution function become the most important entities. However, we must always keep the background in mind.

EXAMPLE 3.2. Consider the situation of Example 3.1. The table at the end of that example gives $F(x) = \Pr[X < x]$ for $x = -1, 0, 1, 2, 3, 4$. Since X must be an integer, the value of $F(x)$ for x not an integer must be the same as the value for the *nearest* integer next below x. Thus the graph of $F(x)$ will be flat between integers, with jumps at $x = 0, 1, 2, 3$. If x is negative, then $X(s) \leq x$ is impossible, $[X \leq x] = \varnothing$, and $F(x) = 0$. If x exceeds 3, then $X(s) \leq x$ is certain, $[X \leq x] = S$, and $F(x) = 1$. Thus the graph of $F(x)$ is as follows (Figure 3.5). Observe that the graph has jumps at $x = 0, 1, 2, 3$, and that at each jump the *upper* value is the correct value for $F(x)$. $\triangle$

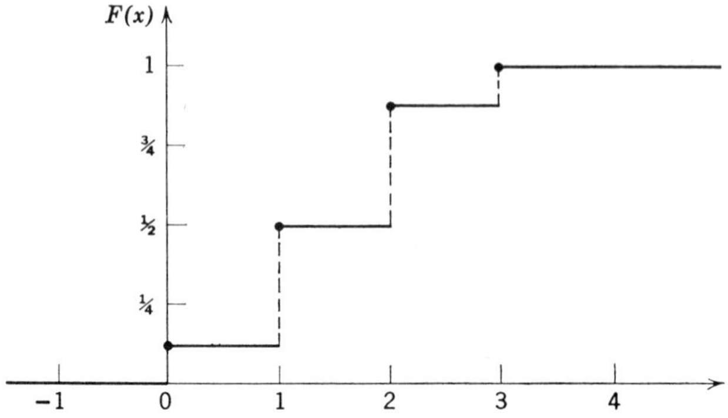

FIGURE 3.5 Distribution function for the number of heads in three coin tosses.

Several properties of $F(x)$ follow directly from its definition (3.4).

$$\text{(a)} \qquad 0 \leq F(x) \leq 1 \qquad \text{for} \qquad -\infty < x < \infty$$

This follows because $F(x)$ is a probability.

If $x_1 \leq x_2$, clearly, $[X \leq x_1] \subset [X \leq x_2]$ since, whenever $X(s) \leq x_1$, necessarily $X(s) \leq x_2$, and it follows that any sample point s in $[X \leq x_1]$ must also be in $[X \leq x_2]$. Thus $\Pr[X \leq x_1] \leq \Pr[X \leq x_2]$, and we have the following property.

$$\text{(b)} \qquad \text{If } x_1 \leq x_2 \qquad \text{then} \qquad F(x_1) \leq F(x_2)$$

This implies that if x increases, then $F(x)$ must also increase (or, at least, not decrease). $F(x)$ is said to be a *nondecreasing* function of x. We observe this phenomenon in Figure 3.5.

In addition

$$\text{(c)} \quad \lim_{x \to +\infty} F(x) = F(\infty) = 1, \qquad \lim_{x \to -\infty} F(x) = F(-\infty) = 0$$

Intuitively these formulas are clear since it would seem that, as $x \to +\infty$, $F(x) = \Pr[X \leq x] \to \Pr[X < +\infty] = \Pr[S] = 1$ and, as $x \to -\infty$, $F(x) = \Pr[X \leq x] \to \Pr[X \leq -\infty] = \Pr[\varnothing] = 0$, since $-\infty < X(s) < \infty$ for every $s \in S$ (see Figure 3.5).

A rigorous proof of (*c*) requires use of axiom iii for probability (Chapter 2, Section 2). Events of the form $[i - 1 < X \leq i]$ are mutually exclusive for all integers i, and

$$S = \cdots \cup [-1 < X \leq 0] \cup [0 < X \leq 1] \cup \cdots \cup [n - 1 < X \leq n] \cup [n < X \leq n + 1] \cup \cdots$$

$$[X \leq n] = \cdots \cup [-1 < X \leq 0] \cup [0 < X \leq 1] \cup \cdots \cup [n - 1 < X \leq n]$$

Thus

$$
\begin{aligned}
1 = \Pr[S] &= \sum_{i=-\infty}^{+\infty} \Pr[i-1 < X \le i] \\
&= \lim_{n \to +\infty} \sum_{i=-\infty}^{n} \Pr[i-1 < X \le i] \\
&= \lim_{n \to +\infty} \Pr[X \le n] \\
&= \lim_{n \to +\infty} F(n)
\end{aligned}
$$

Thus, $\lim_{n \to +\infty} F(n) = 1$. The case where x is not an integer is accounted for by considering the largest integer below x. The proof that $\lim_{x \to -\infty} F(x) = 0$ is similar.

A fourth property of $F(x)$ is more delicate. Suppose that x is made to approach a fixed value x_0 *from the right*; that is, $x \to x_0$ and $x \ge x_0$. We denote this by the symbol $x \to x_0 +$. Then, clearly, any value $X(s)$ of the random variable will be $\le x_0$ if, and only if, it is $\le x$ for each $x \ge x_0$. Using an argument similar to the one used in the preceding paragraph, it follows that $F(x) \to F(x_0)$ as $x \to x_0 +$. Thus we have the property:

(d) $$\lim_{x \to x_0^+} F(x) = F(x_0)$$

We say that $F(x)$ is *continuous on the right*. A similar property for approach from the left does not hold true in general, since we have seen in Figure 3.5 that $F(x)$ may have *jumps* and, thus, may not be continuous from both sides. If a jump does occur at x_0, then $F(x_0)$ is always the *upper* value.

A possible graph for $F(x)$ is shown in Figure 3.6. It can be shown that any function $F(x)$ satisfying properties (a) to (d) is the distribution function of *some* random variable.

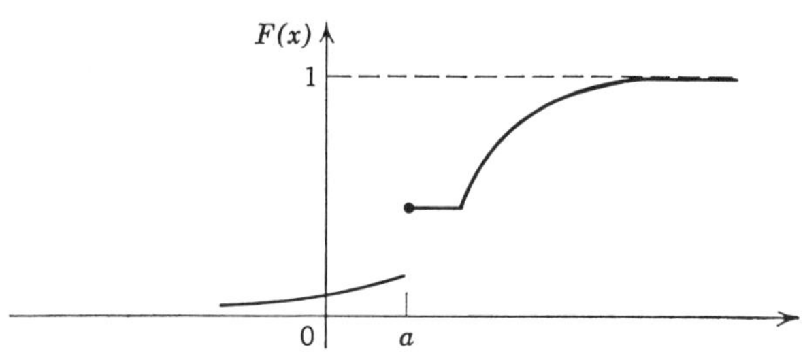

FIGURE 3.6 A possible distribution function.

EXAMPLE 3.3. The function

$$F(x) = \begin{cases} 0 & \text{if} \quad x < 0 \\ x + \dfrac{1}{2} & \text{if} \quad 0 \le x \le \dfrac{1}{2} \\ 1 & \text{if} \quad x > \dfrac{1}{2} \end{cases}$$

can be graphed as shown in Figure 3.7. We easily see that $F(x)$ has the properties given by (a) to (d). Hence $F(x)$ is the distribution function of some random variable X. Notice that $F(x)$ has a jump at $x = 0$, and that $F(0) = \frac{1}{2}$, the *upper* value. This implies that $\Pr[X = 0] = \frac{1}{2}$, $\Pr[X < 0] = 0$. △

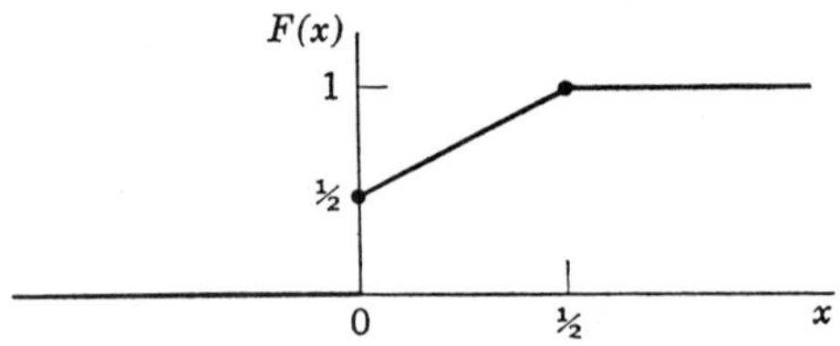

FIGURE 3.7 Graph of $F(x)$ in Example 3.3.

EXAMPLE 3.4. The function

$$F(x) = \begin{cases} 0 & \text{if} \quad x \le 0 \\ 1 - e^{-ax} & \text{if} \quad 0 < x < \infty \end{cases}$$

can be graphed as in Figure 3.8.

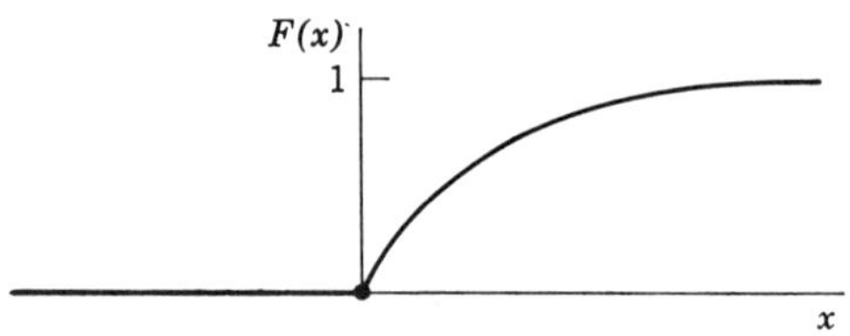

FIGURE 3.8 Graph of $F(x)$ in Example 3.4.

We readily observe that $F(x)$ has the properties given by (a) through (d). This distribution function occurs in many applications. We shall learn in Chapter 7 that, under some reasonable assumption concerning telephone calls, the random variable $X =$ the time between two telephone calls at an exchange will have the *negative exponential distribution* function of this example. △

EXAMPLE 3.5. The function

$$F(x) = \sum_{j \leq x} \binom{n}{j} p^j q^{n-j}$$

where $0 \leq p \leq 1$, $0 \leq q \leq 1$, $p + q = 1$, and n is a fixed positive integer, can be graphed as shown in Figure 3.9. It is clear from this figure that $F(x)$ has the properties given by (1) to (4). We shall learn that this probability distribution function is useful in many applications that involve counting. △

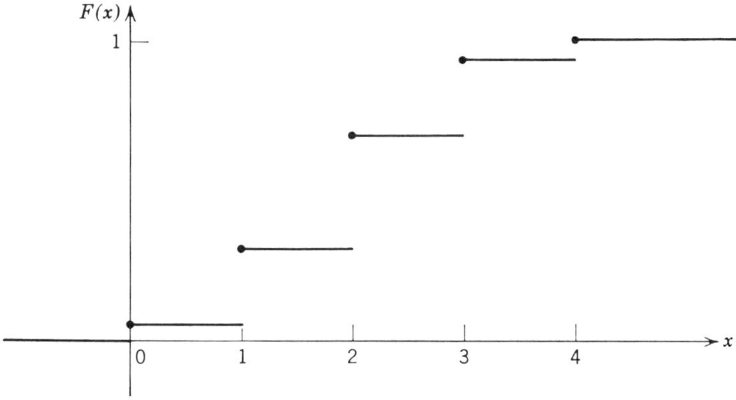

FIGURE 3.9 The distribution function $F(x) = \sum_{j<x} \binom{n}{j} p^j q^{n-j}$; $p = q = \frac{1}{2}$, $n = 4$.

We notice that in Example 3.5 the points of increase of $F(x)$ occur only at integers. If x is not an integer, $F(x)$ has the same value that it had at the integer nearest x on the left. That is, $F(x)$ proceeds from 0 to 1 in a series of jumps. Except at these jumps, $F(x)$ is constant. In Example 3.3 we observed that $F(x)$ had one jump at 0 but, except for that jump, it proceeded "smoothly" from 0 to 1 as x proceeded from left to right. At $x = \frac{1}{2}$, there was a point at which $F(x)$ was continuous but not differentiable. In Example 3.4 we determined that $F(x)$ was continuous everywhere, but that at the point $x = 0$ it was not differentiable. More complicated examples could be constructed, but the ones above illustrate the following point.

It can be shown, by using some deeper analysis, that every probability distribution function can be decomposed into a weighted sum of three parts in the form

$$F(x) = F_1(x) + F_2(x) + F_3(x)$$

where $F_1(x)$ is a function that changes only by jumps (as in Example 3.5, but the jumps need not occur only at the integers), and $F_2(x)$ is continuous (as for Example 3.4). The function $F_3(x)$ is termed *singular*. Singular functions rarely occur in distribution functions of practical interest, and henceforth we assume that the distribution functions discussed have no singular part, that is, we assume that $F_3(x) \equiv 0$. If $F_2(x) \equiv 0$, that is, if the function proceeds only by jumps, we say that the random variable X is a *discrete random variable*. If $F_1(x) \equiv 0$, we say that the random variable X is a *continuous random variable*, although in Section 7 we place additional restrictions on X to make formal manipulations easier. Otherwise, the random variable X is said to be of *mixed type*. In most applied work the random variable is either discrete or continuous. We study only these two cases in detail. There are cases of useful random variables that are of mixed type, but in elementary consideration they arise so infrequently that no special discussion is necessary. Usually, they can be analyzed easily as special cases.

5. Determining Probabilities from the Distribution Function

We observed in Section 4 that $F(x)$ gives probabilities of the form $\Pr[X \le x]$. But we also want to compute other probabilities, such as $\Pr[a < X \le b]$ or $\Pr[X > a]$. We accomplish this as follows (see Example 3.1).

(a) Since

$$[X \le a] \cup [X > a] = S$$

we have

$$\Pr[X \le a] + \Pr[X > a] = \Pr[S] = 1$$

by using axiom iii for probability (Chapter 2, Section 2). Thus

$$\Pr[X > a] = 1 - \Pr[X \le a] = 1 - F(a)$$

(b) Since

$$[X \le a] \cup [a < X \le b] = [X \le b]$$

we have

$$\Pr[X \le a] + \Pr[a < X \le b] = \Pr[X \le b]$$

Thus

$$\Pr[a < X \le b] = \Pr[X \le b] - \Pr[X \le a]$$
$$= F(b) - F(a)$$

Therefore, $F(x)$ determines probabilities for all intervals of the form $[a < X \le b]$. We also need to explore probabilities of the form $\Pr[X = a]$. This is done in Sections 6 and 7 which follow.

EXAMPLE 3.6. Using the probability distribution function given in Example 3.3, we determine that

$$\Pr\left[X \le \frac{1}{4}\right] = F\left(\frac{1}{4}\right) = \frac{3}{4}$$

$$\Pr\left[0 < X \le \frac{1}{4}\right] = F\left(\frac{1}{4}\right) - F(0) = \frac{1}{4}$$

$$\Pr\left[X > \frac{1}{4}\right] = 1 - F\left(\frac{1}{4}\right) = \frac{1}{4}$$

$$\Pr\left[X \le \frac{1}{8}\right] = \frac{5}{8}$$

$$\Pr\left[X > \frac{1}{2}\right] = 1 - F\left(\frac{1}{8}\right) = \frac{3}{8} \quad \triangle$$

6. Discrete Random Variables

In Section 4 we learned that a discrete random variable can be characterized as one whose distribution function $F(x)$ proceeds in jumps (see Figure 3.9 and the discussion that follows). The only nonzero contributions to $F(x)$ occur at some discrete set of points $x_0, x_1, x_2 \dots$ (which, in the example, were integers but, in general, need not be). Therefore, probabilities of nonzero value must all be contributed by the sets of the form:

$$[X = x_i], \qquad i = 0, 1, 2, \dots$$

Thus we can define a function that takes nonzero values only at the x_i. We let the value it takes at each x_i be the amount that $F(x)$ jumps at that point. That is, we define a new function

$$p(x) = \Pr[X = x] \tag{3.5}$$

called the *probability density function* for the discrete random variable X. Clearly, $p(x)$ will have these properties:

(a) $p(x) = 0$ *unless x is one of* $x_0, x_1, x_2, \ldots,$
(b) $0 \le p(x_i) \le 1$, for each x_i in the range, and
(c) $\sum_i p(x_i) = \sum_i \Pr[X = x_i] = 1.$

where the sums are to be taken over the entire range of X.

The functions $F(x)$ and $p(x)$ are related by the formula

$$F(x) = \Pr[X \le x] = \Pr[X = x_i \text{ for some } x_i \le x]$$
$$= \sum_{x_i \le x} \Pr[X = x_i]$$
$$F(x) = \sum_{x_i \le x} p(x_i) \tag{3.6}$$

where the summation is extended over all indices i for which $x_i \le x$.

This relation is illustrated in Figure 3.10. From (3.5) and (3.6) we observe that $p(x)$ completely determines probabilities for events of the form:

$$[X \le a],\ [X < a],\ [X > a],\ [X \ge a]$$
$$[X = a],\ [a < X < b],\ [a \le X < b],\ [a < X \le b],\ [a \le X \le b]$$

We also notice, from (3.6), that the probabilities for these events are defined completely either by the probability distribution function $F(x)$ or by the probability density function $p(x)$, when X is a discrete random variable. Hence, in practical application, we work with either $F(x)$ or $p(x)$, whichever is the easier one to use.

In cases where the range of X consists of a set of integers, we shall frequently use the notation p_x or p_i, in place of $p(x)$ or $p(i)$, for the probability density function. Thus, $p_i = \Pr[X = i]$ for i an integer, $0 \le p_i \le 1$, $\sum p_i = 1$, and $F(x) = \sum_{i \le x} p_i$.

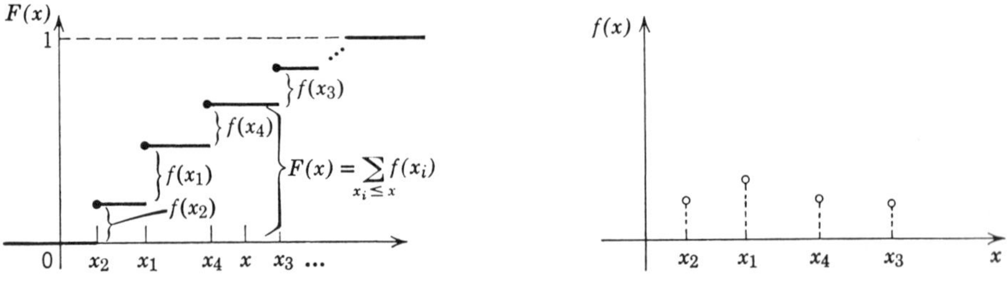

FIGURE 3.10 Relation between the distribution function $F(x)$ and the probability density function $f(x)$ for a discrete random variable.

A useful analogy in studying the probability distribution induced on the real line by a random variable is the analogy of a *mass distribution.* If we assume that a total mass of 1 unit is spread along the real line in such a way that the *probability* associated with any set of points is just the *mass* of the set, then probabilities can be interpreted as masses. With this interpretation the distribution function $F(x)$ is just the mass of the part of the line lying to the left of x (and including x). A *discrete* distribution is one in which the mass distribution consists entirely of *point masses* at the points x_0, x_1, x_2, ... The probability density function $p(x)$, in this case, only gives the mass at the point and, hence, is zero unless x is one of the points x_0, x_1, x_2, ..., and $p(x_i)$ is the mass attached to the point x_i (see Figure 3.11).

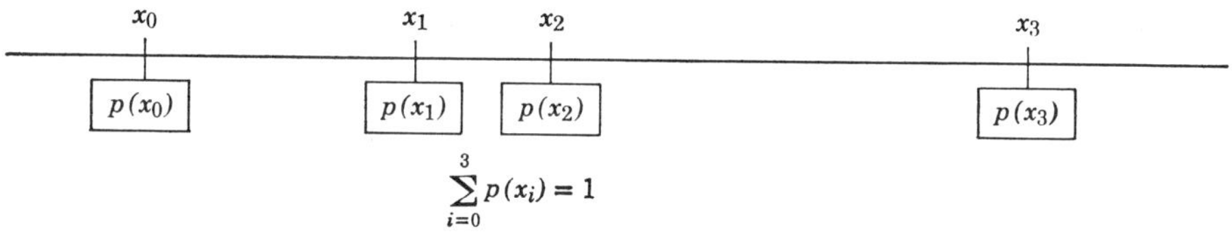

FIGURE 3.11 A physical interpretation of a probability mass function.

We find such a physical model frequently useful for suggesting many operations in probability theory analogous to well-known formulas in mechanics.

EXAMPLE 3.7. In Example 3.2, in which X represents the number of heads in 3 tosses of a coin, the distribution function $F(x)$ is given by Figure 3.5. The corresponding probability density function (see Figure 3.12) is given by

$$p_0 = p(0) = \Pr[X = 0] = \frac{1}{8}$$

$$p_1 = p(1) = \Pr[X = 1] = \frac{3}{8}$$

$$p_2 = p(2) = \Pr[X = 2] = \frac{3}{8}$$

$$p_3 = p(3) = \Pr[X = 3] = \frac{1}{8}$$

and

$$p(x) = 0 \qquad \text{for} \qquad x \neq 0, 1, 2, 3$$ △

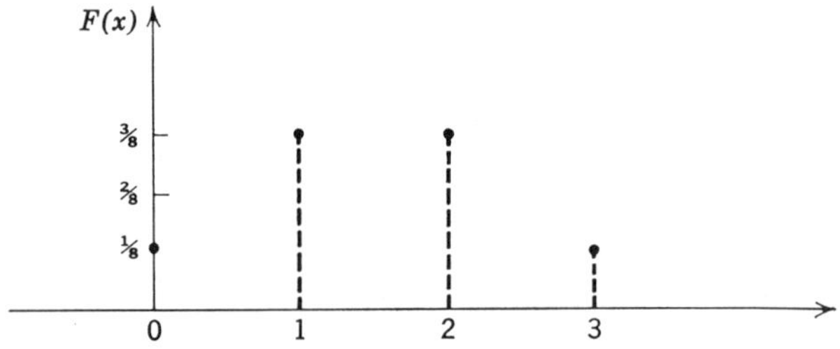

FIGURE 3.12 Probability density function for the number of heads in three coin tosses.

EXAMPLE 3.8. The preceding example can be generalized as follows. Suppose that an honest coin is tossed n times and let $X =$ the number of heads observed. Then X is a discrete random variable with range: 0, 1, 2, ..., n. If i is an integer in this range, $0 \leq i \leq n$, then

$$p_i = p(i) = \Pr[X = i] = \Pr[i \text{ heads in } n \text{ tosses}]$$

By the results of Section 5 of Chapter 2, we know that our sample space S contains 2^n equally likely sample points (sequences of n tosses). The event $[X = i]$ can occur in as many ways as we can choose i of the n tosses to be heads, namely, $\binom{n}{i}$ ways. (The remaining tosses can all be tails in only one way.) Thus

$$p_i = \binom{n}{i} \Big/ 2^n = \binom{n}{i} \frac{1}{2^n} \tag{3.7}$$

for $i = 0, 1, 2, \ldots, n$. If i has any value other than these values, $p_i = 0$.

For each real x,

$$F(x) = \sum_{i \leq x} p_i = \frac{1}{2^n} \sum_{i \leq x} \binom{n}{i}$$

$F(x)$ will be a step-function with jumps at $x = 0, 1, \ldots, n$, and holding constant between jumps. $F(x) = 0$ for $x < 0$, $F(x) = 1$ for $x \geq n$. This function is sketched for $n = 4$ in Figure 3.9. △

EXAMPLE 3.9. In Example 2.3 the example of tossing a coin until the first head appeared was discussed. If the random variable X represents the number of tosses required, then, from the results of that section, we observe that X is a discrete random variable whose range is the set of all positive integers 1, 2, 3, ... (an infinite but countable discrete sequence), and whose probability density function is given by

$$p_i = \Pr[X = i] = \frac{1}{2^i} \qquad \text{for} \qquad i = 1, 2, 3, \ldots$$

Clearly, this set $\{p_i\}$ does define a probability density function, since

$$p_i = 0 \qquad \text{unless} \qquad i = 1, 2, \ldots$$
$$0 \leq p_i \leq 1 \qquad \text{for} \qquad i = 1, 2, \ldots$$
$$\sum_{i=1}^{\infty} p_i = 1$$

The last point can be determined easily by remembering that, for $|r| < 1, \sum_{i=0}^{\infty} ar^i = a/(1-r)$ (the sum formula for an infinite geometric series). Thus

$$\sum p_i = \sum \frac{1}{2^i} = \frac{1/2}{1 - 1/2} = 1 \qquad \triangle$$

7. Continuous Random Variables

We noted in Section 5 that a random variable X is said to be continuous if its distribution function is continuous. This definition, although usually used in the advanced theory of probability, would force us to adopt tools of analysis beyond the scope of this book to study such distributions. Thus we shall restrict our definition of a continuous random variable. In so doing, we shall not seriously restrict the set of random variables available to us for application, but shall simplify the necessary mathematics. Thus, for our purposes, we use the following:

DEFINITION. *A random variable X is* CONTINUOUS *if its distribution function $F(x)$*

(a) *is continuous,*

(b) *has a derivative $f(x) = (d/dx)F(x)$ for all values of x, with possibly a finite number of exceptions, and*

(c) *this derivative is piecewise continuous.* (3.8)

Under these circumstances, the range of X will consist of one or more intervals, finite or infinite.

In this case we readily observe that probabilities of the form $\Pr[X = x]$ are all zero:

$$\Pr[X = x] = 0 \tag{3.9}$$

This follows from the fact that $\Pr[X=x] \le \Pr[x-h < X \le x] = F(x) - F(x-h)$; and this difference approaches 0 as $h \to 0+$ if $F(x)$ is continuous. It follows that formula (3.5) is not an appropriate definition for the probability density function here. It is more useful to try to generalize formula (3.6). The natural generalization of a sum here is an integral. Thus we define the *probability density function*, $f(x)$, in this case to be the derivative given in property (b)

$$f(x) = \frac{d}{dx} F(x)$$

Then formula (3.6) can be generalized by

$$\begin{aligned} F(x) &= F(x) - F(-\infty) \\ &= \int_{-\infty}^{x} \frac{d}{dt} F(t)\, dt \\ &= \int_{-\infty}^{x} f(t)\, dt \end{aligned} \tag{3.10}$$

The analogy with (3.6) is clear, with the sum being replaced by an integral. We notice from (3.10) that the distribution function of a continuous random variable is determined once its density function is given, and conversely.

By using the result of Section 5(b) we have

$$\Pr[a < X \le b] = F(b) - F(a) = \int_a^b f(t)\, dt \tag{3.11}$$

We recall that $\Pr[X = b] = 0$, by (3.9), and thus the left side of (3.11) can be written as $\Pr[a \le X \le b]$ or $\Pr[a < X < b]$ with equal validity.

The function $f(x) = F'(x)$ has these properties:

(i) $f(x) = 0$ *if x is not in the range of X, (since outside of this range $F(x)$ is constant and its derivative is* 0).

(ii) $f(x) \ge 0$ *for all x (since $F(x)$ is a nondecreasing function).*

(iii) $\int_{-\infty}^{\infty} f(x)\, dx = 1$ [*by letting* $x \to +\infty$ *in* (3.10)].

(iv) $f(x)$ *is piecewise continuous.*

Any function that has these properties is the probability density function for *some* continuous random variable.

The connection between the functions f and F is illustrated in Figure 3.13. *Probabilities* are represented by *areas* under the f curve. The total area under the f curve is 1.

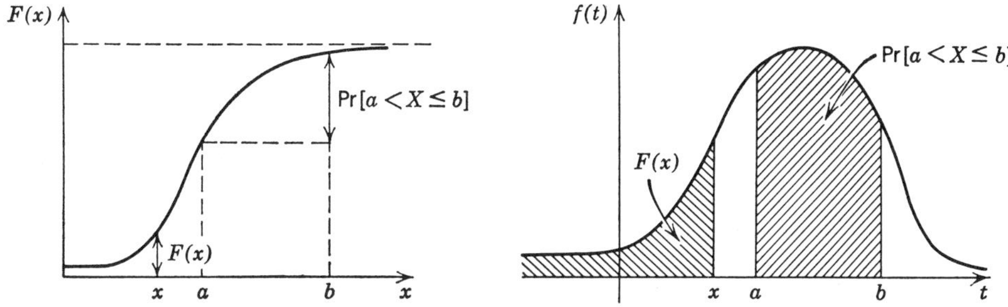

FIGURE 3.13 The determination of probabilities for the event $[a < X \leq b]$ from the distribution function $F(x)$ and the density function $f(t)$.

An alternate interpretation again is to consider the probability distribution as a one-dimensional *mass distribution* of total mass 1 distributed along the x-axis. The probability associated with any interval is then equal to its mass; $F(x)$ is the mass of the interval from $-\infty$ to x; $f(x)$ is the mass density (mass per unit length) at the point x.

Since any actual measuring device can have only a finite number of possible readings, no real-world experiment and measurement can possibly yield a continuous random variable. Nonetheless, even for practical applications, it is convenient to have such random variables in our mathematical model. They are frequently much easier to handle analytically, and we shall observe that often a simple continuous distribution can be used to approximate a complicated discrete distribution (see Figure 3.14).

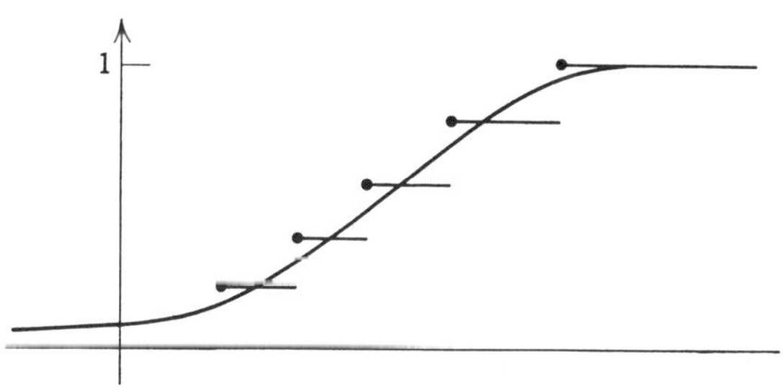

FIGURE 3.14 Continuous approximation to a discrete distribution function.

EXAMPLE 3.10. The function

$$f(x) = \begin{cases} 2x, & \text{if} \quad 0 \leq x \leq 1 \\ 0, & \text{otherwise} \end{cases}$$

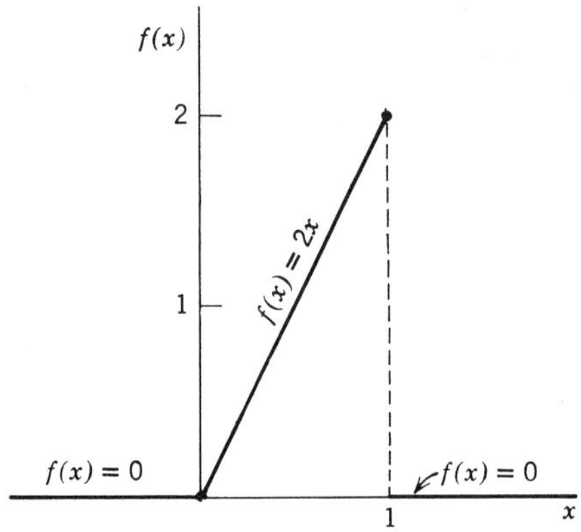

FIGURE 3.15 The probability density function

$$f(x) = \begin{cases} 0 & \text{if} \quad x \leq 0 \\ 2x & \text{if} \quad 0 < x \leq 1 \\ 0 & \text{if} \quad x > 1 \end{cases}$$

is a probability density function satisfying (i) to (iv) above, since $f(x) \geq 0$, for all x, $\int_{-\infty}^{\infty} f(x)\,dx = \int_{-\infty}^{0} 0\,dx + 2\int_{0}^{1} x\,dx + \int_{1}^{\infty} 0\,dx = 1$, and $f(x)$ is piecewise continuous with a discontinuity at $x = 1$.

A graph of $f(x)$ is given in Figure 3.15.

The probability distribution function corresponding to this density function is, by (3.10),

$$F(x) = \int_{-\infty}^{x} f(x)\,dx = \begin{cases} 0, & \text{if} \quad x < 0 \\ x^2, & \text{if} \quad 0 \leq x \leq 1 \\ 1, & \text{if} \quad x > 1 \end{cases}$$

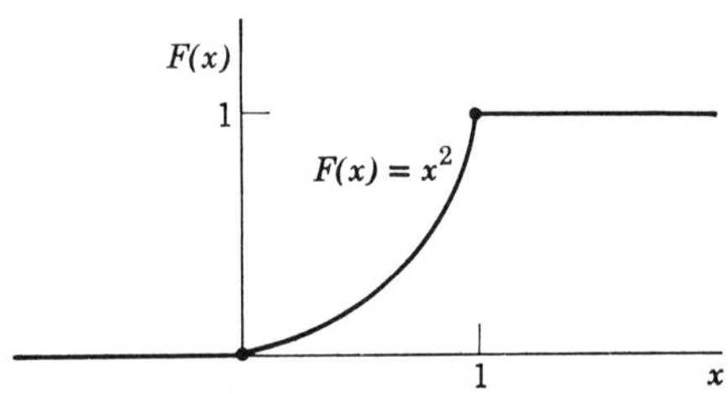

FIGURE 3.16 The probability distribution function

$$F(x) = \begin{cases} 0 & \text{if} \quad x \leq 0 \\ x^2 & \text{if} \quad 0 < x \leq 1 \\ 1 & \text{if} \quad x \geq 1 \end{cases}$$

A graph of $F(x)$ is given in Figure 3.16. From these two functions we can determine probabilities such as

$$\Pr\left[X \leq \frac{1}{2}\right] = F\left(\frac{1}{2}\right) = \int_0^{1/2} 2x\,dx = \frac{1}{4}$$

$$\Pr\left[\frac{1}{4} < X \leq \frac{3}{4}\right] = \int_{1/4}^{3/4} 2x\,dx = F\left(\frac{3}{4}\right) - F\left(\frac{1}{4}\right) = \frac{1}{2}$$

Notice that, since $\Pr[X = b] = 0$, we obtain immediately

$$\Pr\left[\frac{1}{4} < X \leq \frac{3}{4}\right] = \Pr\left[\frac{1}{4} \leq X \leq \frac{3}{4}\right] = \Pr\left[\frac{1}{4} \leq X < \frac{3}{4}\right] = \Pr\left[\frac{1}{4} < X < \frac{3}{4}\right]$$

and in all cases this probability is $\frac{1}{2}$. △

EXAMPLE 3.11. The *negative exponential* distribution function

$$F(x) = \begin{cases} 1 - e^{-ax}, & \text{if} \quad 0 \leq x < \infty \\ 0 & \text{if} \quad x < 0 \end{cases}$$

discussed in Example 3.4, is the distribution function of a continuous random variable. The corresponding probability density function

$$f(x) = F'(x) = \begin{cases} ae^{-ax}, & \text{if} \quad 0 \leq x < \infty \\ 0, & \text{if} \quad x < 0 \end{cases}$$

is graphed in Figure 3.17.

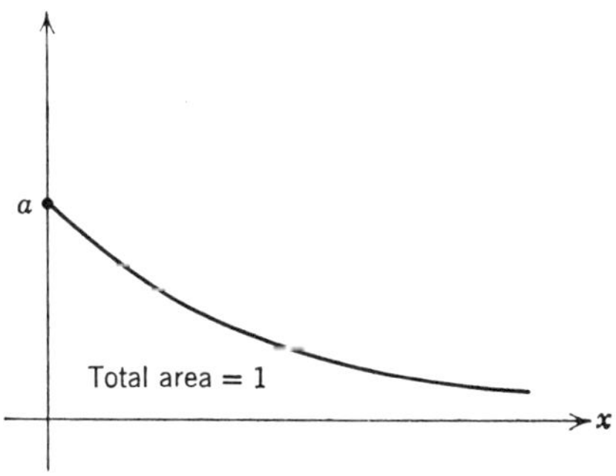

FIGURE 3.17 The negative exponential probability density function

$$f(x) = \begin{cases} ae^{-ax}, & x > 0 \\ 0, & x \leq 0 \end{cases}$$

Observe that, in this case,

$$\Pr[X > x] = \int_x^\infty f(t)\, dt = \int_x^\infty ae^{-at}\, dt$$

$$= -e^{-at}\Big|_x^\infty = e^{-ax}$$

for $x \geq 0$. $\triangle$

EXAMPLE 3.12. The function

$$F(x) = \begin{cases} 1 - e^{-x^b}, & \text{if} \quad x \geq 0 \\ 0, & \text{if} \quad x < 0 \end{cases}$$

where b is a positive constant, is a continuous distribution function. Its derivative

$$f(x) = \begin{cases} bx^{b-1}e^{-x^b}, & \text{if} \quad x > 0 \\ 0, & \text{if} \quad x < 0 \end{cases}$$

exists for all $x \neq 0$ and is at least piecewise continuous. Hence $f(x)$ is the density function, termed a *Weibull density function*, for some continuous random variable. This probability density function has found widespread application in the area of mechanical reliability theory (see Exercises 25 and 26 for a further discussion). $\triangle$

EXAMPLE 3.13. One of the most important probability density functions in the field of statistics is the *normal density function*. We shall discuss it in more detail in Chapter 6. We merely note here that the density function is given by the formula:

$$f(x) = \frac{1}{\sqrt{2\pi}\,\sigma} e^{-(x-\mu)^2/2\sigma^2}, \quad -\infty < x < \infty$$

In Chapter 6 we show that $f(x)$ is a proper probability density function for a continuous random variable. The constants μ, σ^2 represent two parameters that must be specified. In the special case $\mu = 0$, $\sigma = 1$, we speak of the *standard normal* probability density function. The graph of $f(x)$ is a bell-shaped curve symmetric about $x = \mu$ as shown in Figure 3.18.

Because of its wide applicability in describing real-life phenomena and, perhaps more important, because of the Central Limit Theorem (Chapter 7, Section 5), the normal probability density function is one of the most useful continuous densities in probability and statistical theory.

The normal probability *distribution* function cannot be expressed simply in closed form. Of course, it is given by

$$F(t) = \int_{-\infty}^t \frac{1}{\sqrt{2\pi}\,\sigma} e^{-(x-\mu)^2/2\sigma^2}\, dx$$

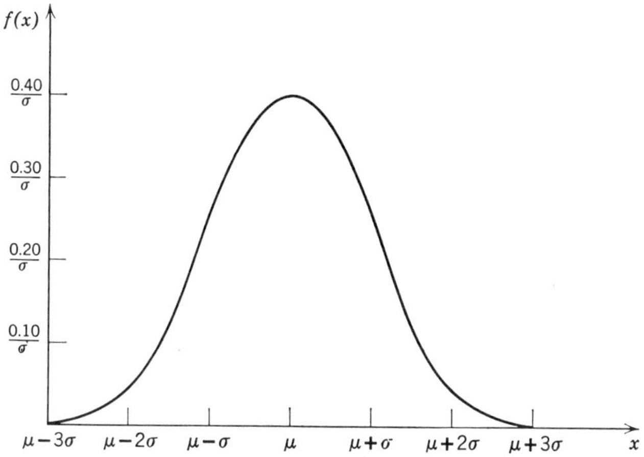

FIGURE 3.18 The normal probability density function

$$f(x) = \frac{1}{\sqrt{2\pi}\,\sigma} e^{-(x-\mu)^2/2\sigma^2}$$

which is a nonelementary integral. By the simple change of variable

$$z = \frac{x-\mu}{\sigma}$$

the computation of $F(t)$ can be transformed to the computation of the distribution function of a *standard* normal distribution ($\mu = 0,\ \sigma = 1$), which is tabulated in every elementary statistics book. A short table (Table 3.3) is given here.

A few examples will illustrate the use of this table in determining probabilities of interest.

First, suppose that $\mu = 0$, $\sigma = 1$, so that X itself is a standard normally distributed random variable. Then

(a) $\Pr[X \le 0] = F(0) = .5000$

Since X is a continuous random variable, we also have

$$\Pr[X < 0] = .5000$$

(b)
$$\begin{aligned}\Pr[X > 1] &= 1 - \Pr[X \le 1]\\ &= 1 - F(1)\\ &= 1 - .8413\\ &= .1587\end{aligned}$$

TABLE 3.3[a]

t	$F(t)$	t	$F(t)$	t	$F(t)$
0.0	0.5000	1.1	0.8643	1.96	0.9750
0.1	0.5398	1.2	0.8849	2.0	0.9772
0.2	0.5793	1.3	0.9032	3.0	0.9987
0.3	0.6179	1.333	0.9088	4.0	0.9999+
0.333	0.6306	1.4	0.9192		
0.4	0.6554	1.5	0.9332		
0.5	0.6915	1.6	0.9452		
0.6	0.7257	1.645	0.9500		
0.667	0.7475	1.7	0.9554		
0.7	0.7580	1.8	0.9641		
0.8	0.7881	1.9	0.9713		
0.9	0.8159				
1.0	0.8413				

[a]Standard Normal Distribution Function $F(t) = \frac{1}{\sqrt{2\pi}} \int_{-\infty}^{t} e^{-z^2/2}\, dz$

(Notice that $\Pr[X \geq 1]$ will also be .1587.)

$$\text{(c)} \quad \Pr[1 < X \leq 2] = \Pr[X \leq 2] - \Pr[X \leq 1] = F(2) - F(1) = .9772 - .8413 = .1359$$

Since this density function is symmetric about 0, we have,

$$F(-x) = 1 - F(x)$$

Consequently,

$$\text{(d)} \quad \Pr[X \leq -1] = F(-1) = 1 - F(1) = .1587$$

$$\text{(e)} \quad \Pr[X > -1] = 1 - \Pr[X \leq -1] = \Pr[X \leq 1] = .8413$$

$$\text{(f)} \quad \Pr\left[X > -\frac{1}{2}\right] = 1 - \Pr\left[X \leq -\frac{1}{2}\right] = \Pr\left[X \leq \frac{1}{2}\right] = .6915$$

$$\text{(g)}\quad \Pr\left[-1 < X < -\frac{1}{2}\right] = \Pr\left[X \leq -\frac{1}{2}\right] - \Pr[X \leq -1]$$
$$= .4528$$

$$\text{(h)}\quad \Pr\left[-\frac{1}{2} < X \leq \frac{1}{3}\right] = F\left(\frac{1}{3}\right) - F\left(-\frac{1}{2}\right)$$
$$= F\left(\frac{1}{3}\right) - \left[1 - F\left(\frac{1}{2}\right)\right]$$
$$= F\left(\frac{1}{3}\right) - 1 + F\left(\frac{1}{2}\right)$$
$$= .6306 - 1 + .6915$$
$$= .3221$$

Suppose next that $\mu = 60$, $\sigma = 3$, so that X is normal but not standard normal. The substitution $z = (x - 60)/3$ will transform the integrals needed to compute probabilities for X into integrals that can be evaluated using the standard normal table.
For example

$$\text{(i)}\quad \Pr[X \leq 60] = \int_{-\infty}^{60} f(x)\, dx$$
$$= \frac{1}{3\sqrt{2\pi}} \int_{-\infty}^{60} e^{-(x-60)^2/18}\, dx$$
$$= \frac{1}{\sqrt{2\pi}} \int_{-\infty}^{0} e^{-z^2/2}\, dz$$

if the substitution $z = (x - 60)/3$ is made. Thus we determine from the standard normal table that $\Pr[X \leq 60] = .5000$.

$$\text{(j)}\quad \Pr[57 \leq X \leq 64] = \int_{57}^{64} f(x)\, dx$$
$$= \frac{1}{3\sqrt{2\pi}} \int_{57}^{64} e^{-(x-60)^2/18}\, dx$$
$$= \frac{1}{\sqrt{2\pi}} \int_{-1}^{4/3} e^{\ z^2/2}\, dz$$
$$= F(4/3) - F(-1)$$
$$= .9088 - (1 - .8413)$$
$$= .7501$$

using the substitution $z = (x - 60)/3$ and the standard normal table. Here F refers to the standard normal distribution function given in the table. $\triangle$

Many other special probability distribution and density functions arise in applications. A more thorough discussion of the most important of these special distributions will be deferred until Chapter 6.

EXERCISES

1. Which of the following correspondence rules define random variables?

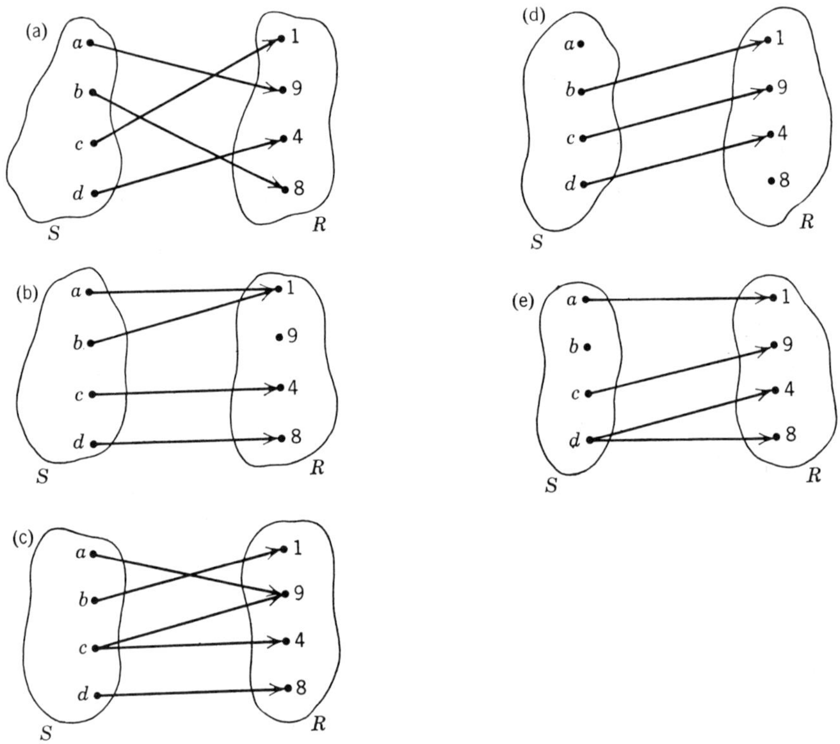

2. For each of the random variables found in Exercise 1, assume that the sample points are equally likely and determine
   (a) $\Pr[X \leq 4]$
   (b) $\Pr[X \geq 8]$
   (c) $\Pr[4 \leq X \leq 8]$
   (d) $F(x)$
   (e) p_i for $i = 0, 1, 2, \ldots$
   (f) Sketch $F(x)$ and p_i

3. A piece of electronic equipment is selected at random and its resistance measured. A typical sample point s is then a measured resistance (in ohms). If $X(s) = s$, is X a random variable? Describe its domain and range.

4. In Exercise 3, suppose that

$$S = \{s: 5 \leq s \leq 6\}$$

and that probabilities are assigned to the subintervals $c < s < d$ of the interval from 5 to 6 so that intervals (c, d) of equal length are equally likely. Determine

(**a**) $\Pr[X \leq 5\frac{1}{4}]$

(**b**) $\Pr[X > 5\frac{3}{4}]$

(**c**) $\Pr[5\frac{1}{8} < X \leq 5\frac{2}{3}]$

(**d**) $F(x)$

(**e**) Is X a continuous random variable?

(**f**) $f(x)$

(**g**) Sketch $F(x), f(x)$

(*Note*. The probability density function here is termed a *uniform density function*.)

5. Consider the relation $y^2 = x$ where y and x are real numbers, $x > 0$. Does this relation define y as a function of x? Does $y = x^2$ define such a function?

6. In Example 3.9 an honest coin is tossed until the first head appears. Define the sample space for this experiment. Let X be the random variable that assigns to each sample point the number of tails until the first head appears. Sketch the relation between points in the sample space and the integers 0, 1, 2, ... Is X a discrete or continuous random variable?

7. Sketch $F(x)$ for the probability density function given in Example 3.9. [*Hint*. Use the formula for the sum of a finite geometric series to determine $F(x)$ for x an integer.]

8. A piece of equipment is checked once every hour. Each hour it has probability $\frac{1}{5}$ of breaking down, and different hours are assumed independent as far as breakdowns are concerned. The equipment starts in operating condition. Let $X =$ the number of hours until it is observed to have broken down. (**a**) Find the range, probability density function, and distribution function of the random variable X. (**b**) Sketch these functions. (**c**) What is the probability that the equipment will last at least 3 hours?

9. Find the value of the constant k so that

$$f(x) = \begin{cases} kx^2(1 - x^3), & 0 < x < 1 \\ 0 \quad , & \text{otherwise} \end{cases}$$

is a proper probability density function of a continuous random variable, and sketch this function.

10. The lifetime (in hours) of a certain piece of equipment is a continuous random variable X having range $0 < x < \infty$ and probability density function:

$$f(x) = \begin{cases} \dfrac{xe^{-x/10}}{100}, & 0 < x < \infty \\ 0, & \text{otherwise} \end{cases}$$

(a) Show that this is a probability density function.
(b) Compute the probability that the lifetime exceeds 20 hours, $\Pr[X > 20]$.

11. A continuous random variable X is said to be *triangularly distributed* on the interval $[a, b]$ if it has probability density function:

$$f(x) = \begin{cases} 0, & \text{if } x < a \\ C(x-a), & a \leq x \leq \dfrac{a+b}{2} \\ C(b-x), & \dfrac{a+b}{2} \leq x \leq b \\ 0, & x > b \end{cases}$$

Sketch $f(x)$. What value must the constant C have in order that $f(x)$ be a probability density function? Determine

(a) $\Pr\left[X \geq \dfrac{a+b}{2}\right]$

(b) $\Pr\left[\dfrac{2a+b}{3} \leq X \leq \dfrac{a+2b}{3}\right]$

(c) $\Pr\left[X \leq \dfrac{2a+b}{3}\right]$

(d) $\Pr\left[\dfrac{3a+b}{4} \leq X \leq \dfrac{2a+b}{3}\right]$

(e) $F(x)$, and sketch $F(x)$

12. Is the function

$$F(x) = \begin{cases} C \ln x, & 0 \leq x \leq 2 \\ 0, & x < 0 \\ 1, & x > 2 \end{cases}$$

a probability distribution for any constant C? Give reasons.

13. Is the function

$$F(x) = \begin{cases} C \ln x, & 1 \le x \le 2 \\ 0 \quad , & x < 1 \\ 1 \quad , & x > 2 \end{cases}$$

a probability distribution for some constant C? Give reasons.

14. For what values of λ is

$$F(x) = 1 - e^{-\lambda x}, \qquad 0 \le x \le \infty$$

a probability distribution function? ($F(x) = 0$ for $x < 0$.)

15. In Exercise 14 find the corresponding $f(x)$ and determine

(a) $\Pr[X > 1]$

(b) $\Pr[2 \le X \le 3]$

(c) $\Pr[2 < X < 3]$

(d) $\Pr[X > y]$, for any $y \ge 0$

16. Several important probability distributions are related to the *gamma function* $\Gamma(\alpha)$ defined by

$$\Gamma(\alpha) = \int_0^\infty t^{\alpha-1} e^{-t} \, dt$$

for $\alpha > 0$.

(a) Using integration by parts derive the formula

$$\Gamma(\alpha) = (\alpha - 1)\Gamma(\alpha - 1)$$

for $\alpha > 1$.

(b) Using the result of (a) show that $\Gamma(n) = (n-1)!$ whenever n is a positive integer.

17. A random variable X having range $0 < x < \infty$ and probability density function of the form

$$f(x) = \begin{cases} Cx^{\alpha-1} e^{-\beta x}, & 0 < x < \infty \\ 0 \quad , & \text{otherwise} \end{cases}$$

for some positive constants α, β, and C is said to have a *gamma density.*

(a) Whenever $\alpha > 0$, $\beta > 0$, show that $f(x)$ is a probability density function provided that the constant C has the value:

$$C = \frac{\beta^\alpha}{\Gamma(\alpha)}$$

(b) Sketch $f(x)$ in the cases $\alpha = 2, 4$, $\beta = 1$. Notice that the case $\alpha = 1$ gives the negative exponential distribution, and that the probability density in Exercise 10 is a special case of the gamma density with $\alpha = 2$, $\beta = \frac{1}{10}$.

18. In a recent investigation of work times of a hospital admittance desk, it has been found that the admittance nurse (an RN) performs 2 jobs depending on the type of patient she is admitting. Usually, she merely types out a history of the patient. Sometimes the patient needs immediate assistance, and she conducts him to his room. Measurements taken on the time to do her job lead to the conjecture that the density function of her work time is given by

$$f(x) = 3p^2e^{-3px} + \frac{(1-p)^2(2)^2}{3\,!}\,x^2\,e^{-2(1-p)x}, \qquad 0 \le x < \infty$$

for some constant p. For what values of p is this a probability density function?

19. A study at a certain bank led to the conjecture that the number of arrivals in a 15-minute period is a discrete random variable X with probability density function:

$$p_i = \begin{cases} \dfrac{\lambda^i\, e^{-\lambda}}{i!}, & i = 0, 1, 2, \ldots \\ 0 \quad , & \text{otherwise} \end{cases}$$

For what values of λ is this a proper probability density function? This density function is very important in many situations and is called the *Poisson density function.* We will encounter it again in later chapters.

20. Using the Poisson density function in Exercise 19, determine

(a) $\Pr[X \ge 1]$
(b) $\Pr[X < 2]$
(c) $\Pr[2 \le X \le 4]$

Interpret these probabilities in terms of bank arrivals.

21. If the random variable X has a probability density function of the form

$$f(x) = \begin{cases} Cx^{\alpha-1}(1-x)^{\beta-1}, & 0 < x < 1 \\ 0 \quad , & \text{otherwise} \end{cases}$$

where α, β, and C are positive constants, X is said to have a *beta density.*

(a) In case $\alpha = 2$, $\beta = 2$, show that $C = 6$, and sketch $f(x)$ in this case
(b) Compute C in case $\alpha = 1/2$ and $\beta = 2$

22. Consider a sequence of independent experiments each having only two possible outcomes, "success" or "failure," so that Pr["success"] $= p$, Pr["failure"] $= q$, $p + q = 1$, on each trial. Define the random variable X to be the *number* of trials required until the *first* "success" occurs. If $p > 0$, show that X has range all positive integers 1, 2, 3, ..., and probability density function

$$p_i = pq^{i-1} = (1-q)q^{i-1}$$

for $i = 1, 2, 3, \ldots$. Check that $\sum_{i=1}^{\infty} p_i = 1$ by using the formula for the sum of a geometric progression. X is said to have a *geometric* probability density function. (See Example 3.9 and Exercises 6, 7, and 8.)

23. Is the function

$$f(x) = \begin{cases} \sin x, & 0 \le x \le \pi/2 \\ 0 \quad , & \text{otherwise} \end{cases}$$

a proper probability density function? Determine

(a) $\Pr[X < x]$

(b) $\Pr[X \le x]$

(c) $\Pr\left[\frac{\pi}{6} \le X < \frac{\pi}{4}\right]$

(d) $\Pr\left[X > \frac{\pi}{3}\right]$

24. In the process of manufacturing a part, one finds that it is over-sized, within tolerance, or undersized. The manufacturing process manufactures parts that are at least a units in size, and no more than b units. Define a sample space for this manufacturing process. Define three events of interest for the quality control of this process if an oversized part is defined to be one whose dimension is more than $(a + 3b)/4$ and an undersized part is one whose dimension is less than $(2a + b)/3$. If the size is a triangularly distributed random variable on $[a, b]$ (see Exercise 11), find the probabilities of the three events $E_1 =$ the part is oversized; $E_2 =$ the part is within tolerance; $E_3 =$ the part is undersized. Are the events E_1, E_2, E_3 mutually exclusive? If the part is undersized then it must be scrapped at an extra cost of C_1. If it is within tolerance, it has no extra cost. If it is oversized, it can be corrected at a cost of $C_3[x - (a + 3b)/4]$ for $(a + 3b)/4 \le x \le b$. Define the random variable Y to be the extra cost of the product. What is the density function for Y? Sketch the density and distribution functions for Y. Is Y a continuous random variable?

25. In the study of the reliability of electrical and mechanical parts, the lifetime X of a given part can be considered as a random variable. If X has the density function

$$f(x) = \frac{b}{\theta - x_0}\left(\frac{x - x_0}{\theta - x_0}\right)^{b-1} \exp\left[-\left(\frac{x - x_0}{\theta - x_0}\right)^{b}\right]; \qquad x_0 < x < \infty$$

then X is said to have a *Weibull* distribution. By using the substitution $u = [(x - x_0)/(\theta - x_0)]^b$, show that $f(x)$ is a probability density function for $b > 0$, $0 < x_0 < \theta$. Find $F(x)$ for such a random variable. Show that $\ln\{\ln[1/(1 - F(x)]\}$ is a linear function of $\ln(x - x_0)$ with slope b. One calls b the *Weibull slope*. Show that $\Pr[X \le \theta]$ does not depend on b or x_0. θ is called the *characteristic life* of a part.

26. If X has a Weibull distribution, as in Exercise 25, determine

(a) $\Pr[X \leq a]$, if $a > x_0$
(b) $\Pr[X \geq a]$, if $a > x_0$
(c) $\Pr[a \leq X \leq c]$, if $c > a > x_0$

27. Show that if X is a continuous random variable whose density function is symmetric about 0 [that is, $f(-x) = f(x)$ for every x], then

(a) $F(0) = 1/2$
(b) $\Pr[X > a] = 1/2 - \int_0^a f(x)\,dx,\ a > 0$
(c) $\Pr[-a < X < a] = 2F(a) - 1, a > 0$

Hence we need only compute tables of $F(x)$ for positive values of x for such a random variable (for example, standard normal).

28. Using the table of the normal distribution function, determine these probabilities for a random variable having a standard normal distribution.

(a) $\Pr[X \leq 3]$
(b) $\Pr[X \leq -2]$
(c) $\Pr[X \geq 2]$
(d) $\Pr[X \geq 3]$
(e) $\Pr[2 < X < 3]$
(f) $\Pr[-2 < X < 2]$
(g) $\Pr[1/2 < X < 3/4]$
(h) $\Pr[-1/2 < X < 3/4]$
(i) $\Pr[-3/4 < X < -1/2]$
(j) $\Pr[-3 < X < 3]$

29. Suppose that X is a normally distributed random variable with $\mu = 7$, $\sigma = 2$. Determine the following probabilities and give some reason for the computation you make

(a) $\Pr[X \leq 6]$
(b) $\Pr[X < 6]$
(c) $\Pr[X \geq 10]$
(d) $\Pr[6 \leq X < 10]$
(e) $\Pr[X > 13]$
(f) $\Pr[10 < X \leq 13]$
(g) $\Pr[4 \leq X \leq 6]$
(h) $\Pr[4 \leq X \leq 10]$
(i) $\Pr[0 \leq X \leq 13]$
(j) $\Pr[X \leq 15]$

30. Suppose X is a normally distributed random variable with parameters μ and σ. Using the standard normal table, find

(a) $\Pr[X < \mu]$
(b) $\Pr[\mu - \sigma < X < \mu + \sigma]$
(c) $\Pr[\mu - 2\sigma < X < \mu + 2\sigma]$
(d) $\Pr[\mu - 1.96\sigma < X < \mu + 1.96\sigma]$
(e) $\Pr[X > \mu - 1.96\sigma]$
(f) $\Pr[X < \mu + 1.96\sigma]$
(g) $\Pr[\mu - 1.645\sigma < X < \mu + 1.645\sigma]$
(h) $\Pr[\mu - \sigma < X < \mu + 2\sigma]$
(i) $\Pr[\mu - 2\sigma < X < \mu + \sigma]$
(j) $\Pr[|X - \mu| > 3\sigma]$

4 • Sets of Random Variables and Random Sequences

1. Introduction

In Chapter 3 we were concerned with the properties of a single random variable defined on a given sample space. In many applications it is important to study two or more random variables defined on the same sample space. Thus, for example, if we consider the experiment of selecting a radio tube from a set S of tubes, and if X represents the lifetime (in hours) of the tube selected, and Y represents the gas pressure inside the tube, then X and Y are both random variables defined on the *same* sample space S. An electrical engineer might be interested in relations between these two quantities; that is, he would be interested in the *pair* (X, Y) of random variables.

In this chapter, we consider the problem of two random variables, their associated distribution and density functions, and some properties, such as independence of the random variables. These concepts are extended easily in later sections to the case of many random variables defined on the same sample space. The last three sections of the chapter show how these ideas lead naturally to the concept of a *random sequence* or a *discrete parameter random process*, and give some preliminary discussion of such processes. These concepts are extended and elaborated in Chapters 7 to 10.

2. Pairs of Random Variables

Let X and Y be two random variables defined on some given sample space S (for example, X = height and Y = weight of a randomly selected individual). Each random variable is a function whose domain is S and whose range is some set of real numbers (see Figure 4.1).

Suppose now that we consider X and Y to have values on real axes which are coordinate axes of the ordinary two-dimensional plane. In this manner, each sample point s defines a *pair* of real numbers, $x = X(s)$ and $y = Y(s)$, where (x, y) can be treated as the co-ordinates of some point in the plane. Thus the pair (X, Y) of random variables can be considered as a function that to each

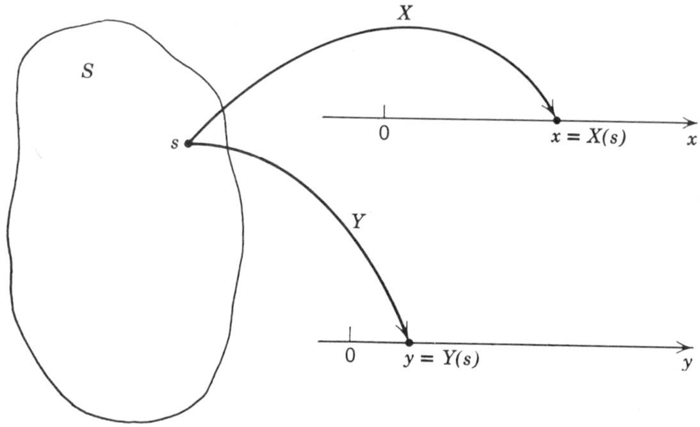

FIGURE 4.1 Two random variables X and Y considered as real-valued functions on the sample space.

point s in the sample space assigns a point (x, y) in the plane (see Figure 4.2). We shall find this geometric interpretation very valuable.

By using this idea, we observe that we can construct correspondences between certain *events* (subsets of S) and certain *regions* in the (x, y) plane. For example, for fixed x_1 and y_1, the event $A = [X = x_1 \text{ and } Y = y_1]$ is the set of all sample points s having X-value x_1 and Y-value y_1, $\{s : X(s) = x_1 \text{ and } Y(s) = y_1\}$. The corresponding region in the (x, y) plane, therefore, will consist only of the single point (x_1, y_1). [See Figure 4.3(a).]

The event $B = [X \leq x_1] = \{s : X(s) \leq x_1\}$ will correspond to the region consisting of all points (x, y) whose x-coordinate is less than or equal to x_1, $\{(x, y) : x \leq x_1\}$; that is, the region to the *left* of the line $x = x_1$ [see Figure 4.3(b)].

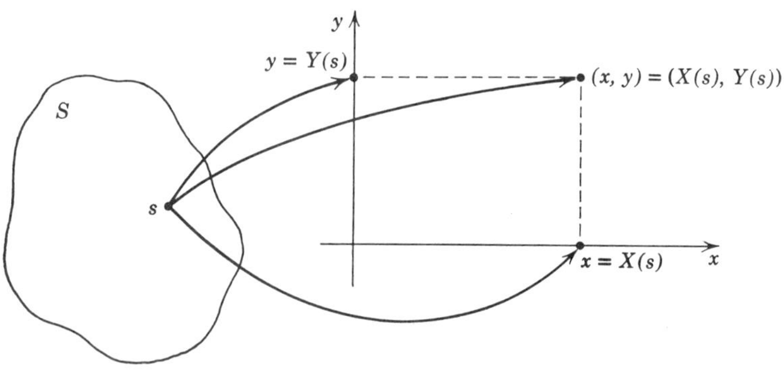

FIGURE 4.2 (X, Y) as a function from S to the plane.

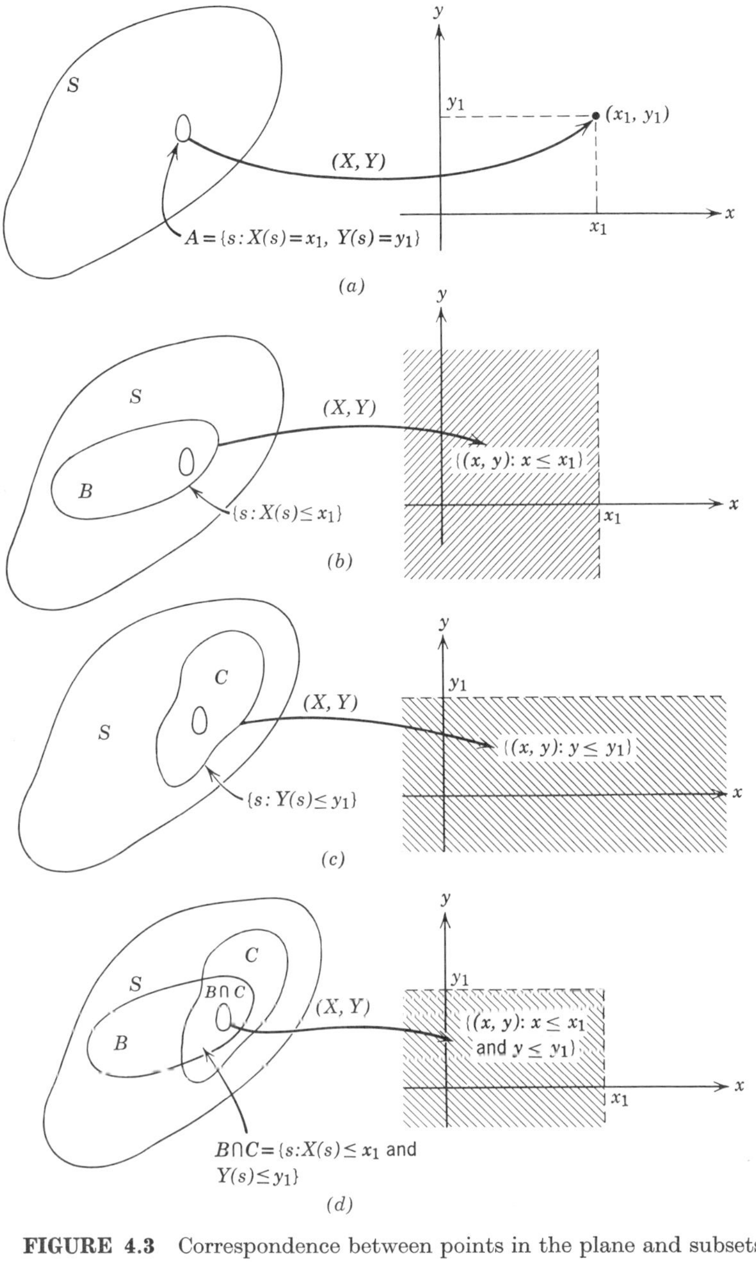

FIGURE 4.3 Correspondence between points in the plane and subsets of S.

Similarly, the event $C = [Y \leq y_1] = \{s : Y(s) \leq y_1\}$ will correspond to the region *below* the line $y = y_1$ [see Figure 4.3(c)].

If we consider the event consisting of all sample points, s, for which *both* $X(s) \leq x_1$ *and* $Y(s) \leq y_1$,

$$[X \leq x_1 \text{ and } Y \leq y_1] = [X \leq x_1] \cap [Y \leq y_1] = B \cap C \tag{4.1}$$

then this will correspond to the *intersection* of the two shaded regions in Figure 4.3 (b) and (c). [See Figure 4.3(d).]

Events of these forms, and the corresponding regions in the (x, y) plane, will be the basic building blocks needed to analyse the probability structure of pairs of random variables.

3. Distribution Functions

All of the events discussed in Section 2, being subsets of the sample space, will have probabilities. Thus, we can talk about probabilities of the form:

$$\Pr[X \leq x_1], \qquad \Pr[Y \leq y_1], \qquad \Pr[X \leq x_1 \text{ and } Y \leq y_1]$$

The first two, considered as functions of x_1 and y_1, are simply the *distribution functions* of X and Y, as defined in Chapter 3. The third probability will depend on *both* x_1 *and* y_1 and, hence, is a function of two variables termed the *joint distribution function of* X *and* Y. For simplicity we drop the subscripts from the x and y, and define

$$F_X(x) = \Pr[X \leq x] = \Pr[A_1] \tag{4.2}$$

$$F_Y(y) = \Pr[Y \leq y] = \Pr[A_2] \tag{4.3}$$

$$F(x, y) = \Pr[X \leq x \text{ and } Y \leq y] = \Pr[A_1 \cap A_2] \tag{4.4}$$

where

$$A_1 = \{s : X(s) \leq x\}, \qquad A_2 = \{s : Y(s) \leq y\} \tag{4.5}$$

$F(x, y)$ is the *joint distribution function* of X and Y. If we must specify exactly which random variables are involved, we shall write $F_{X,Y}(x, y)$ in place of $F(x, y)$. Thus, to each region in the (x, y)-plane of the form shaded in Figure 4.3(d), $F(x, y)$ assigns a number, a *probability*.

Recall that two events, A and B, were termed *independent* if $\Pr[A \cap B] = \Pr[A] \Pr[B]$. Extending these ideas, we define the random variables X and Y to be *independent* if the events A_1 and A_2 are independent *for every* x *and* y. Thus, X and Y are independent random variables if

$$\Pr[A_1 \cap A_2] = \Pr[A_1]\Pr[A_2] \tag{4.6}$$

or

$$\Pr[X \leq x \text{ and } Y \leq y] = \Pr[X \leq x]\Pr[Y \leq y] \tag{4.7}$$

or, in terms of the distribution functions,

$$F(x, y) = F_X(x)F_Y(y) \tag{4.8}$$

for every value of x and y.

Thus the condition for independence is that $F(x, y)$ factor in the form (4.8). Actually, it is easy to show that if $F(x, y)$ factors into any product of two nonnegative functions, one of x only and the other of y only,

$$F(x, y) = u(x)v(y) \tag{4.9}$$

then, for some nonzero constant c, $c\,u(x) = F_X(x)$ and $(1/c)v(y) = F_Y(y)$, so that $F(x, y) = c\,u(x) \cdot (1/c)v(y) = F_X(x)F_Y(y)$ and we can conclude that X and Y are independent.

EXAMPLE 4.1. A Geiger counter is being bombarded by subatomic praticles. Let X and Y represent the time intervals between the first and second, and between the second and third clicks of the counter. Thus, X and Y are random variables with ranges $0 < x < \infty$, $0 < y < \infty$. It can be shown that they have joint distribution function

$$F(x, y) = \begin{cases} 1 - e^{-ax} - e^{-ay} + e^{-a(x+y)}, & 0 < x < \infty;\ 0 < y < \infty \\ 0, & \text{otherwise} \end{cases}$$

where a is some positive constant (Figure 4.4). Thus the probability that the first time interval does not exceed 2 seconds and that the second time interval does not exceed 1 second is

$$\Pr[X \leq 2 \text{ and } Y \leq 1] = 1 - e^{-2a} - e^{-a} + e^{-3a}$$

In general, we observe that

$$F(x, y) = \begin{cases} (1 - e^{-ax})(1 - e^{-ay}), & 0 < x < \infty,\ 0 < y < \infty \\ 0, & \text{otherwise} \end{cases}$$

and it follows that $F(x, y)$ factors into the form (4.9) where

$$u(x) = \begin{cases} 1 - e^{-ax}, & 0 < x < \infty; \\ 0, & \text{otherwise}\ ; \end{cases} \qquad v(y) = \begin{cases} 1 - e^{-ay}, & 0 < y < \infty \\ 0, & \text{otherwise} \end{cases}$$

Hence, X and Y are independent in this case. We can show that $u(x) = F_X(x)$ and $v(y) = F_Y(y)$ here. △

4. Properties of Joint Distribution Functions

Since $F(x, y)$ is a function of two variables, its graphical representation would require three dimensions [one for x, one for y, and one for $F(x, y)$]. (See Figure 4.4.) We notice the following properties that correspond to the

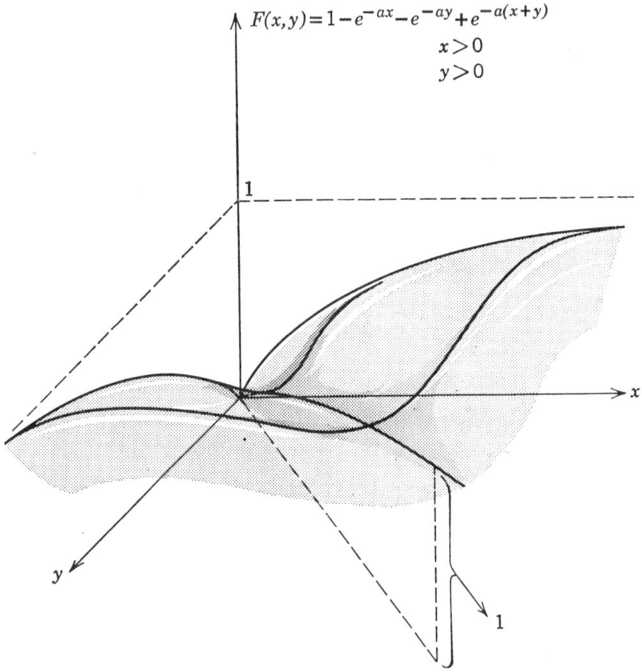

FIGURE 4.4 A continuous distribution function.

properties of the distribution function of a single random variable discussed in Section 4 of Chapter 3.

(a) $0 \leq F(x, y) \leq 1$, *for* $-\infty < x < \infty$, $-\infty < y < \infty$. This follows since $F(x, y)$ is a probability.

(b) *If* $x_1 \leq x_2$ *and* $y_1 \leq y_2$, *then* $F(x_1, y_1) \leq F(x_2, y_2)$. This follows since the region consisting of all points (x, y) satisfying $x \leq x_1$ and $y \leq y_1$, will lie *inside* the region $x \leq x_2$ and $y \leq y_2$ (see Figure 4.5). It follows that $[X \leq x_1$ and $Y \leq y_1] \subset [X \leq x_2$ and $Y \leq y_2]$ and, hence, that $F(x_1, y_1) \leq F(x_2, y_2)$.

(c)
$$\lim_{\substack{x \to \infty \\ y \to \infty}} F(x, y) = F(\infty, \infty) = 1$$

$$\lim_{x \to -\infty} F(x, y) = F(-\infty, y) = 0$$

$$\lim_{y \to -\infty} F(x, y) = F(x, -\infty) = 0$$

These properties seem reasonable since the event $[X \leq x$ and $Y \leq y]$ approaches a certain event if $x \to \infty$, $\to y\ \infty$, and approaches an impossible event if $x \to -\infty$, or if $y \to -\infty$. A rigorous proof would follow the lines of the proof of property (c) of Section 4, Chapter 3.

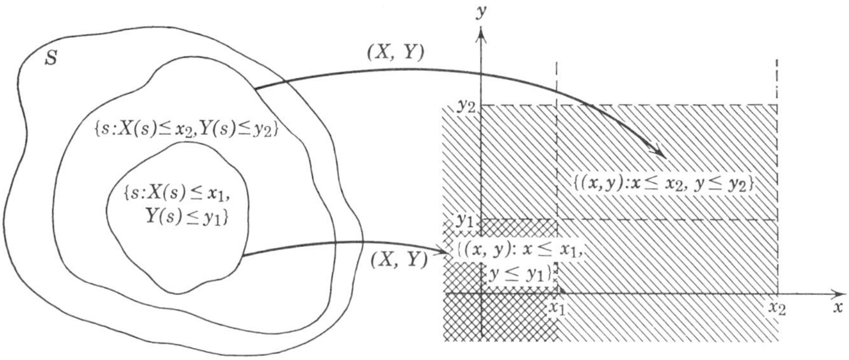

FIGURE 4.5 Graph of sets $[X \leq x_1;\ Y \leq y_1] \subset [X \leq x_2;\ Y \leq y_2]$ for $x_1 \leq x_2$, $y_1 \leq y_2$.

(d)
$$\lim_{x \to a+} F(x, y) = F(a, y)$$

$$\lim_{y \to b+} F(x, y) = F(x, b)$$

Again, this seems reasonable, and a proof would be similar to the proof of property (d) of Section 4, Chapter 3.

We notice that the function $F(x, y)$ in Example 4.1 satisfies properties (a) to (d). However, consider the following example.

EXAMPLE 4.2. Let

$$F(x, y) = \begin{cases} 1 - e^{-(x+y)}, & 0 < x < \infty, \quad 0 < y < \infty \\ 0, & \text{otherwise} \end{cases}$$

We easily check that this function satisfies properties (a) to (d). In spite of this fact, $F(x, y)$ is *not* a joint distribution function. To determine this, assume that there are random variables X and Y having $F(x, y)$ as joint distribution function. Consider the events:

$$A = [X \leq 1 \text{ and } Y \leq 2]$$
$$B = [X \leq 2 \text{ and } Y \leq 1]$$
$$A \cap B = [X \leq 1 \text{ and } Y \leq 1]$$

and

$$C = [X \leq 2 \text{ and } Y \leq 2]$$

Clearly, $A \subset C$, $B \subset C$ and, hence, $A \cup B \subset C$. Thus, $\Pr[A \cup B] \leq \Pr[C]$, or $\Pr[C] - \Pr[A \cup B] \geq 0$. But

$$\Pr[C] - \Pr[A \cup B] = \Pr[C] - \{\Pr[A] + \Pr[B] - \Pr[A \cap B]\}$$

using property (c) of Section 3 in Chapter 2. If these probabilities are all expressed in terms of $F(x, y)$, we have

$$\begin{aligned}\Pr[C] - \Pr[A \cup B] &= F(2, 2) - \{F(1, 2) + F(2, 1) - F(1, 1)\} \\ &= F(2, 2) - F(1, 2) - F(2, 1) + F(1, 1) \\ &= (1 - e^{-4}) - (1 - e^{-3}) - (1 - e^{-3}) + (1 - e^{-2}) \\ &= -(e^{-4} - 2e^{-3} + e^{-2}) \\ &= -(e^{-2} - e^{-1})^2 < 0\end{aligned}$$

contradicting the fact that this probability must be ≥ 0. Hence $F(x, y)$ is *not* a joint distribution function. △

This shows that properties (a) to (d) are not sufficient to guarantee that $F(x, y)$ is a distribution function. An additional property is necessary. It is the following.

(e) *Whenever $a \leq b$ and $c \leq d$, then*

$$F(a, c) - F(a, d) - F(b, c) + F(b, d) \geq 0.$$

To demonstrate this, an argument similar to the one used in Example 4.2 shows that

$$F(a, c) - F(a, d) - F(b, c) + F(b, d) = \Pr[a < X \leq b \text{ and } c < Y \leq d]$$

(see Figure 4.8 and the derivation of formula (4.12) in the next section). Property (e) follows since a probability must be nonnegative.

The calculation in Example 4.2 shows that (e) fails in case $a = c = 1$,

$b = d = 2$ and, hence, that the function of this example cannot be a distribution function.

If we let $y \to \infty$, then the event $[X \leq x \text{ and } Y \leq y]$ approaches the event $[X \leq x \text{ and } Y < \infty]$ which is the same as the event $[X \leq x]$, since the condition $Y < \infty$ is *always* satisfied (every real number is less than ∞). It follows that, as $y \to \infty$,

$$\begin{aligned} F(x, y) &= \Pr[X \leq x \text{ and } Y \leq y] \\ &\to \Pr[X \leq x] = F_X(x) \end{aligned}$$

or

$$F(x, \infty) = F_X(x) \tag{4.10}$$

This argument is not quite rigorous, but it can be justified in a manner similar to the one for property (c).

Similarly

$$F(\infty, y) = F_Y(y) \tag{4.11}$$

Formulas (4.10) and (4.11) show how the *individual* distribution functions of X and Y can be found from their *joint* distribution function. When the individual distribution functions are obtained in this way, they are sometimes termed *marginal* distribution functions.

The preceding discussion applies equally to discrete as well as to continuous random variables. Consider the following discrete example.

EXAMPLE 4.3. An honest coin is tossed three times, and the following random variables are defined.

$X =$ the number of heads on the first two tosses

$$(X = 0, 1, \text{ or } 2)$$

$Y =$ the number of heads on the third toss

$$(Y = 0 \text{ or } 1)$$

The sample space S here consists of the eight sample points:

$$S = \begin{Bmatrix} s_1 & s_2 & s_3 & s_4 & s_5 & s_6 & s_7 & s_8 \\ HHH, & HHT, & HTH, & THH, & THT, & TTH, & HTT, & TTT \end{Bmatrix}$$

All are equally likely, each having probability $\frac{1}{8}$.

The corresponding values of X and Y are given in Table 4.1.

TABLE 4.1

S	$X(s)$	$Y(s)$
$s_1 = HHH$	2	1
$s_2 = HHT$	2	0
$s_3 = HTH$	1	1
$s_4 = THH$	1	1
$s_5 = THT$	1	0
$s_6 = TTH$	0	1
$s_7 = HTT$	1	0
$s_8 = TTT$	0	0

We then can determine the following events:

$$[X \le 2 \text{ and } Y \le 1] = \{s_1, s_2, s_3, s_4, s_5, s_6, s_7, s_8\} = S$$
$$[X \le 2 \text{ and } Y \le 0] = \{s_2, s_5, s_7, s_8\}$$
$$[X \le 1 \text{ and } Y \le 1] = \{s_3, s_4, s_5, s_6, s_7, s_8\}$$
$$[X \le 1 \text{ and } Y \le 0] = \{s_5, s_7, s_8\}$$
$$[X \le 0 \text{ and } Y \le 1] = \{s_6, s_8\}$$
$$[X \le 0 \text{ and } Y \le 0] = \{s_8\}$$

Because the above events are all combinations of mutually exclusive, equally likely elementary events of the form $\{s_i\}$, we can construct Table 4.2 for the joint distribution function $F(x, y)$.

TABLE 4.2[a]

x \ y	0	1
0	1/8	1/4
1	3/8	6/8
2	1/2	1

[a] The Joint Distribution Function $F(x, y)$ of X, Y.

This table gives the value of $F(x, y)$ only for certain values of x and y; namely, $x = 0, 1, 2$ and $y = 0, 1$. However, the distribution function

$F(x, y)$ is defined for *all* values of x and y. For values not in this table we need only use the fact that X and Y must take only these integer values. Thus, for example,

$$F(3/2, 1/2) = \Pr\left[X \leq \frac{3}{2} \text{ and } Y \leq \frac{1}{2}\right]$$

$$= \Pr[X \leq 1 \text{ and } Y \leq 0]$$

$$= F(1, 0) = \frac{3}{8}$$

$$F(7, 0) = \Pr[X \leq 7 \text{ and } Y \leq 0]$$

$$= \Pr[X \leq 2 \text{ and } Y \leq 0]$$

$$= F(2, 0) = \frac{1}{2}$$

The marginal distribution functions $F_X(x)$ and $F_Y(y)$ can be obtained from (4.10) and (4.11), and are given in Table 4.3 for these integer values of x and y. Again these functions are extended to other values of X and Y by the fact that X and Y can take only these integer values.

TABLE 4.3

x	$F_X(x) = \Pr[X \leq x]$	y	$F_Y(y) = \Pr[Y \leq y]$
0	1/4	0	1/2
1	3/4	1	1
2	1		

The graphs of $F_X(x)$ and $F_Y(y)$ are, therefore, the step-functions sketched in Figure 4.6.

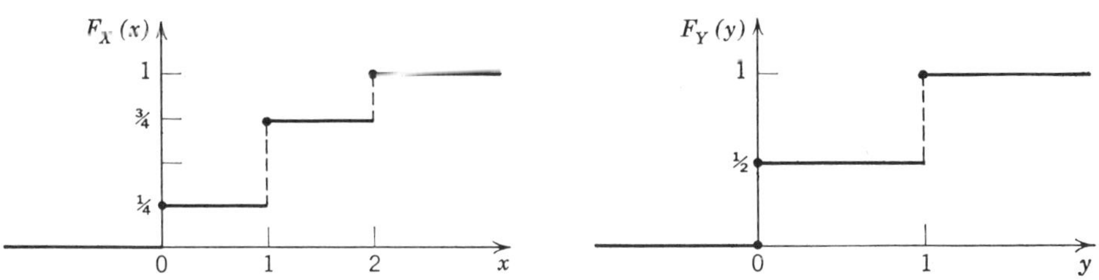

FIGURE 4.6 Marginal distribution functions in Example 4.3.

Notice that

$$F(0, 0) = \frac{1}{4} \cdot \frac{1}{2} = F_X(0)F_Y(0)$$

$$F(0, 1) = \frac{1}{4} \cdot 1 = F_X(0)F_Y(1)$$

$$F(1, 0) = \frac{3}{4} \cdot \frac{1}{2} = F_X(1)F_Y(0)$$

$$F(1, 1) = \frac{3}{4} \cdot 1 = F_X(1)F_Y(1)$$

$$F(2, 0) = 1 \cdot \frac{1}{2} = F_X(2)F_Y(0)$$

$$F(2, 1) = 1 \cdot 1 = F_X(2)F_Y(1)$$

and, in general,

$$F(x, y) = F_X(x) \cdot F_Y(y)$$

Thus, we observe that these random variables X, Y, are independent. Their joint distribution is given in Table 4.2, and their marginal distributions are given in Table 4.3 and are graphed in Figure 4.6. Notice that $F(x, y)$ does, indeed, satisfy properties (a) to (e). △

5. Determining Probabilities from a Joint Distribution Function

If X and Y are given random variables, we frequently want to find the probabilities of certain events defined in terms of X and Y, for which we know the joint distribution function. Thus, for example, any event of the form $[X \leq x_1$ and $Y \leq y_1]$ has probability simply $F(x_1, y_1)$. For more complicated events, it is useful to sketch the region in the (x, y)-plane in which the values of X, Y must lie, in order to define the required event. [For example, the region corresponding to the event $[X \leq x_1$ and $Y \leq y_1]$ is shaded in Figure 4.3(d).]

An event of the form $[a < X \leq b$ and $Y \leq y_1]$, for fixed $a < b$ and y_1, would correspond to the region R sketched in Figure 4.7.

In Figure 4.7 the event $[X \leq b$ and $Y \leq y_1]$ corresponds to the region shaded (\\\\\), and the event $[X \leq a$ and $Y \leq y_1]$ corresponds to the region shaded (/////). These two events will have probabilities $F(b, y_1)$ and $F(a, y_1)$.

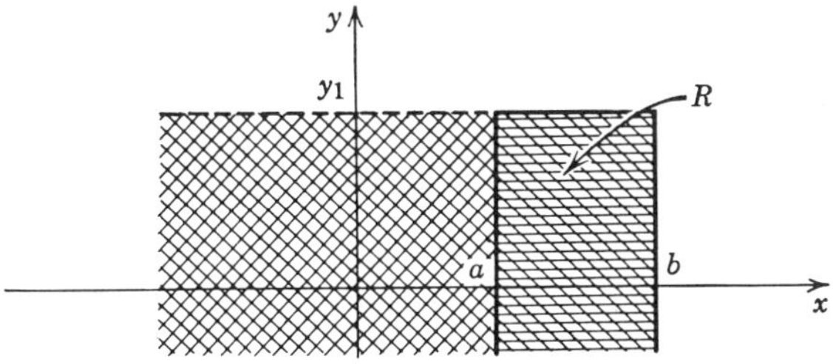

FIGURE 4.7 Region R defining the event $[a < X \leq b \text{ and } Y \leq y_1]$.

Clearly, the region R corresponding to the event $[a < X \leq b$ and $Y \leq y_1]$ will be the difference between these two regions [shaded ($\equiv$) in the figure]. Hence

$$\Pr[a < X \leq b \text{ and } Y \leq y_1] = F(b, y_1) - F(a, y_1)$$

Analogously, we observe that

$$\Pr[X \leq x_1 \text{ and } a < Y \leq b] = F(x_1, b) - F(x_1, a)$$

Next, we wish to determine the probability of a more general event of the form $R = [a < X \leq b$ and $c < Y \leq d]$, for fixed $a < b$, $c < d$. Consider these events:

$$A = [X \leq b \text{ and } Y \leq d] \supset R$$
$$B_1 = [X \leq a \text{ and } Y \leq d]$$
$$B_2 = [X \leq b \text{ and } Y \leq c]$$

We notice that any point of A which does not lie in R must lie in B_1 or B_2 (Figure 4.8). Thus

$$R \cup (B_1 \cup B_2) = A$$

where, clearly, R and $B_1 \cup B_2$ are mutually exclusive.

Hence

$$\Pr[R] + \Pr[B_1 \cup B_2] = \Pr[A]$$
$$\Pr[R] = \Pr[A] - \Pr[B_1 \cup B_2]$$
$$= \Pr[A] - (\Pr[B_1] + \Pr[B_2] - \Pr[B_1 \cap B_2])$$
$$= \Pr[A] - \Pr[B_1] - \Pr[B_2] + \Pr[B_1 \cap B_2]$$

But

$$\Pr[A] = F(b, d), \ \Pr[B_1] = F(a, d), \ \Pr[B_2] = F(b, c)$$

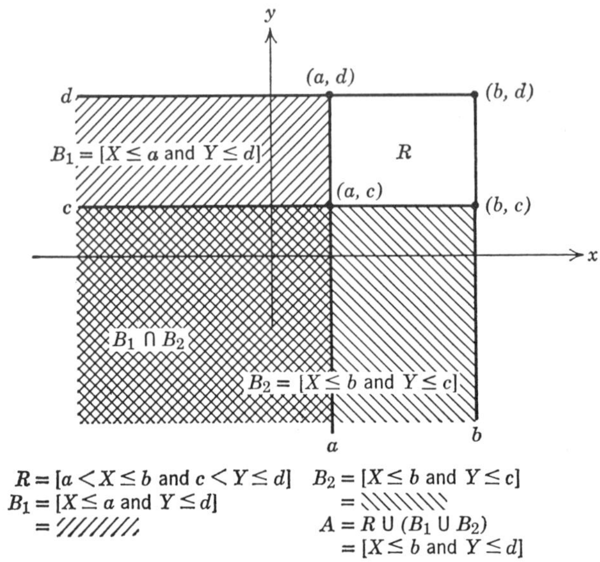

FIGURE 4.8 Regions corresponding to certain joint events.

and

$$\Pr[B_1 \cap B_2] = \Pr[X \leq a \text{ and } Y \leq c] = F(a, c)$$

Thus

$$\Pr[a < X \leq b \text{ and } c \leq Y \leq d] = F(b, d) - F(a, d) - F(b, c) + F(a, c) \tag{4.12}$$

Intuitively, to determine the probability of the event R, we must deduct from the probability of the larger event A the probabilities of B_1 and B_2. But (Figure 4.8) this will deduct the probability of the intersection $B_1 \cap B_2$ *twice*; hence, it must be added back in once to obtain the required probability.

In case X and Y are independent, $F(x, y) = F_X(x)F_Y(y)$, formula (4.12) can be rewritten as follows.

$$\begin{aligned}\Pr[a < X \leq b \text{ and } c < Y \leq d] &= F_X(b)F_Y(d) - F_X(a)F_Y(d) \\ &\quad - F_X(b)F_Y(c) + F_X(a)F_Y(c) \\ &= [F_X(b) - F_X(a)][F_Y(d) - F_Y(c)] \\ &= \Pr[a < X \leq b]\Pr[c < Y \leq d] \end{aligned} \tag{4.13}$$

In other words, independence of X and Y implies that every event of the form $[a < X \leq b]$ is independent of every event of the form $[c < Y \leq d]$.

EXAMPLE 4.4. Consider the distribution function in Example 4.1.

$$F(x, y) = \begin{cases} 1 - e^{-ax} - e^{-ay} + e^{-a(x+y)}, & 0 \leq y < \infty, 0 \leq x < \infty \\ 0, & \text{otherwise} \end{cases}$$

and, for simplicity, assume that $a = 1$.
Thus

$$\begin{aligned} F(1, 1) &= 1 - e^{-1} - e^{-1} + e^{-2} = 0.399 \\ F(1, 2) &= 1 - e^{-1} - e^{-2} + e^{-3} = 0.447 \\ F(2, 1) &= 1 - e^{-2} - e^{-1} + e^{-3} = 0.447 \\ F(2, 2) &= 1 - e^{-2} - e^{-2} + e^{-4} = 0.748 \end{aligned}$$

and, by (4.12),

$$\begin{aligned} \Pr[1 < X \leq 2 \text{ and } 1 < Y \leq 2] &= F(2, 2) - F(1, 2) - F(2, 1) + F(1, 1) \\ &= 0.748 - 0.447 - 0.447 + 0.399 \\ &= 0.253 \end{aligned}$$

△

EXAMPLE 4.5. Let X and Y represent the lifetimes of two similar tubes in a piece of electronic equipment. Assume that X and Y are independent with the same distribution function, and with joint distribution function:

$$F(x, y) = F_X(x)F_Y(y) = F(x)F(y)$$

It is frequently desirable to determine the joint distribution of the time U until the *first* tube failure and the time V until the *second* tube failure. Thus in any experiment $U =$ the smaller of X and Y, while $V =$ the larger of X and Y:

$$U = \text{minimum}\,\{X, Y\}, \qquad V = \text{maximum}\,\{X, Y\}$$

Clearly

$$\begin{aligned} F_V(v) &= \Pr[V \leq v] = \Pr[X \leq v \text{ and } Y \leq v] \\ &= F_X(v)F_Y(v) = [F(v)]^2 \\ F_U(u) &= \Pr[U \leq u] = 1 - \Pr[U > u] \\ &= 1 - \Pr[X > u \text{ and } Y > u] \\ &= 1 - [1 - F_X(w)][1 - F_Y(u)] \\ &\quad 1 - [1 - F(u)]^2 = F(u)[2 - F(u)] \end{aligned}$$

For $u \leq v$,

$$\begin{aligned}
\Pr[U \leq u \text{ and } V > v] &= \Pr[(X \leq u \text{ and } Y > v) \text{ or } (Y \leq u \text{ and } X > v)] \\
&= \Pr[X \leq u \text{ and } Y > v] + \Pr[Y \leq u \text{ and } X > v] \\
&= \{\Pr[X \leq u] - \Pr[X \leq u \text{ and } Y \leq v]\} \\
&\quad + \{\Pr[Y \leq u] - \Pr[Y \leq u \text{ and } X > v]\} \\
&= \{F(u) - F(u, v)\} + \{F(u) - F(u, v)\} \\
&= 2\{F(u) - F(u, v)\} \\
&= 2F(u)[1 - F(v)]
\end{aligned}$$

Thus the joint distribution function of U and V is given by

$$\begin{aligned}
G(u, v) &= \Pr[U \leq u \text{ and } V \leq v] \\
&= \Pr[U \leq u] - \Pr[U \leq u \text{ and } V > v] \\
&= F_U(u) - 2F(u)[1 - F(v)] \\
&= F(u)[2 - F(u)] - 2F(u)[1 - F(v)] \\
&= F(u)[2F(v) - F(u)]
\end{aligned}$$

for $u \leq v$. If $u > v$, then

$$G(u, v) = \Pr[U \leq u \text{ and } V \leq v] = \Pr[V \leq v] = [F(v)]^2$$

since $U \leq V$ always.

Suppose for explicitness that X and Y have ranges $0 < X < 1$, $0 < Y < 1$, and distribution functions;

$$F(x) = \begin{cases} 0, & \text{for } x \leq 0 \\ x, & \text{for } 0 < x < 1 \\ 1, & \text{for } x \geq 1 \end{cases} \qquad F(y) = \begin{cases} 0, & \text{for } y \leq 0 \\ y, & \text{for } 0 < y < 1 \\ 1, & \text{for } y \geq 1 \end{cases}$$

Then

$$F_V(v) = [F(v)]^2 = v^2, \qquad \text{for} \qquad 0 < v < 1$$

$$F_U(u) = F(u)[2 - F(u)] = u(2 - u), \qquad \text{for} \qquad 0 < u < 1$$

If A is the event "only one tube is operating at time $\frac{1}{2}$," then $A = [U < \frac{1}{2} < V]$, and

$$\Pr[A] = \Pr\left[U \leq \frac{1}{2} \text{ and } V > \frac{1}{2}\right]$$

$$= 2F\left(\frac{1}{2}\right)\left[1 - F\left(\frac{1}{2}\right)\right]$$

$$= 2 \cdot \frac{1}{2} \cdot \frac{1}{2} = \frac{1}{2} \qquad \triangle$$

6. Density Functions: Discrete Case

A pair of random variables (X, Y) are said to have a *discrete* joint distribution if X and Y individually are *discrete* random variables in the sense discussed in Section 6, Chapter 3. If X and Y individually are *continuous* random variables, as discussed in Section 7, Chapter 3, then the pair (X, Y) is said to have a *continuous* joint distribution. The *mixed* case occurs when one of the random variables is discrete and the other continuous. When this happens, we usually can easily modify our discussion to take care of it (see Example 4.18).

Assume now that X and Y are discrete random variables. The discrete random variables that will be most important in our later discussions all will have ranges consisting of a set of *integers*. Thus, let us now assume that X and Y are *integer-valued* random variables. Whenever (i, j) is a pair of integers, define

$$p(i, j) = \Pr[X = i \text{ and } Y = j] \tag{4.14}$$

The fact that X and Y are integer-valued random variables implies that the sum of $p(i, j)$ over all integer pairs (i, j) is 1. If we interpret probability as a two-dimensional mass distribution, $p(i, j)$ represents the point mass attached to the point with coordinates (i, j). If the pair (i, j) is *not* a point in the joint range of (X, Y), then $p(i, j) = 0$.

Clearly, $p(i, j)$ represents a function that assigns a number to each *pair* of integers (i, j). This function is termed the *joint probability density function* of X and Y. It has these properties:

(a) $0 \leq p(i, j) \leq 1$ *for each* i, j

(b) $\sum_{i,j} p(i, j) = \sum_i \sum_j p(i, j) = 1$

(c) $F(x, y) = \sum_{i \leq x} \sum_{j \leq y} p(i, j) = \sum_{\substack{i \leq x \\ j \leq y}} p(i, j)$

If $p(i) = p_i$ and $q(j) = q_j$ represent the individual probability density functions of X and Y, $p_i = \Pr[X = i]$, $q_j = \Pr[Y = j]$, then

$$p_i = \Pr[X = i] = \sum_j \Pr[X = i \text{ and } Y = j] = \sum_i p(i, j)$$

with a similar formula for q_j. Thus we have

$$p_i = \sum_j p(i, j), \qquad q_j = \sum_i p(i, j) \tag{4.15}$$

The functions p_i and q_j are sometimes referred to as the *marginal probability density functions* of X and Y.

If X and Y are *independent* random variables, formula (4.13) can be applied as follows:

$$\begin{aligned} p(i, j) &= \Pr[X = i \text{ and } Y = j] \\ &= \Pr[i - 1 < X \le i \text{ and } j - 1 < Y \le j] \text{ (since } X \text{ and } Y \text{ are integers)} \\ &= \Pr[i - 1 < X \le i] \Pr[j - 1 < Y \le j] \\ &= \Pr[X = i] \Pr[Y = j] \\ &= p_i q_j \end{aligned}$$

Thus, if X and Y are independent,

$$p(i, j) = p_i q_j \tag{4.16}$$

and we easily show that, conversely, the identity (4.16) implies that X and Y are independent.

EXAMPLE 4.6. Consider the coin-tossing Example 4.3 of Section 4. A coin is tossed three times. X = the number of heads on the first two tosses, and Y = the number of heads on the third toss. The values of $p(i, j)$ are easily computed from Table 4.1 and are given in Table 4.4.

TABLE 4.4[a]

i \ j	0	1	p_i
0	1/8	1/8	1/4
1	1/4	1/4	1/2
2	1/8	1/8	1/4
q_j	1/2	1/2	

[a] Joint Density $p(i, j)$ for X = Number of Heads on 2 Coin Tosses; Y = Number of Heads on Third Toss.

Identity (4.16) can be checked, in this case, from Table 4.4. Since X and Y are independent here, (4.16) must hold true. The distribution function $F(x, y)$ can also be determined from (4.8) and can be compared with the values in Table 4.2. Thus, for example,

$$\begin{aligned} F(1, 1) &= \sum_{i \le 1} \sum_{j \le 1} p(i, j) \\ &= p(0, 0) + p(0, 1) + p(1, 0) + p(1, 1) \\ &= \frac{3}{4} = \left(\frac{3}{4}\right) \cdot (1) = F_X(1) F_Y(1) \end{aligned}$$ △

EXAMPLE 4.7. In many experimental studies we want to know if two random variables are independent. One method used to study this problem is the following. Suppose that we have collected data on 1000 rats, some of which have been subjected to doses of a drug suspected of causing a particular type of tumor. Of the 1000 rats, some have been given a heavy dosage of the drug, some a light dosage, and some none of the drug. We find these results (Table 4.5):

TABLE 4.5

	Tumor Present		
	Yes	No	
Heavy dosage	10	90	100
Light dosage	15	485	500
No drug	12	388	400
	37	963	1000

Suppose we let

$$X = \begin{cases} 0 & \text{if rat received heavy dosage} \\ 1 & \text{if rat received light dosage} \\ 2 & \text{if rat received no dosage} \end{cases}$$

$$Y = \begin{cases} 0 & \text{if rat has tumor} \\ 1 & \text{if rat does not have tumor} \end{cases}$$

and assume that the proportions obtained from the above table can be used as estimates of the probabilities of events, such as

$$p(0, 0) = \Pr[X = 0,\ Y = 0] \approx 10/1000 = 0.01$$

If these assumptions are valid, we can summarize our experimental results in Table 4.5 below.

TABLE 4.6[a]

$i \backslash j$	0	1	$p_i \approx$
0	0.010	0.090	0.10
1	0.015	0.485	0.50
2	0.012	0.388	0.40
$q_j \approx$	0.037	0.963	1.00

[a] Estimated Joint Probability Density $p(i, j)$ from Experimental Results

The marginal probabilities p_i and q_j are also estimated in this Table. Now we argue as follows.

If X and Y are independent (for example, if there is no relation between the level of dosage and the presence or absence of the tumor), then we must have $p(i, j) = p_i q_j$. For example, $p(0, 0) \approx (.10)(.037) = .0037$.

That is, *if we assume that the* random variables X and Y are independent, then the joint probability density function should be as given in Table 4.7.

TABLE 4.7[a]

$i \backslash j$	0	1
0	0.0037	0.0963
1	0.0185	0.4815
2	0.0148	0.3852

[a] Estimated Joint Probability Density Function if X, Y are Independent, $p(i, j) = p_i q_j$.

Clearly the entries in Table 4.6 and 4.7 differ. But this could have been expected even if X and Y are independent, since the densities are only approximated. The question then is not whether these results agree exactly but whether they are in disagreement enough that a person would doubt the assumption of independence used to obtain Table 4.7. The detailed analysis of such situations can be found in works on *statistics*, specifically *contingency tables*. △

EXAMPLE 4.8. TRINOMIAL DISTRIBUTION. Suppose that a certain random experiment has three possible outcomes: outcomes of Types 1, 2, or 3 (for example, win, lose, or draw), with probabilities p, q, and r, $p+q+r=1$. Assume that this experiment is repeated independently n times, where n is some fixed integer. Let X and Y represent the number of Type 1 outcomes and Type 2 outcomes, respectively, in this sequence of trials. Therefore, $n-X-Y$ represents the number of Type 3 outcomes. The random variables X, Y are said to have a *trinomial distribution.* Their joint range consists of all pairs (i, j) of nonnegative integers satisfying $0 \leq i+j \leq n$.

To compute the probability $p(i, j)$ that $X=i$ and $Y=j$, we must first count the number of ways that i of the trials could be selected to be of Type 1, $\binom{n}{i}$ ways, must multiply this by the number of ways that j of the remaining trials could be selected to be of Type 2, $\binom{n-i}{j}$ ways, and must multiply the result by the probability of a given set of i outcomes of Type 1, j of Type 2, and $n-i-j$ of Type 3, which is $p^i q^j r^{n-i-j}$. Thus we have

$$p(i, j) = \binom{n}{i}\binom{n-i}{j} p^i q^j r^{n-i-j}$$

$$= \frac{n!}{i!j!(n-i-j)!} p^i q^j r^{n-i-j} \tag{4.17}$$

for $0 \leq i+j \leq n$.

This distribution is an extension of the *binomial* distribution discussed in Example 2.23 of Chapter 2. If formula (4.15), together with the binomial theorem, is used, we find that, individually, X and Y have binomial distributions

$$p_i = \binom{n}{i} p^i (1-p)^{n-i}, \qquad q_j = \binom{n}{j} q^j (1-q)^{n-j} \tag{4.18}$$

Notice that, in this case, $p(i, j) \neq p_i q_j$ and, hence, that X and Y are *not* independent. △

To illustrate one use of the multinomial distribution, suppose that an inspection plan calls for inspecting 5 parts and for either accepting each part, rejecting each part, or submitting the part to further inspection, with probabilities $p = .70$, $q = .20$, $r = .10$, respectively. What is the probability that all 5 parts must be reinspected? What is the probability that none of the parts must be reinspected? What is the probability that at least one of the parts must be reinspected?

To answer the first question, let $X =$ the number accepted and $Y =$ the number rejected. If all 5 must be reinspected, $X = 0$, $Y = 0$, and $n - X - Y = 5$. Thus we observe that

$$\begin{aligned} p(0, 0) &= \binom{5}{0}\binom{5}{0} p^0 q^0 r^5 \\ &= r^5 = (\cdot 10)^5 = 10^{-5} \end{aligned}$$

To answer the second question we agree that if none of the 5 must be reinspected, $n - X - Y = 0$, $X + Y = n = 5$. Thus, to compute the required probability we must sum the $p(i, j)$ over all pairs i, j satisfying $i + j = n$. This gives

$$\begin{aligned} \sum_{i+j=n} p(i, j) &= \sum_{i+j=n} \frac{n!}{i!\,j!\,0!} p^i q^j r^0 = \sum_{i=0}^{n} \binom{n}{i} p^i q^{n-i} \\ &= \binom{5}{0} p^0 q^5 + \binom{5}{1} p^1 q^4 + \cdots + \binom{5}{5} p^5 q^0 \\ &= (p + q)^5 = (.90)^5 = 0.59 \end{aligned}$$

using the binomial formula.

Finally, to determine the probability that at least one part must be reinspected, we have

$$1 - (\text{probability that none must be reinspected}) = 1 - 0.59 = 0.41$$

7. Density Functions: Continuous Case

As in the single variable case, formula (4.14) cannot be extended directly to define a density function in the continuous case since, normally, in this case, $\Pr[X = x \text{ and } Y = y] = 0$. Instead we seek to generalize property (c) of the previous section that relates the distribution and density functions. Since the natural continuous analogue of summation is integration, we define the *joint probability density function*, $f(x, y)$, of the continuous random variables X, Y as a function which, when integrated, yields $F(x, y)$; that is, so that

$$F(x, y) = \int_{-\infty}^{x} \int_{-\infty}^{y} f(s, t)\, dt\, ds \tag{4.19}$$

To guarantee the existence and uniqueness of this function, we confine our attention here to pairs of continuous random variables, (X, Y), whose distribution function, $F(x, y)$, is continuous for all (x, y) and has a continuous mixed partial derivative $\partial^2 F/\partial x\, \partial y$ for all but possibly a finite number of values of x and of y. In this case, we can differentiate (4.19) twice, and can express $f(x, y)$ in terms of $F(x, y)$ as

$$f(x, y) = \frac{\partial^2 F(x, y)}{\partial x\, \partial y} \tag{4.20}$$

Conversely, if (4.20) holds true, and $f(x, y)$ is sufficiently smooth, we can integrate and show that (4.19) must hold true.

The joint probability density function $f(x, y)$ satisfies the following properties.

(a) $f(x, y) \geq 0$ *for all* x, y.

(b) $\int_{-\infty}^{\infty} \int_{-\infty}^{\infty} f(x, y)\, dy\, dx = 1$.

(c) $f(x, y)$ *is continuous for all except possibly finitely many values of* x *or of* y.

Property (b) follows from (4.19) by letting $x \to \infty$, $y \to \infty$, and property (c) holds true by our assumption concerning $\partial^2 F/\partial x\, \partial y$. Property (a) seems reasonable if we interpret joint probability as a two-dimensional mass distribution, with $f(x, y)$ being the *mass density* (mass per unit area) at the point (x, y). A rigorous proof of (a) would involve expressing the partial derivative of $F(x, y)$ as a limit of a difference quotient and the use of property (e) of Section 4.

If R is any region of the two-dimensional plane, then the probability that (X, Y) lies in this region is obtained by integrating $f(x, y)$ over the region. In particular,

$$\Pr[a < X \leq b \text{ and } c < Y \leq d] = \int_{a}^{b} \int_{c}^{d} f(x, y)\, dy\, dx \tag{4.21}$$

Since $\Pr[X = a] = 0 = \Pr[Y = c]$, it follows that the integral in (4.21) also represents $\Pr[a \leq X \leq b \text{ and } c \leq Y \leq d]$ and $\Pr[a < X < b \text{ and } c < Y < d]$.

Thus, if $f(x, y)$ is represented geometrically by a surface, the volumes under this surface give probabilities (see Figure 4.9).

If we interpret $f(x, y)$ as a mass density, then integration over a plane region amounts to determining the total mass in that region.

By (4.10),

$$F_X(x) = F(x, \infty) = \int_{-\infty}^{x} \int_{-\infty}^{\infty} f(s, t)\, dt\, ds$$

and hence

$$f_X(x) = F'_X(x) = \int_{-\infty}^{\infty} f(x, t)\, dt$$

or

$$f_X(x) = \int_{-\infty}^{\infty} f(x, y)\, dy \tag{4.22}$$

where $f_X(x)$ is the individual probability density function of the random variable X. Similarly

$$f_Y(y) = \int_{-\infty}^{\infty} f(x, y)\, dx \tag{4.23}$$

When obtained in this way, $f_X(x)$ and $f_Y(y)$ are sometimes termed the *marginal probability density functions* of X and Y.

If X and Y are independent, the differentiation of formula (4.8) yields

$$f(x, y) = \frac{\partial^2}{\partial x\, \partial y}[F_X(x)F_Y(y)] = F'_X(x)F'_Y(y)$$

$$f(x, y) = f_X(x)f_Y(y) \tag{4.24}$$

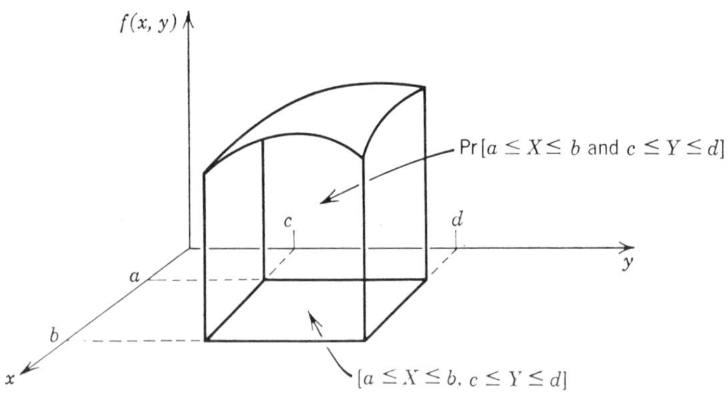

FIGURE 4.9 Probability as volume under the probability density surface $f(x,y)$.

Thus, if X and Y are independent, their joint probability density function factors into the product of the individual density functions. Conversely, if (4.24) holds true, then integration yields (4.8), and we would conclude that X and Y are independent.

Notice that if $f(x, y)$ is known, then $f_X(x)$ and $f_Y(y)$ can be determined by (4.22) and (4.23), but that, conversely, knowledge of $f_X(x)$ and $f_Y(y)$ does *not* in general determine $f(x, y)$. In the *independent* case, however, $f_X(x)$ and $f_Y(y)$ *do* determine $f(x, y)$, by (4.24).

EXAMPLE 4.9. Let X and Y be continuous random variables with ranges $0 < x < \infty$, $0 < y < \infty$, and joint probability density function:

$$f(x, y) = \begin{cases} e^{-(x+y)}, & 0 < x < \infty, 0 < y < \infty \\ 0, & \text{otherwise} \end{cases}$$

In this case

$$\begin{aligned} f_X(x) &= \int_{-\infty}^{\infty} f(x, y)\, dy \\ &= \int_0^{\infty} e^{-(x+y)}\, dy, \qquad \text{since} \qquad f(x, y) = 0 \qquad \text{for} \qquad y < 0 \\ &= e^{-x} \int_0^{\infty} e^{-y}\, dy \\ &= \begin{cases} e^{-x}, & \text{for} \quad 0 < x < \infty \\ 0, & \text{for} \quad -\infty < x \le 0 \end{cases} \end{aligned}$$

Similarly

$$f_Y(y) = \begin{cases} e^{-y}, & 0 < y < \infty \\ 0, & \text{otherwise} \end{cases}$$

Notice that $f(x, y) = f_X(x) f_Y(y)$, so that X and Y are independent here.

To compute a joint probability, we must integrate $f(x, y)$. For example

$$\begin{aligned} \Pr[0 < X < 1 \quad \text{and} \quad 1 < Y < 2] &= \int_0^1 \int_1^2 f(x, y)\, dy\, dx \\ &= \int_0^1 \int_1^2 e^{-(x+y)}\, dy\, dx \\ &= \left(\int_0^1 e^{-x}\, dx\right)\left(\int_1^2 e^{-y}\, dy\right) \\ &= (1 - e^{-1})(e^{-1} - e^{-2}) = 0.148. \quad \triangle \end{aligned}$$

EXAMPLE 4.10. Suppose that, in the previous example, X represents the lifetime (in years) of a certain electronic tube, and that Y represents the lifetime of a replacement tube installed after the failure of the first tube. The time until the second tube fails is given by $Z = X + Y$.

Z clearly has range $0 < z < \infty$, and its distribution and density functions can be calculated as follows.

$$\begin{aligned} F_Z(z) &= \Pr[Z \leq z] = \Pr[X + Y \leq z] \\ &= \iint_{x+y \leq z} f(x, y)\, dx\, dy \\ &= \int_0^z \int_0^{z-y} e^{-(x+y)}\, dx\, dy \\ &= 1 - (1 + z)\, e^{-z}, \qquad \text{for} \qquad z > 0 \end{aligned}$$

$$f_Z(z) = F'_Z(z) = ze^{-z}, \qquad \text{for} \qquad z > 0$$

The probability that the total lifetime exceeds 2 years is

$$\Pr[Z > 2] = 1 - F_Z(2) = 3e^{-2} = .405 \qquad \triangle$$

EXAMPLE 4.11. Let X and Y be continuous random variables with ranges $0 < x < \infty$, $0 < y < \infty$, and joint probability density function:

$$f(x, y) = \begin{cases} xe^{-x(y+1)}, & 0 < y < \infty, \quad 0 < x < \infty \\ 0, & \text{otherwise} \end{cases}$$

$$\begin{aligned} f_X(x) &= \int_0^\infty xe^{-x(y+1)}\, dy = xe^{-x} \int_0^\infty e^{-xy}\, dy \\ &= e^{-x}, \qquad 0 < x < \infty \end{aligned}$$

$$\begin{aligned} f_Y(y) &= \int_0^\infty xe^{-x(y+1)}\, dx \\ &= \frac{1}{(y+1)^2}, \qquad 0 < y < \infty \end{aligned}$$

We observe that $f(x, y) \neq f_X(x) f_Y(y)$, so that X and Y are *not* independent here. $\triangle$

EXAMPLE 4.12. THE BIVARIATE NORMAL DISTRIBUTION. The natural generalization to bivariate distributions of the normal distribution, discussed in Example 3.13 of Chapter 3, is the *bivariate normal distribution*. Two random variables, X and Y, are said to have this distribution if they have ranges $-\infty < x < \infty$, $-\infty < y < \infty$, and joint probability density function of the form:

$$f(x, y) = \frac{1}{2\pi\sigma_X\sigma_Y\sqrt{1-\rho^2}} \exp\left\{\frac{-1}{2(1-\rho^2)}\left[\left(\frac{x-\mu_X}{\sigma_X}\right)^2 - 2\rho\left(\frac{x-\mu_X}{\sigma_X}\right)\left(\frac{y-\mu_Y}{\sigma_Y}\right) + \left(\frac{y-\mu_Y}{\sigma_Y}\right)^2\right]\right\} \quad (4.25)$$

The quantities μ_X, μ_Y, σ_X, σ_Y, ρ are constants with $\sigma_X > 0$, $\sigma_Y > 0$ and $-1 \leq \rho \leq 1$. Routine calculus computations show (i) that this is indeed a probability density function; (ii) that the marginal densities $f_X(x)$, $f_Y(y)$ are normal; and (iii) that X and Y are independent if, and only if, $\rho = 0$. $\triangle$

The bivariate normal occurs rather often in sampling-type problems. Another interesting area of application is concerned with gunfire. Here, it is supposed that, because of barrel whip, aiming errors, windage, projectile defects, and many other things, a gun aimed to hit a point (x', y') will, in general, hit a different point, and the probability distribution for the point of impact can be modeled as a pair of random variables, (X, Y), where X is the horizontal deviation of the hit point and Y is the range deviation of the hit point, measured from some given origin. The probability density function for the impact point is given by (4.25) with $\mu_X = x'$ and $\mu_Y = y'$. Depending on the type of fire and the characteristics of the weapon, ρ may be practically 0, or it may be large.

EXAMPLE 4.13. The function

$$f(x, y) = \begin{cases} 2e^{-x-y}, & 0 \leq x \leq y, \quad 0 \leq y < \infty \\ 0, & \text{otherwise} \end{cases}$$

satisfies all of the properties of a joint probability density function. Furthermore

$$f(x, y) = (2e^{-x})(e^{-y})$$

within the joint range. But

$$\begin{aligned} f_X(x) &= 2\int_x^\infty e^{-x-y}\,dy \\ &= 2e^{-x}\int_x^\infty e^{-y}\,dy \\ &= 2e^{-2x} \end{aligned}$$

$$f_Y(y) = 2\int_0^y e^{-x-y}\,dx$$
$$= 2e^{-y}[1 - e^{-y}]$$

Clearly

$$f(x, y) \neq f_X(x) f_Y(y)$$

and thus X, Y are not independent random variables. This illustrates the principle that for X and Y to be independent, $f(x, y)$ must factor into a function of x multiplied by a function of y for *all x and y, not just within the joint range of X and Y.* △

8. Conditional Distributions

If two random variables are independent, all of the properties of their joint distribution are determined once their marginal density or distribution functions are known. In practical applications, this provides considerable simplification, since these marginal distributions can be estimated separately, and from them the joint probability distribution and density functions can be determined using (4.8) and (4.16) or (4.24). In the case where the random variables are not independent, the joint probabilities can no longer be determined in this simple fashion. Then the question is: What must we know to determine the joint probability density and distribution function of two random variables that are not independent?

We recall that if A and B are two events not necessarily independent, the joint probability of A and B is

$$\Pr[A \cap B] = \Pr[A \mid B] \Pr[B]$$
$$= \Pr[B \mid A] \Pr[A]$$

where $\Pr[A \mid B]$ ($\Pr[B \mid A]$) is the condition probability of A given that B has occurred (of B given that A has occurred).

In particular, if A and B are events of the form $A = [Y \le y]$, $B = [X \le x]$, we write

$$\Pr[Y \le y \mid X \le x] = \Pr[A \mid B]$$

If we hold x fixed, and consider this probability as a function of y, it is termed the *conditional distribution function of Y given that $X \le x$*:

$$G(y \mid x) = \Pr[Y \le y \mid X \le x]$$

If X and Y are *discrete* random variables, then for any pair (i, j) in the joint range of (X, Y) let us define the following conditional probability:

$$p_{ij} = \Pr[Y = j \mid X = i]$$
$$= \frac{\Pr[X = i \text{ and } Y = j]}{\Pr[X = i]} = \frac{p(i, j)}{p_i} \tag{4.26}$$

If p_{ij} is considered as a function of j, for fixed i, it is termed the *conditional probability density function* of Y given that $X = i$. For fixed i, p_{ij} clearly *is* a discrete density function since, using (4.15),

$$\sum_j p_{ij} = \sum_j \frac{p(i, j)}{p_i} = \left(\frac{\sum_j p(i, j)}{p_i}\right)$$
$$= \frac{p_i}{p_i} = 1$$

The notation p_{ij} does not display which random variable is the Y-variable whose conditional probabilities are involved, and which is the X-variable on which the condition is placed. When using this notation, therefore, we must make the distinction clear from the discussion.

In case X and Y are continuous random variables, in analogy with (4.26) we define the quantity:

$$f(y \mid x) = \frac{f(x, y)}{f_X(x)} \tag{4.27}$$

to be the *conditional probability density function of Y given that $X = x$* (provided $f_X(x) \neq 0$].

For fixed x, $f(y \mid x)$ clearly is a probability density function, since, using (4.22),

$$\int_{-\infty}^{\infty} f(y \mid x)\, dy = \int_{-\infty}^{\infty} \frac{f(x, y)\, dy}{f_X(x)} = \frac{f_X(x)}{f_X(x)} = 1$$

To interpret $f(y \mid x)$ as the probability density function of a certain conditional distribution, a limiting argument is necessary: For given x, y, and h, with $h > 0$ and $f_X(x) \neq 0$, define

$$F(y; x, h) = \Pr[Y \leq y \mid x \leq X \leq x + h] \tag{4.28}$$
$$= \frac{\Pr[x \leq X \leq x + h \text{ and } Y \leq y]}{\Pr[x \leq X \leq x + h]}$$
$$= \frac{\displaystyle\int_x^{x+h} \int_{-\infty}^{y} f(s, t)\, dt\, ds}{\displaystyle\int_x^{x+h} f_X(s)\, ds}$$

(See Figure 4.10.)

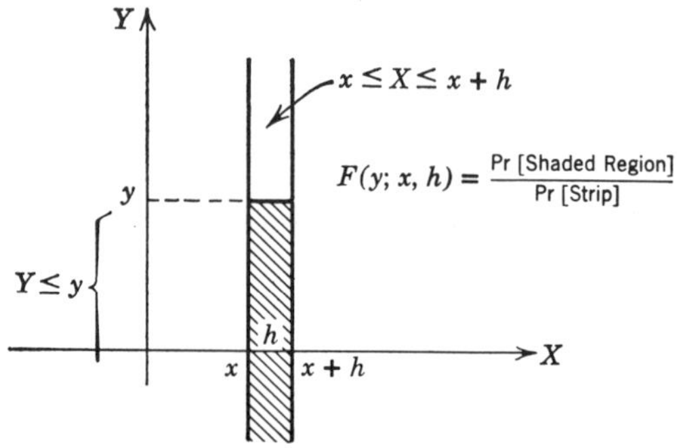

FIGURE 4.10 Evaluation of the conditional distribution function for continuous random variables.

The conditional probability $\Pr[Y \leq y \mid X = x]$ normally will not exist according to the definition of conditional probability that we have been using, since $\Pr[X = x] = 0$ in the continuous case. Nonetheless, it is clear from (4.28) that if we let $h \to 0$, $F(y; x, h)$ might approach a function that could be interpreted as the conditional distribution function of Y given $X = x$. Thus we *define* the *conditional distribution function of* Y, *given* $X = x$, by

$$F(y \mid x) = \lim_{h \to 0} F(y; x, h)$$

To evaluate this limit we use l'Hôpital's rule which, we recall, states that in computing the limit of a quotient which reduces to 0/0, one should compute instead the limit of the quotient of derivatives. We easily check that the hypotheses necessary to apply this rule are satisfied here. Thus

$$F(y \mid x) = \lim_{h \to 0} F(y; x, h) = \lim_{h \to 0} \frac{\int_x^{x+h} \int_{-\infty}^{y} f(s, t)\, dt\, ds}{\int_x^{x+h} f_X(s)\, ds}$$

$$= \lim_{h \to 0} \frac{\int_{-\infty}^{y} f(x+h, t)\, dt}{f_X(x+h)} \; *$$

* Using the fact that $(d/dh) \int_x^{x+h} \phi(s) ds = \phi(x+h)$ by the Fundamental Theorem of Calculus.

$$= \frac{\int_{-\infty}^{y} \lim_{h \to 0} f(x+h, t)\, dt}{\lim_{h \to 0} f_X(x+h)}$$

$$= \frac{\int_{-\infty}^{y} f(x, t)\, dt}{f_X(x)} = \int_{-\infty}^{y} \frac{f(x, t)}{f_X(x)}\, dt$$

$$= \int_{-\infty}^{y} f(t \mid x)\, dt \tag{4.29}$$

where the interchange of limit and integral is also easily justified. (The same conclusion results if $h < 0$.) Comparison of (4.29) with (3.10) shows that the conditional probability density function, $f(y \mid x)$, bears the same relation to the conditional distribution function, $F(y \mid x)$, as the ordinary density function bears to the ordinary distribution function.

If (4.26) and (4.27) are written in the form:

$$p(i, j) = p_i p_{ij}, \qquad f(x, y) = f_X(x) f(y \mid x) \tag{4.30}$$

clearly, one natural way of determining the joint density function of two given random variables is to find the density function of one of those variables and the conditional density of the other, given the first, and multiply.

Notice also that if X and Y are *independent* $[p(i, j) = p_i q_j$ or $f(x, y) = f_X(x) f_Y(y)]$, then

$$p_{ij} = q_j, \quad \text{or} \quad f(y \mid x) = f_Y(y) \tag{4.31}$$

where q_j represents the discrete density of Y. In the independent case, the conditional density of Y is the same as the unconditional density of Y.

EXAMPLE 4.14. In Example 4.7, we observed that we had reason to believe that the random variables X and Y were not independent and that we had reason to believe that a heavy dosage of a drug affected the probability of a tumor. To pursue this point further, let us assume that the data given in Table 4.6 are in fact the joint probabilities for the random variables X, Y. We would want to ask of this data the following kind of question: If the rat has been given a heavy dosage, what is his probability of getting the tumor? In terms of the random variables of Example 4.7, we want to know $\Pr[Y=0 \mid X=0]$. From (4.26) this is given by

$$p_{00} = \frac{\Pr[X=0,\ Y=0]}{\Pr[X=0]} = \frac{.010}{.100} = 0.10$$

Since $q_0 = \Pr[Y=0] = 0.037$, we notice that the conditional probability is considerably larger than the corresponding marginal probability. Roughly speaking, the information given by the conditioning variable significantly (in this case) changes the probability of the other random variable. Table 4.8 gives the entire conditional probability density function of Y given X for Example 4.7.

TABLE 4.8[a]

i \ j	0	1
0	0.100	0.900
1	0.030	0.970
2	0.030	0.970

[a] The Conditional Density Function $p_{ij} = \Pr[Y=j \mid X=i]$

△

EXAMPLE 4.15. Consider the situation discussed in Example 4.11 with $f(x, y) = xe^{-x(y+1)}$ for $0 < y < \infty$, $0 < x < \infty$. In this case

$$f(y \mid x) = \frac{xe^{-x(y+1)}}{e^{-x}} = xe^{-xy}$$

for $0 < y < \infty$, $0 < x < \infty$. If x has a value for which $f_X(x) \neq 0$ while $f(x, y) = 0$, that is, if $x > 0$ but $y \leq 0$, then $f(y \mid x) = 0$. If $x \leq 0$, $f(y \mid x)$ is undefined.

This situation is sketched in Figure 4.11. Observe that, for a fixed value of x, the graph of $f(y \mid x)$ is related to the curve on the $f(x, y)$-surface of points with fixed x-coordinate. If we denote this curve by $v = v(y) = xe^{-x(y+1)}$ (x fixed), then $f(y \mid x) = e^x v(y)$. The factor e^x is required to adjust

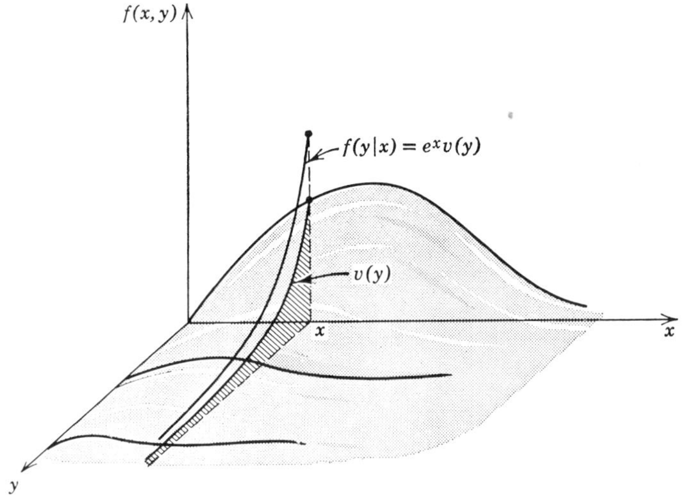

FIGURE 4.11 Cross sections of joint density surface $f(x,y) = xe^{-x(y+1)}$, with relation to $f(y|x)$.

the curve so that the total area under it is 1, a necessary requirement for any probability density function.

Notice also that we can compute the opposite conditional density function

$$f(x \mid y) = \frac{f(x, y)}{f_Y(y)} = \frac{xe^{-x(y+1)}}{1/(y+1)^2} = x(y+1)^2 \, e^{-x(y+1)}$$

for $0 < y < \infty$, $0 < x < \infty$. The student should make a sketch similar to Figure 4.11 in which this curve is indicated. △

EXAMPLE 4.16. In Example 4.9, observe that

$$f(y \mid x) = \frac{e^{-(x+y)}}{e^{-x}} = e^{-y} = f_Y(y)$$

as must be the case when X and Y are independent. △

EXAMPLE 4.17. The following kind of argument has been used to model the length of time that occurs between successive demands for airplane spare parts. Suppose that the random variable X represents the time between successive calls for spare parts and

$$f(x \mid \lambda) = \lambda e^{-\lambda x}, \qquad 0 \le x < \infty$$

The value λ is itself a value of a random variable Λ whose density function is given by

$$f_\Lambda(\lambda) = \begin{cases} \lambda e^{-\lambda}, & 0 \le \lambda < \infty \\ 0, & \text{otherwise} \end{cases}$$

That is, we suppose that the time between demands depends on the value taken by another random variable Λ. Given that $\Lambda = \lambda$, the time between demands is a negative exponentially distributed random variable. The random variable Λ has the given marginal density function. Physically Λ might be interpreted as a characteristic of the particular item in the airplane that has broken down.

We want to know the marginal density function of X, the inter-demand time. We notice immediately, using (4.30), that

$$f(x, \lambda) = \lambda^2 e^{-\lambda(x+1)}, \qquad 0 \le x < \infty, \qquad 0 \le \lambda < \infty$$

We easily check that $f(x, \lambda)$ is a proper joint probability density function. From it, we obtain

$$f_X(x) = \int_0^\infty \lambda^2 e^{-\lambda(x+1)}\, d\lambda$$

$$= \frac{2}{(x+1)^3}, \qquad 0 \le x < \infty$$

$$F_X(x) = \int_0^x f_X(s)\, ds$$

$$= \begin{cases} 1 - (1 + x)^{-2}, & 0 \le x < \infty \\ 0, & \text{otherwise.} \end{cases}$$

$f_X(x)$ and $F_X(x)$ are shown in Figure 4.12. We can then determine probabilities for various levels of demand as follows

$$\Pr[X \le 1] = F_X(1) = 1 - (1 + 1)^{-2} = \frac{3}{4}$$

$$\Pr[X > 2] = 1 - F_X(2) = 1 - [1 - (1 + 2)^{-2}] = \frac{1}{9}$$

$$\Pr[1 \le X \le 3] = F_X(3) - F_X(1) = [1 - (1 + 3)^{-2}] - [1 - (1 + 1)^{-2}]$$

$$= \frac{1}{4} - \frac{1}{16} = \frac{3}{16}$$

Notice that these probabilities are quite different from those obtained from $f(x \mid \lambda)$ for a given λ (for example, for a given item). $\triangle$

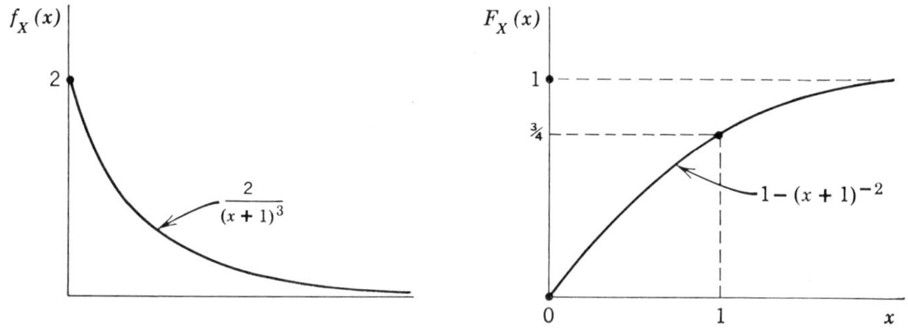

FIGURE 4.12 The marginal density and distribution function for X in Example 4.17.

EXAMPLE 4.18. In the previous examples, we required that both random variables be continuous or discrete. All of our results carry over to the mixed case in which one of the variables is discrete and the other is continuous. To illustrate these ideas, we consider the following problem, which is of importance in the area of waiting-line theory (see Chapter 13).

Suppose that people arrive singly and at random at a theatre ticket booth, line up for service, are served, and depart. The number M_s of individuals arriving in a time interval of fixed length s is one random variable of interest here. This is a discrete random variable with range $m = 0, 1, 2, \ldots$. Let us assume that it has a probability density function of the form

$$\Pr[M_s = m] = \frac{e^{-\lambda s}(\lambda s)^m}{m!} \tag{4.32}$$

for $m = 0, 1, 2, \ldots$, where λ is a given positive constant. The justification and realism of such an assumption about M_s will be discussed in Chapter 7, Section 7. Another random variable of interest is the time T that it takes the ticket seller to service a single customer. This is a continuous random variable that we assume has probability density function:

$$f_T(t) = \begin{cases} \lambda e^{-\lambda t}, & 0 < t < \infty \\ 0, & \text{otherwise} \end{cases} \tag{4.33}$$

Again, the reason for this assumption will be discussed later. Still another random variable of interest is the number N of arrivals during the time T that one customer is being served. We wish to find the probability density function of this random variable N.

Let us start by considering the pair (T, N). Formula (4.32) tells us what the probability density of N would be if T had a fixed value, $T = t$. In other words (4.32) can be used to give the *conditional* probability density of N given that $T = t$:

$$f(n \mid t) = \Pr[N = n \mid T = t] = \frac{e^{-\lambda t}(\lambda t)^n}{n!} \tag{4.34}$$

for $n = 0, 1, 2, \ldots$. The *joint* probability density function of (T, N) is then obtained by multiplying (4.33) and (4.34), using (4.30):

$$\begin{aligned} f(t, n) &= f_T(t) f(n \mid t) \\ &= \frac{\lambda e^{-\lambda t}\, e^{-\lambda t}(\lambda t)^n}{n!} \\ &= \frac{\lambda e^{-2\lambda t}(\lambda t)^n}{n!} \end{aligned}$$

for $n = 0, 1, 2, \ldots,$ and $0 < t < \infty$. Notice that all these manipulations carry through in spite of the fact that T is continuous and N is discrete. To find the probability density function of N alone, we apply (4.23), integrating $f(t, n)$, since T is continuous:

$$\begin{aligned} p_n = \Pr[N = n] &= \int_0^\infty f(t, n)\, dt \\ &= \int_0^\infty \frac{\lambda e^{-2\lambda t}(\lambda t)^n}{n!}\, dt \\ &= \frac{1}{n!} \int_0^\infty e^{-2\lambda t}(\lambda t)^n\, d(\lambda t) \end{aligned}$$

This integral can be evaluated in various ways. If, for instance, we substitute $\lambda t = u$ and use a table of integrals, we obtain, for $n = 0, 1, 2, \ldots,$

$$\begin{aligned} p_n &= \frac{1}{n!} \frac{n!}{2^{n+1}} \\ &= \left(\frac{1}{2}\right)^{n+1} \end{aligned} \tag{4.35}$$ △

Clearly this example has many analogues, such as orders coming in to a factory, particles being counted in a Geiger counter, demands coming to a computer facility, breakdown and repair of electronic equipment, etc. It is interesting to note that here the resulting density, p_n, does not involve the constant λ. Our discussion is also easily generalized to cover the case when the value of λ in the density of M_s differs from that in the density of T (see Exercise 21).

In Example 2.3 in Chapter 2, a distribution almost identical to the distribution of (4.35) occurred when we discussed the experiment of tossing a coin until the first head appeared. Coin-tossing, card-playing, etc., at first might appear to have little connection with the real world of the scientist or engineer. Yet in this problem, which is considerably more complicated and "real" than a simple coin-tossing experiment, we obtain the same density function. Hence, all of the analysis for the coin-tossing experiment is valid for the random variable N. In fact, we could *simulate* the arrival process during a service time by the simple expedient of flipping a coin. These relations abound in scientific and engineering studies, and are studied under categories such as *similitude* or *simulation*. When those relations can be found they provide the scientist with a powerful tool for studying the behavior of things that are difficult to observe (for example, N) by watching the behavior of a much more mundane and easily observed phenomena (coin-tossing). Thus we would be mistaken if we concluded that seemingly unrealistic ideas such as coin-tossing and card-playing are devoid of important, useful, science- and engineering-oriented content.

9. Many Random Variables

In the previous sections we have restricted our attention to two random variables. These concepts extend immediately to more than two random variables defined on the same sample space. In this section we indicate some of the extensions.

Let X_1, X_2, ... X_n be n random variables defined on the same sample space. Then their *joint distribution function* is defined as

$$F(x_1, x_2 \ldots x_n) = \Pr[X_1 \leq x_1 \text{ and } X_2 \leq x_2 \text{ and} \ldots \text{and } X_n \leq x_n]$$

For brevity of notation, we write

$$F(x_1, x_2 \ldots x_n) = \Pr[X_1 \leq x_1; X_2 \leq x_2; \ldots; X_n \leq x_n] \tag{4.36}$$

Again, we can consider the discrete or continuous cases. If the random variables are discrete (with integer ranges), then their joint probability density function is

$$p(i_1, i_2, \ldots, i_n) = \Pr[X_1 = i_1; X_2 = i_2; \ldots; X_n = i_n] \tag{4.37}$$

The marginal densities of one or more of the random variables are obtained by summing (4.37) appropriately. For example

$$p_{X_n}(i_n) = \Pr[X_n = i_n] = \sum_{i_1} \sum_{i_2} \cdots \sum_{i_{n-1}} p(i_1, i_2, \ldots, i_n) \tag{4.38}$$

$$\begin{aligned} p_{X_1, \cdots, X_{n-1}}(i_1, \ldots, i_{n-1}) &= \Pr[X_1 = i_1; \ldots; X_{n-1} = i_{n-1}] \\ &= \sum_{i_n} p(i_1, i_2, \ldots, i_n) \end{aligned} \tag{4.39}$$

Conditional densities are defined similarly:

$$\begin{aligned} p(i_n \mid i_1, \ldots, i_{n-1}) &= \Pr[X_n = i_n \mid X_1 = i_1; \ldots; X_{n-1} = i_{n-1}] \\ &= \frac{p(i_1, i_2, \ldots, i_n)}{p_{X_1, \cdots, X_{n-1}}(i_1, \ldots, i_{n-1})} \end{aligned} \tag{4.40}$$

or

$$p(i_1, i_2, \ldots, i_n) = p_{X_1 \, X_{n-1}}(i_1, \ldots, i_{n-1}) p(i_n \mid i_1, \ldots, i_{n-1}) \tag{4.41}$$

In the continuous case, assuming that $F(x_1, x_2, \ldots, x_n)$ has a continuous mixed partial derivative, we define the *joint probability density function* as

$$f(x_1, x_2, \ldots, x_n) = \frac{\partial^n F(x_1, x_2, \ldots, x_n)}{\partial x_1 \, \partial x_2 \ldots \partial x_n} \tag{4.42}$$

Integration of this function gives probabilities, such as

$$\begin{aligned} &\Pr[a_1 \le X_1 \le b_1; \ldots; a_n \le X_n \le b_n] \\ &\qquad = \int_{a_n}^{b_n} \cdots \int_{a_1}^{b_1} f(x_1, x_2, \ldots, x_n) \, dx_1 \ldots dx_n \end{aligned} \tag{4.43}$$

Again, marginal densities are obtained by integration. For example

$$f_{X_n}(x_n) = \int_{-\infty}^{\infty} \cdots \int_{-\infty}^{\infty} f(x_1, x_2, \ldots, x_n) \, dx_1 \ldots dx_{n-1} \tag{4.44}$$

$$f_{X_1, \cdots, X_{n-1}}(x_1, \ldots, x_{n-1}) = \int_{-\infty}^{\infty} f(x_1, x_2, \ldots, x_n) \, dx_n \tag{4.45}$$

Conditional densities can also be defined:

$$f(x_n \mid x_1, \ldots, x_{n-1}) = \frac{f(x_1, x_2, \ldots, x_n)}{f_{X_1, \cdots, X_{n-1}}(x_1, \ldots, x_{n-1})} \tag{4.46}$$

or

$$f(x_1, x_2, \ldots, x_n) = f_{X_1, \cdots, X_{n-1}}(x_1, \ldots, x_{n-1}) f(x_n \mid x_1, \ldots, x_{n-1}) \tag{4.47}$$

The random variables $X_1, X_2, \ldots, X_n$ are said to be *mutually independent* if

$$p(i_1, i_2, \ldots, i_n) = p_{i_1}^{(1)} p_{i_2}^{(2)} \ldots p_{i_n}^{(n)} \tag{4.48}$$

where $p_i^{(j)}$ denotes the probability density function of X_j, $p_i^{(j)} = p_{X_j}(i) = \Pr[X_j = i]$. In the continuous case *mutual independence* means

$$f(x_1, x_2, \ldots, x_n) = f_{X_1}(x_1) f_{X_2}(x_2) \ldots f_{X_n}(x_n) \tag{4.49}$$

If the variables are mutually independent, then

$$p(i_n \mid i_1, \ldots, i_{n-1}) = p_{i_n}^{(n)} \tag{4.50}$$

or

$$f(x_n \mid x_1, \ldots, x_{n-1}) = f_{X_n}(x_n) \tag{4.51}$$

Notice that it is possible for every pair of random variables in the set to be independent (*pairwise independent*) without the entire set being mutually independent.

EXAMPLE 4.19. Suppose two coins are tossed, and let X_1 represent the number of heads on the first coin tossed (0 or 1), let X_2 represent the number of heads on the second coin tossed (0 or 1), and let X_3 represent the number of matches (0 or 1). By the result of Example 2.21 in Chapter 2 we observe that the *pairs* (X_1, X_2), (X_1, X_3), and (X_2, X_3) are each independent, but that the set (X_1, X_2, X_3) is not mutually independent. $\triangle$

Problems involving several random variables frequently can be simplified by representing the n-tuples $(X_1, X_2, \ldots, X_n)$ and $(x_1, x_2, \ldots, x_n)$ as *vectors*

$$\mathbf{X} = (X_1, X_2, \ldots, X_n), \qquad \mathbf{x} = (x_1, x_2, \ldots, x_n)$$

We make the convention that the inequality $\mathbf{X} \leq \mathbf{x}$ means that $X_i \leq x_i$ for $i = 1, 2, \ldots, n$. The joint distribution function is then simply

$$F(\mathbf{x}) = \Pr[\mathbf{X} \leq \mathbf{x}]$$

The use of vector and matrix operations when working with sets of several random variables greatly simplifies many computations.

EXAMPLE 4.20. *The multivariate normal distribution.* The single variable normal distribution defined in Example 3.13 and the bivariate normal of Example 4.12 can be generalized to many variables most conveniently by using vector and matrix terminology. Random variables $X_1, X_2, \ldots, X_n$

are said to have a *multivariate normal distribution* if they have ranges $-\infty < x_i < \infty$ and joint probability density function

$$f(\mathbf{x}) = f(x_1, \ldots, x_n) = \frac{1}{(2\pi)^{n/2}\sqrt{|\mathbf{A}|}} \exp\{-\tfrac{1}{2}(\mathbf{x} - \boldsymbol{\mu})\mathbf{A}^{-1}(\mathbf{x} - \boldsymbol{\mu})'\} \tag{4.52}$$

where

(i) $\boldsymbol{\mu} = (\mu_1, \ldots, \mu_n)$ is a fixed vector;
(ii) "prime" denotes "transpose";
(iii) $\mathbf{A}$ is a fixed square matrix satisfying $\mathbf{xAx}' > 0$ for every nonzero vector $\mathbf{x}$;
(iv) $|\mathbf{A}|$ represents the determinant of $\mathbf{A}$ [which can be shown to be positive, using (iii)].

This distribution is particularly important in applications to statistics. One crucial property for statistical theory is that if $X_1, \ldots, X_n$ have a multivariate normal distribution and that if $Y_1, Y_2, \ldots, Y_k$ are obtained from $X_1, \ldots, X_n$ by a *linear* transformation

$$Y_i = \sum_{j=1}^{n} b_{ij} X_j \qquad \text{for} \qquad i = 1, 2, \ldots, k$$

then $Y_1, Y_2, \ldots, Y_k$ also have a multivariate normal (or degenerate normal) distribution. △

10. Random Sequences or Discrete Parameter Random Processes

One of the basic purposes of this book is to study the situation when a random experiment leads not to a single random variable, but to an entire *sequence*

$$\{X_i : i = 1, 2, 3, \ldots\} = \{X_1, X_2, X_3, \ldots\}$$

of random variables. We have already encountered these sequences when we discussed sequences of coin-tossings or parts inspections. Sequences of random variables have wide applicability to diverse fields such as business orders, breakdowns of machinery, lifetimes of electronic equipment, subatomic particle counting, computer operation, communication systems, epidemics, genetic linkages, and many others. Any system that varies in a random fashion with time and is observed at some sequence of times falls into this pattern.

A sequence of random variables defined on the same sample space is termed a *random sequence* or a *discrete parameter random process*. (The term *discrete parameter* refers to the subscript i in the sequence $\{X_i : i = 1, 2, 3, \ldots\}$.) Sometimes the phrase *stochastic process* is used instead of *random process*.

The ranges of the random variables may be either discrete or continuous. The discrete case is somewhat simpler to handle and is sufficient for a wide variety of applications. We will concentrate most heavily on it. For our present introduction, assume that the random variables in the sequence $\{X_1, X_2, X_3, \ldots\}$ are discrete, with ranges consisting of sets of *integers* ("*states* of the *system*"). The statement that the jth random variable has value m, $X_j = m$, is expressed by saying that "the system is in *state m* at the jth *stage*," or that "the system is in *state m* at *time j*."

Basically our problem consists in determining, or in answering certain questions about, the joint probability density function or distribution function of $X_1, X_2, \ldots, X_n$,

$$p(i_1, i_2, \ldots, i_n) = \Pr[X_1 = i_1; X_2 = i_2; \ldots; X_n = i_n]$$

or

$$F(x_1, x_2, \ldots, x_n) = \Pr[X_1 \leq x_1; X_2 \leq x_2; \ldots; X_n \leq x_n]$$

We may wish to know these functions for a certain value of n, for all n, for all sufficiently large n, or, possibly, we may want to answer questions that involve limits as $n \to \infty$.

One way of computing these probabilities is by the repeated use of formula (4.41):

$$\begin{aligned} p(i_1, i_2, \ldots, i_n) &= p_{X_1, \ldots, X_{n-1}}(i_1, \ldots, i_{n-1}) p(i_n \mid i_1, \ldots, i_{n-1}) \\ &= \ldots \\ &\vdots \end{aligned}$$

$$p(i_1, i_2, \ldots, i_n) = p_{i_1}{}^{(1)}\, p(i_2 \mid i_1)\, p(i_3 \mid i_1, i_2) \ldots p(i_n \mid i_1, \ldots, i_{n-1}) \tag{4.53}$$

Here $p_{i_1}{}^{(1)}$, as previously, stands for the density function of X_1, $p_{i_1}{}^{(1)} = \Pr[X_1 = i_1]$, and the meanings of the other conditional probabilities are obvious. In terms of direct probabilities, (4.53) can be written as

$$\begin{aligned} \Pr[X_1 = i_1; \ldots; X_n = i_n] = \Pr[X_1 = i_1] \Pr[X_2 = i_2 \mid X_1 = i_1] \ldots \\ \Pr[X_n = i_n \mid X_1 = i_1; \ldots; X_{n-1} = i_{n-1}] \end{aligned}$$

In complex problems the determination of the individual density function $p_{i_1}{}^{(1)}$ of X_1 and the various conditional probability density functions appearing in (4.53) may be very difficult. In surprisingly many practical problems, however, certain general simplifications are possible that greatly facilitate the discussion. The two most important simplifications—the *independent* case and the *Markov dependent* case—are discussed briefly in the next two sections. More extensive discussions, particularly of the second case, will occur throughout the remainder of the book.

11. Random Sequences: Independent Case

The simplest possible random sequence $\{X_1, X_2, \ldots\}$ is a sequence of *mutually independent* random variables (every finite subset is mutually independent). In this case

$$p(i_n \mid i_1, \ldots, i_{n-1}) = p_{i_n}^{(n)}$$

and

$$p(i_1, i_2, \ldots, i_n) = p_{i_1}^{(1)} p_{i_2}^{(2)} \cdots p_{i_n}^{(n)} \tag{4.54}$$

In this case, only the individual probability density functions $p_i^{(1)}, p_i^{(2)}, \ldots$ of $X_1, X_2, \ldots$ must be determined to compute the joint probabilities.

If, in addition, all the random variables have the *same* distribution, $p_i^{(j)} = p_i$ for each j, the sequence is said to be a sequence of *independent, identically-distributed* random variables. In this case (4.54) becomes

$$p(i_1, i_2, \ldots, i_n) = p_{i_1} p_{i_2} \cdots p_{i_n} \tag{4.55}$$

Simple examples of these processes are easy to construct. Given a sequence of tosses of an honest coin as the random experiment, the sequence $\{X_1, X_2, \ldots\}$, where X_n is taken to be 1 or 0 depending on whether the nth toss is a head or a tail, will be a sequence of independent, identically distributed random variables. The process would have states $\{0, 1\}$ and common probability density function $p_0 = \frac{1}{2}$, $p_1 = \frac{1}{2}$. A slightly more useful model is the following.

EXAMPLE 4.21. Consider a sequence $\{X_1, X_2, \ldots\}$ of independent random variables, each of which can take only the values $+1$ and -1, with probabilities p and q, respectively ($p + q = 1$). Thus $p_1{}^{(n)} = p$, $p_{-1}{}^{(n)} = q$, for each n. Thus the joint probability density function is

$$p(i_1, i_2, \ldots, i_n) = p^k q^{n-k}$$

where $i_j = \pm 1$ for each j, and k represents the number of $+1$'s in the set $(i_1, i_2, \ldots, i_n)$.

This simple model has many important applications to fields such as variations in inventory size, gambling wins and losses, Brownian motion, switching circuits ($+1 =$ closed switch, $-1 =$ open switch), binary communication codes, and many others (see Chapter 7, Section 5). △

12. Random Sequences: Markov Dependent Case

The random sequence $\{X_1, X_2, \ldots\}$ is said to be a *Markov dependent sequence* or a *Markov process* if the conditional probability

$$p(i_n \mid i_1, \ldots, i_{n-1}) = \Pr[X_n = i_n \mid X_1 = i_1; \ldots; X_{n-1} = i_{n-1}]$$

depends only on the values of i_n and i_{n-1}. This means that

$$p(i_n \mid i_1, \ldots, i_{n-1}) = p(i_n \mid i_{n-1}) \tag{4.56}$$

$$\Pr[X_n = i_n \mid X_1 = i_1; \ldots; X_{n-1} = i_{n-1}] = \Pr[X_n = i_n \mid X_{n-1} = i_{n-1}]$$

Generally speaking, identity (4.56) means that, if the system is known to have reached state i_{n-1} at the $(n-1)$st stage, it is immaterial what chain of states it passed through on the way to this stage as far as predicting (giving probabilities for) the state it will enter at the *next* stage is concerned. In other words, the system has no "memory" that would allow it to use information about its behavior *before* a known state was reached to modify the probabilities for the next stage. If the state of the system is known at the "present time" (stage $n-1$), then the "future" (stage n) behavior of the system does not depend on the "past" (stages $1, 2, \ldots, n-2$). This property is frequently stated somewhat vaguely as: "In a Markov dependent sequence, knowledge of the present makes the future independent of the past."

For such a sequence (4.53) becomes

$$p(i_1, i_2, \ldots, i_n) = p_{i_1}^{(1)}\, p(i_2 \mid i_1) p(i_3 \mid i_2) \ldots p(i_n \mid i_{n-1}) \tag{4.57}$$

The probabilities $p(i_n \mid i_{n-1})$ are termed *transition probabilities*. The notation $p(i_n \mid i_{n-1})$ is somewhat ambiguous here since this function depends on *three* variables, i_n, i_{n-1} and n, not just two. A better notation will be introduced later.

Markov dependant random sequences are of wide applicability. In general, any process which is "memoryless" in the sense discussed, and which keeps a running total, or lists net position or effect, will fall into this category. In Chapters 7, 8, 10, and 13 discuss these numerous processes in detail.

As a simple example, consider a sequence of tosses of an honest coin, and let X_n denote the total number of heads in the first n tosses. The sequence $\{X_1, X_2, \ldots\}$ is not an independent sequence since knowledge of the number of heads in the first $n-1$ tosses definitely affects the number of heads in the first n tosses. However, if X_{n-1} is known, say $X_{n-1}=k$ (k heads in the first $n-1$ tosses), then X_n must take one of the values k or $k+1$ with equal probability, and any further information about the exact *order* in which the k heads occurred in the first $n-1$ tosses (information about $X_1, X_2, \ldots, X_{n-2}$) will not change these probabilities. Thus

$$p(i_n \mid i_1, \ldots, i_{n-1}=k) = p(i_n \mid i_{n-1}=k) = \begin{cases} \dfrac{1}{2}, & \text{if } \quad i_n = k \quad \text{or} \quad k+1 \\ 0, & \text{otherwise} \end{cases}$$

We conclude that this *is* a Markov dependent sequence.

Notice that in the preceding illustration, the conditional probability $p(i_n \mid i_{n-1})$ did not, in fact, involve the value of n. When this happens, we say that the sequence has *stationary transition probabilities*. In this case, we will adopt a somewhat different notation for the transition probabilities:

$$p_{ij} = p(i_n = j \mid i_{n-1} = i) = \Pr[X_n = j \mid X_{n-1} = i] \tag{4.58}$$

where the *stationarity* is indicated by the fact that p_{ij} does not depend on n. Thus p_{ij} represents the probability that a transition from state i to state j takes place at any given stage of the sequence. For such a sequence, formula (4.57) can be written

$$p(i_1, i_2, \ldots, i_n) = p_{i_1}^{(1)} p_{i_1 i_2} p_{i_2 i_3} \cdots p_{i_{n-1} i_n} \tag{4.59}$$

and we observe that all joint probability density functions are determined once the *initial* probability density function $p_i^{(1)}$ and the *transition* function p_{ij} are known.

EXAMPLE 4.22. In Example 4.21 we considered an independent sequence $\{X_1, X_2, \ldots\}$ where each $X_i = +1$ or -1, with probabilities p and q. Suppose we now define

$$Y_n = X_1 + X_2 + \cdots + X_n$$

for $n = 1, 2, \ldots,$ and consider the sequence:

$$\{Y_1, Y_2, \ldots\}$$

If we interpret the X-sequence as a sequence of wins-and-losses, then Y_n represents a player's net winnings after n trials. If the X-sequence represents an independent sequence of unit steps of $+1$ or -1 on the real axis, then Y_n will represent the position after n steps. For this reason the sequence $\{Y_1, Y_n, \ldots\}$ is termed a *random walk* (see Chapter 7, Section 5). This particular random sequence is extremely important both theoretically and in many practical problems. The study of its properties, and certain generalizations, occupies a large part of probability theory. Here we only point out that $\{Y_1, Y_2, \ldots\}$ is clearly *not* an independent sequence in spite of the fact that it is derived from the independent sequence $\{X_1, X_2, \ldots\}$. It *is* a Markov dependent sequence since

$$Y_n = Y_{n-1} + X_n$$

and, hence, if Y_{n-1} is given, Y_n will depend only on X_n, which is independent of all previous X's and Y's. Thus

$$\begin{aligned} p(i_n \mid i_{n-1}) &= \Pr[Y_n = i_n \mid Y_{n-1} = i_{n-1}] \\ &= \Pr[Y_{n-1} + X_n = i_n \mid Y_{n-1} = i_{n-1}] \\ &= \Pr[i_{n-1} + X_n = i_n \mid Y_{n-1} = i_{n-1}] \\ &= \Pr[X_n = i_n - i_{n-1}] \end{aligned}$$

since X_n is independent of $X_1, \ldots, X_{n-1}$ and, hence, of Y_{n-1}. It follows that

$$p(i_n \mid i_{n-1}) = \begin{cases} p, & \text{if} \quad i_n = i_{n-1} + 1 \\ q, & \text{if} \quad i_n = i_{n-1} - 1 \\ 0, & \text{otherwise} \end{cases}$$

Thus the sequence is observed to have stationary transition probabilities

$$p_{ij} = \begin{cases} p, & \text{if} \quad j = i + 1 \\ q, & \text{if} \quad j = i - 1 \\ 0, & \text{otherwise} \end{cases}$$

for $i, j = 0, \pm 1, \pm 2, \ldots$.

In Figure 4.13 we show graphically two runs of a simulated experiment to yield an X-sequence with $p = q = \frac{1}{2}$. In Figure 4.14 the corresponding realizations of the random walk Y-sequence are shown. $\triangle$

EXAMPLE 4.23. Consider this inventory problem. Items arrive at a retail outlet singly and at random in such a way that the number M_s arriving in time s has probability density function:

$$\Pr[M_s = n] = e^{-\lambda s}(\lambda s)^n / n!, \qquad n = 0, 1, 2, \ldots$$

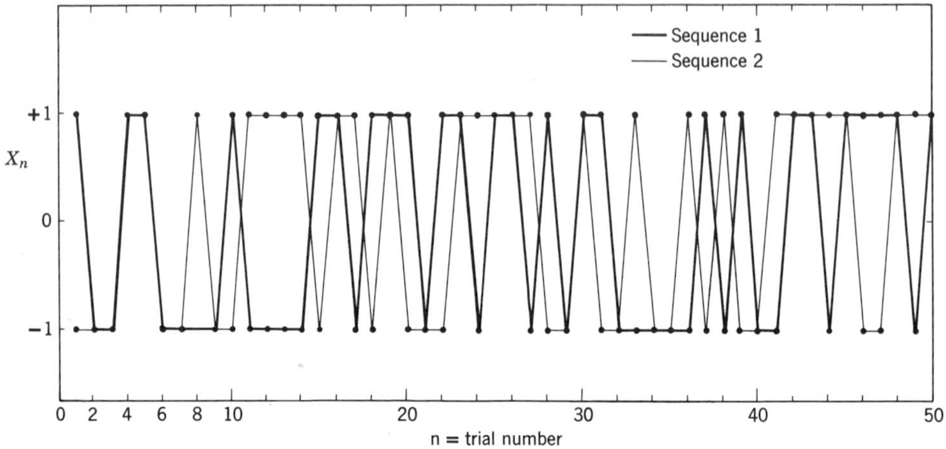

FIGURE 4.13 Two experimental realizations of an independent sequence $X_1, X_2, \ldots,$ where $X_n = \pm 1$ with probability 1/2.

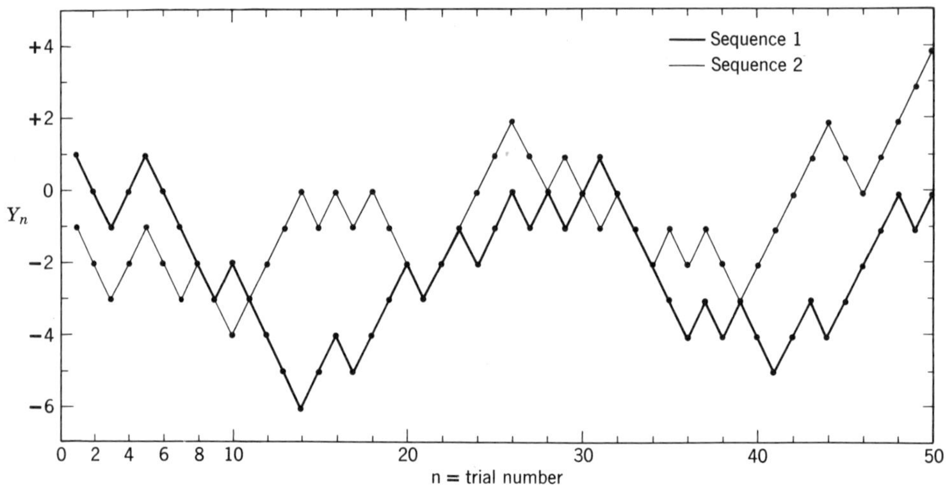

FIGURE 4.14 The corresponding experimental realizations of the random walk sequence $Y_1, Y_2, \ldots,$ where $Y_n = X_1 + \ldots + X_n$.

Customers arrive independently in such a way that the time interval T between two buying customers has probability density function:

$$\lambda e^{-\lambda t}, \qquad 0 < t < \infty$$

If the stock is exhausted when a customer arrives, he leaves without buying.

The analogy with Example 4.18 is clear, except that here customers may continue to arrive even when the stock is empty, unlike the working of the ticket seller.

As in Example 4.18 we can show that the number N of item arrivals between two successive customers is a random variable having probability density function:

$$\left(\frac{1}{2}\right)^{n+1}, \qquad n = 0, 1, 2, \ldots$$

We can also show that the sequence $\{N_1, N_2, \ldots\}$ of item arrivals between successive customers is an independent sequence. If $M_n = N_1 + \ldots + N_n$, then the sequence $\{M_1, M_2, \ldots\}$ is a Markov dependent sequence in which M_n represents the total number of items that have arrived up to the arrival of the nth customer. Let us define a new sequence $\{Z_1, Z_2, \ldots\}$ recursively:

$$Z_1 = \begin{cases} N_1 - 1, & \text{if } N_1 > 0 \\ 0, & \text{if } N_1 = 0 \end{cases}$$

$$Z_{i+1} = \begin{cases} Z_i + N_{i+1} - 1, & \text{if } Z_i + N_{i+1} > 0 \\ Z_i + N_{i+1}, & \text{if } Z_i + N_{i+1} = 0 \end{cases}$$

We see that Z_n represents the number of items remaining in stock after the departure of the nth customer. The sequence $\{Z_1, Z_2, \ldots\}$ can also be shown to be Markov dependent. The properties of this sequence are important in the field of *waiting-line theory*, and are studied in Chapter 13. △

EXAMPLE 4.24. In reliability theory we suppose that a component has a physical lifetime that is a random variable. Assume that whenever a component fails it is immediately replaced by another, and let $\{X_1, X_2, \ldots\}$ represent the sequence of lifetimes of successive components. It is usual (and seems reasonable in many real systems) to assume that this sequence is a sequence of independent, identically distributed random variables. They are continuous rather than discrete random variables, but the discussion of the previous sections can be extended to cover this case. The quantity

$$Y_n = X_1 + X_2 + \cdots + X_n$$

would represent the time until the failure of the nth component. The sequence $\{Y_1, Y_2, \ldots\}$ is a Markov dependent sequence that uses an extension of definition (4.56) to cover the continuous case. △

EXERCISES

1. Two random variables X and Y have ranges $X=0, 1$; $Y=0, 1, 2$. For points (x, y) in the joint range, the joint distribution function $F(x, y)$ is given by the following table, with the function extended to points outside the range in the usual way.

$x \backslash y$	0	1	2
0	1/12	1/3	2/3
1	1/6	7/12	1

Compute

(a) $\Pr[0 < X < 2 \quad \text{and} \quad 0 < Y < 2]$

(b) $F_X(x) \quad \text{and} \quad F_Y(y)$

(c) $\Pr[X = 1 \quad \text{and} \quad Y = 1]$

(d) $\Pr[X = 0 \quad \text{and} \quad Y < 1\tfrac{1}{2}]$

2. Two random variables X and Y have ranges $0 < X < 1$, $0 < Y < 1$, and joint distribution function:

$$F(x, y) = \begin{cases} 0 \ \text{ for } x \leq 0 \quad \text{or} \quad y \leq 0 \\ xy \ \text{for } 0 < x < 1 \quad \text{and} \quad 0 < y < 1 \\ x \ \text{ for } 0 < x < 1 \quad \text{and} \quad y \geq 1 \\ y \ \text{ for } 0 < y < 1 \quad \text{and} \quad x \geq 1 \\ 1 \ \text{ for } x \geq 1 \quad \text{and} \quad y \geq 1 \end{cases}$$

(a) Sketch the joint range of X and Y.

(b) Check that $F(x, y)$ satisfies properties (a) to (e) of Section 4.

(c) Compute $F_X(x)$ and $F_Y(y)$.

(d) Compute

(i) $\Pr[X > \tfrac{1}{2} \quad \text{and} \quad Y > \tfrac{1}{2}]$

(ii) $\Pr[X > \tfrac{1}{2} \quad \text{and} \quad Y < 2]$

(iii) $\Pr[X = \tfrac{1}{2} \quad \text{and} \quad Y = \tfrac{1}{2}]$

3. For each of the following determine for what value (or values) of the constant k is $F(x, y)$ the joint distribution function of some pair of random variables. Justify your answers.

(a) $$F(x, y) = \begin{cases} k(1 - e^{-x})(1 - e^{-y}), & \text{for} \quad x > 0, y > 0 \\ 0 \quad , & \text{otherwise} \end{cases}$$

(b) $$F(x, y) = \begin{cases} (1 - ke^{-x})(1 - e^{-y}), & \text{for} \quad x \geq 0, y \geq 0 \\ 0 \quad , & \text{otherwise} \end{cases}$$

(c) $$F(x, y) = \begin{cases} (1 - ke^{-x})(1 - e^{-y}), & \text{for} \quad x > 0, y > 0 \\ 0 \quad , & \text{otherwise} \end{cases}$$

4. Two parts of the same machine have lifetimes X and Y, where these are independent random variables each having the negative exponential distribution of Example 3.4,

$$F(x) = \begin{cases} 1 - e^{-ax}, & \text{for} \quad x > 0 \\ 0 \quad , & \text{for} \quad x \leq 0 \end{cases}$$

where a is a positive constant.

(a) Compute the joint distribution function $F(x, y)$.

(b) Compute $\Pr[X < a \text{ and } Y > 2a]$.

(c) As in Example 4.5, let U and V represent the time until the first failure and the second failure, $U = \min\{X, Y\}$, $V = \max\{X, Y\}$.

(i) Compute $F_U(u)$, $F_V(v)$, and the joint distribution function of U and V.

(ii) Using (*i*) compute the probability that the first failure occurs before time a and the second after time $2a$, $\Pr[U < a \text{ and } V > 2a]$. Compare your result with the answer to part (b) and comment.

5. Let X and Y be two discrete random variables with joint probability density function given by

x \ y	1	2	3
1	1/12	1/6	1/12
2	1/6	1/4	1/12
3	1/12	1/12	0

Compute the probability of these events:

(a) X is less than $2\frac{1}{2}$

(b) X is even

(c) XY is even

(d) Y is odd, given that X is odd

6. Two cards are drawn from a deck of 52 (without replacement). Let X be the number of aces and Y the number of kings obtained.

(a) Find the joint range and joint probability density function of X and Y.

(b) Determine the individual density and distribution functions of X and Y.

(c) Compute $\Pr[X \geq Y]$.

7. A company manufactures items, each of which can have 0, 1, or 2 defects with probabilities 0.7, 0.2, 0.1. If an item has 2 defects, however, it is caught by the inspectors and is replaced by a perfect item before delivery. Let X represent the original number of defects in an item produced and Y the number of defects in the corresponding delivered item. Give the joint probability density function of X and Y and the density function of Y alone, in tabular form. If a delivered item is observed to have 0 defects, what is the probability that it is a replacement?

8. Derive formulas (4.18) for the trinomial distribution.

9. In a long-range radio communication system, errors in transmission result from either human error in sending or receiving, or natural interference (static). Suppose that a series of 20 symbols are sent. Each has probability 0.10 of being in error because of natural interference, and probability 0.05 of being in error because of human factors only. Assume independence of the error possibilities of different symbols.
   - **(a)** What is the probability of 2 natural interference errors and 1 human error in 20 symbols?
   - **(b)** If 2 out of 20 symbols are found to be in error, what is the probability that both were the result of human error?

10. Let X and Y be random variables having joint probability density functon: $f(x, y) = e^{-(x+y)}$ for $x > 0$, $y > 0$; $f(x, y) = 0$ otherwise. Compute
   - **(a)** $\Pr[X > 1 \text{ and } Y > 1]$
   - **(b)** $\Pr[X > 1 \text{ or } Y > 1]$
   - **(c)** $\Pr[X > 1]$
   - **(d)** $\Pr[X + Y \leq 4]$
   - **(e)** $\Pr[X = Y]$
   - **(f)** $\Pr[X > 1 \mid Y > 1]$
   - **(g)** $\Pr[X > Y \mid Y > 1]$

11. In Exercise 10, let $Z = X + Y$.
   - **(a)** Compute $F_Z(z) = \Pr[Z \leq z]$.
   - **(b)** Compute $f_Z(z)$.
   - **(c)** Compute the joint distribution function $F_{X,Z}(x, z)$ of X and Z.
   - **(d)** Compute $f_{X,Z}(x, z)$.
   - **(e)** Are X and Z independent? Explain.
   - **(f)** If X and Y represent the lifetimes of a light bulb and a replacement bulb, interpret the meaning of Z.

12. Two random variables X and Y have joint density function $f(x, y) = cxy$ for $0 < x < y < 1$; $f(x, y) = 0$ otherwise.
   - **(a)** Sketch the joint range of X and Y.
   - **(b)** Find the constant c.

(c) Compute $f_X(x)$ and $f_Y(y)$.

(d) Are X and Y independent? Explain.

13. A gun is aimed at a certain point (taken as the origin of a coordinate system). Because of random factors the actual hit-point can be any point (X, Y) in a circle of radius R about the origin. Assume that the joint density of X and Y is constant in this circle:

$$f(x, y) = C \text{ for } x^2 + y^2 \leq R^2, \quad f(x, y) = 0, \text{ otherwise.}$$

(a) Compute C.

(b) Show that $f_X(x) = (2/\pi R)\sqrt{1 - (x/R)^2}$ for $-R \leq x \leq R$

$f_X(x) = 0$ otherwise

(c) If the target area is a square with diagonal $2R$, centered at the origin, what is the probability of hitting the target?

(d) If the target is a highway of width $R/2$ centered along the x-axis, what is the probability of hitting the highway?

(e) Are X and Y independent? Explain.

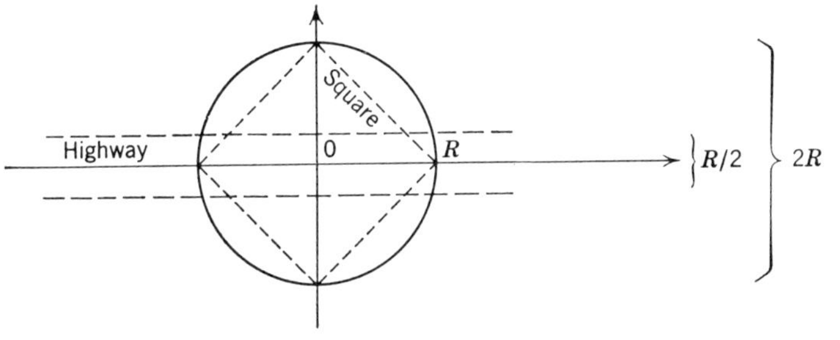

14. Let X, Y, and Z be random variables with joint density function:

$f(x, y, z) = 12x^2yz$ for $0 < x < 1$, $0 < y < 1$, $0 < z < 1$; $f(x, y, z) = 0$ otherwise.

Compute

$$\Pr[X < Y < Z]$$

15. In Exercise 7, compute (in tabular form) the conditional density

$$p_{ij} = \Pr[Y = j \mid X = i] \qquad \text{for} \qquad i, j = 0, 1, 2$$

16. Let X and Y be continuous random variables with joint density function: $f(x, y) = e^{-y}$ for $0 < x < y$; $f(x, y) = 0$ otherwise.

(a) Compute the individual density functions of X and Y.

(b) Are X and Y independent? Explain.

(c) Determine the condition density $f(x \mid y)$.

(d) Compute $\Pr[X \leq 1 \mid Y = 2]$.

17. If X and Y are continuous random variables with joint density function that is given by the following, compute the constant c, the individual density and distribution functions of X and Y, $f(y \mid x)$ and $f(x \mid y)$:

(a) $f(x, y) = cxy$ for $0 < x < 1,\ 0 < y < 1$

(b) $f(x, y) = cx(1 - y)$ for $0 < x < 1,\ 0 < y < 1$

(c) $f(x, y) = cxe^{-y}$ for $0 < x < y$

(d) $f(x, y) = c(1 - x^2 - y^2)$ for $x^2 + y^2 \le 1$

18. In Example 4.15, make a sketch similar to Figure 4.11 but indicate the surface cross-sections for fixed y, and their relation to $f(x \mid y)$.

19. If X represents the time interval between successive automobiles that pass a given intersection, then the event $[X > a]$ means that the next car has not arrived a time units after the first car has passed. If we observe that this event has occurred, then $Y = X - a$ represents the remaining time that we must wait until the next car passes. Suppose X has probability density function: $f(x) = e^{-x}$ for $x > 0$; $f(x) = 0$ otherwise.

(a) Compute the conditional distribution function of Y
$G(y) = \Pr[Y \le y \mid X > a] = \Pr[X - a \le y \mid X > a]$.

(b) Compute the conditional density $g(y) = G'(y)$.

(c) Compare $g(y)$ to $f(y)$ and comment concerning the relation between the conditional distribution of Y and the distribution of the original X.

20. In Example 4.18 compute the conditional density function $f(t \mid n)$, and interpret its meaning.

21. In Example 4.18, assume that M_s has discrete density $e^{-\lambda s}(\lambda s)^m/m!$ for $m = 0$, $1, 2, \ldots$, and T has density $\mu e^{-\mu t}$ for $t > 0$, where λ and μ are not necessarily the same.

(a) Compute the joint density of T and N.

(b) Show that N has density $p_n = \mu\lambda^n/(\lambda + \mu)^{n+1}$ for $n = 0, 1, 2, \ldots$.

(c) Show that $\Pr[N \ge 1] < \frac{1}{2}$ if and only if $\lambda < \mu$.

22. Let $\{X_1, X_2, \ldots\}$ be an independent random sequence where each X_i takes only the values 1 and 0 with probabilities p and q, $p + q = 1$. Show that $X_1, X_2, \ldots,$ X_n has joint density $p(i_1, i_2, \ldots, i_n) = p^t q^{n-t}$ where $t = \sum_{k=1}^{n} i_k$.

23. In Exercise 22, let $S_k = \sum_{i=1}^{k} X_i$ for $k = 1, 2, \ldots$.

(a) Show that the sequence $\{S_1, S_2, \ldots\}$ is a Markov dependent sequence.

(b) Show that the transition probabilities are given by the following, where $\alpha = i_n - i_{n-1}$:

$$p(i_n \mid i_{n-1}) = \begin{cases} p^\alpha q^{1-\alpha} & \text{for} \quad \alpha = 0 \text{ or } 1 \\ 0 & \text{otherwise} \end{cases}$$

24. A communication system has the property that, if a symbol is correctly transmitted, the probability is 0.9 that the next symbol is correct. While, if a symbol is incorrectly transmitted, the probability is 0.5 that the next symbol is incorrect. The transmission can be modeled by the Markov dependent random sequence $\{X_1, X_2, \ldots\}$ where $X_i = 1$ if the ith symbol is correctly transmitted, $X_i = 0$ if the ith symbol is incorrect. Assume the probability is 0.7 that the first symbol is correctly transmitted.

 (a) Compute the transition probabilities $p(i_n \mid i_{n-1})$.

 (b) Compute $p(i_1, i_2, \ldots, i_n)$.

 (c) Compute $\Pr[X_3 = 1]$.

 (d) If the kth symbol is correct, $X_k = 1$, what is the probability that the $k + 2$nd symbol is correct, that is, $X_{k+2} = 1$.

25. Let $\{X_1, X_2, \ldots\}$ be an independent sequence of discrete random variables. Let

$$S_k = \sum_{i=1}^{k} X_i \text{ for } k = 1, 2, \ldots$$

Show that $\{S_1, S_2, \ldots\}$ is a Markov dependent sequence.

5 • Expectation

1. Introduction

If X is a random variable, we have observed that, to describe completely the probability behavior of X, we must specify an entire *function*, the distribution function $F(x)$, or the probability density function $f(x)$ or $p(x_i)$. Frequently, we want a single *number*, instead of a function, that gives us as much information about the random variable as a single number can, namely, where the "center" of the distribution of X lies. Many possible ways of constructing this number exist. The *median*, which is defined as the number t for which $\Pr[X \leq t) = \Pr[X \geq t]$, (if such a number exists), or the *mode*, which is defined as the number t, for which $f(t)$ or $p(x_i)$ is a maximum (again provided such a number exists uniquely), are two values sometimes used for this purpose.

2. Expectation

The most important and useful number for the location of the "center" of the distribution of X is the *expectation* or *mean*, usually denoted by $E[X]$. (See Figure 5.1.)

If X is a random variable, then the expectation, $E[X]$, of X is defined by

$$E[X] = \begin{cases} \sum_i x_i p(x_i), & \text{in the discrete case} \\ \int_{-\infty}^{\infty} x f(x)\, dx, & \text{in the continuous case} \end{cases} \tag{5.1}$$

$E[X]$ is defined by (5.1) provided that the sum or integral involved converges *absolutely*:

$$\left(\sum_i |x_i|\, p(x_i) < \infty \qquad \text{or} \qquad \int_{-\infty}^{\infty} |x| f(x)\, dx < \infty \right)$$

Otherwise, we say that $E[X]$ does not exist.

If X is a discrete random variable with range a set of integers and probability density function $\{p_i\}$, then (5.1) takes the form:

$$E[X] = \sum_i ip_i \tag{5.2}$$

It is important to realize that $E[X]$ is a fixed number, a *constant*. If we interpret the probability distribution of X as a mass distribution of total mass 1, then $E[X]$ represents the *center of gravity* of that mass distribution. Alternately, $E[X]$ can be interpreted as the *weighted average* of the values of X, using the probability density function, $f(x)$ or $p(x_i)$, as the weight function.

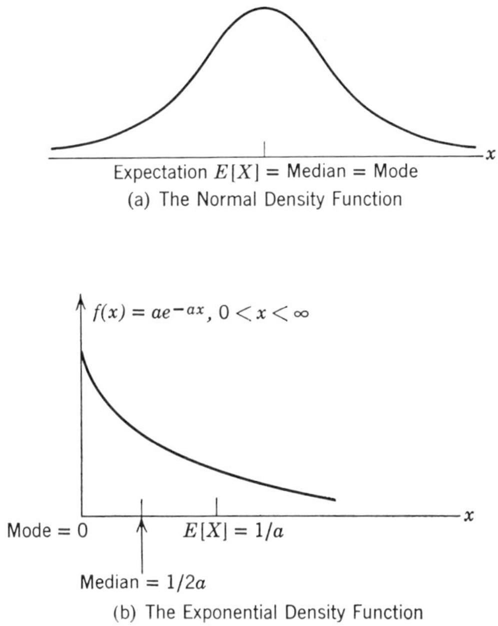

FIGURE 5.1 Mean, median, and mode for two density functions.

EXAMPLE 5.1. Consider an experiment that has two possible outcomes, "success" and "failure," with probabilities p and q, $p + q = 1$. For instance, toss a weighted coin and designate "heads" as a "success," and "tails" as a "failure." Let the random variable X assign the number 1 to a "success" and 0 to a "failure." Thus, X has range $\{0, 1\}$ and probability density function: $p_0 = q$, $p_1 = p$. Such a random variable is said to have a *Bernoulli* distribution. In this case

$$E[X] = \sum ip_i = 0p_0 + 1p_1 = 0q + 1p = p.$$

In particular, if $p = q = \frac{1}{2}$, then $E[X] = \frac{1}{2}$. △

This example shows that $E[X]$ should not be interpreted as a value that we would "expect" X to take. If X has range $\{0, 1\}$, then we could never expect X to take the value $\frac{1}{2}$.

EXAMPLE 5.2. Suppose that X has the negative exponential density function discussed in Example 3.11 of Chapter 3.

$$f(x) = \begin{cases} ae^{-ax}, & \text{for} \quad 0 \le x < \infty \\ 0, & \text{otherwise} \end{cases}$$

Then, using integration by parts,

$$\begin{aligned} E[X] &= \int_{-\infty}^{\infty} x\,f(x)\,dx = \int_0^{\infty} x(ae^{-ax})\,dx \\ &= a\int_0^{\infty} xe^{-ax}\,dx \\ &= \frac{1}{a}\int_0^{\infty} ue^{-u}\,du, \qquad \text{where} \quad u = ax \\ &= \frac{1}{a} \; \triangle \end{aligned}$$

EXAMPLE 5.3. Let X be a continuous random variable having range $-\infty < x < \infty$, and probability density function:

$$f(x) = \frac{1}{\pi(1 + x^2)}, \qquad -\infty < x < \infty$$

The graph of $f(x)$ is the symmetric, bell-shaped curve sketched in Figure 5.2. To show that $f(x)$ is a probability density function, we must show that $\int_{-\infty}^{\infty} f(x)\,dx = 1$. This can be done directly:

$$\int_{-\infty}^{\infty} \frac{dx}{\pi(1 + x^2)} = \lim_{\substack{a \to -\infty \\ b \to +\infty}} \frac{1}{\pi} \operatorname{Arctan} x \Big|_a^b = \frac{1}{\pi}\left[\frac{\pi}{2} - \left(\frac{-\pi}{2}\right)\right] = 1$$

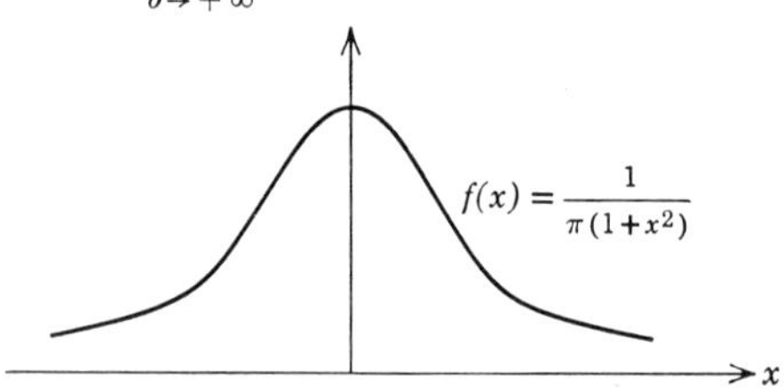

FIGURE 5.2 The Cauchy density function.

A random variable having this probability density function is said to have a *Cauchy distribution*. For such a random variable, definition (5.1) gives

$$E[X] = \frac{1}{\pi} \int_{-\infty}^{\infty} \frac{x}{1+x^2}\, dx$$

Now

$$\frac{1}{\pi} \int_{-\infty}^{\infty} \left| \frac{x}{1+x^2} \right| dx = \frac{2}{\pi} \int_{0}^{\infty} \frac{x}{1+x^2}\, dx = \lim_{t \to \infty} \frac{2}{\pi} \ln(1+t^2) = \infty$$

Thus the integral does not converge absolutely, and $E[X]$ does not exist in this case. △

3. Functions of a Random Variable

In many applications a random variable of interest Y is determined directly by another random variable X, whose distribution is known. This means that there is some function ϕ that can be used to determine the values of Y directly from the values of X,

$$Y = \phi(X)$$

More precisely, for each sample point s, $Y(s) = \phi[X(s)]$. The determination of the distribution of Y from the distribution of X depends on how complicated the function ϕ is. We shall not attempt to give a general method; instead we discuss certain important special cases. The methods used in these cases cover most applications of interest.

EXAMPLE 5.4. LINEAR FUNCTIONS. The simplest and most important situation occurs when the function ϕ is *linear*

$$Y = aX + b$$

This can be interpreted by saying that Y differs from X only in origin and scale of measurement. Thus, if X represents a randomly selected temperature in degrees Centegrade, $Y = \frac{9}{5}X + 32$ will represent the same temperature in degrees Fahrenheit. If X represents a dimensional measurement in inches, $Y = 25.4\ X$ will represent the same measurement in millimeters.

The range of Y clearly will consist of all numbers y of the form $y = ax + b$ where x is in the range of X.

The distribution function of Y is

$$\begin{aligned} F_Y(y) &= \Pr[Y \leq y] = \Pr[aX + b \leq y] \\ &= \Pr\left[X \leq \frac{y-b}{a}\right], \qquad \text{provided } a \text{ is positive} \\ &= F_X\left(\frac{y-b}{a}\right) \end{aligned}$$

Thus the probability density function will be

$$p_Y(y) = p_X\left(\frac{y-b}{a}\right) \qquad \text{in the discrete case}$$

$$f_Y(y) = \frac{dF_Y(y)}{dy} = \frac{1}{a} f_X\left(\frac{y-b}{a}\right) \qquad \text{in the continuous case}$$

$(a > 0)$. $\triangle$

EXAMPLE 5.5. Let X have the standard normal probability density function discussed in Example 3.13

$$f_X(x) = \frac{1}{\sqrt{2\pi}} e^{-x^2/2}, \qquad -\infty < x < \infty$$

and let

$$Y = X^2$$

This example differs fundamentally from the previous one in that two different values of X will normally correspond to the same value of Y; for example, $X = +2$ and -2 both yield $Y = 4$. The range of Y is clearly $0 \leq y < \infty$. Its distribution function is

$$\begin{aligned} F_Y(y) &= \Pr[Y \leq y] = \Pr[X^2 \leq y] \\ &= \Pr[-\sqrt{y} \leq X \leq \sqrt{y}], \qquad \text{for} \qquad y > 0 \\ &= F_X(\sqrt{y}) - F_X(-\sqrt{y}) \\ &= 2F_X(\sqrt{y}) - 1, \qquad \text{using the symmetry of } X \end{aligned}$$

Its probability density function is

$$
\begin{aligned}
f_Y(y) &= \frac{dF_Y(y)}{dy} = 2F'_X(\sqrt{y})\,\frac{1}{2\sqrt{y}} \\
&= \frac{1}{\sqrt{y}}\,f_X(\sqrt{y}) \\
&= \frac{1}{\sqrt{2\pi y}}\,e^{-y/2}, \qquad \text{for} \quad y > 0 \\
f_Y(y) &= 0, \qquad\qquad\qquad\quad \text{for} \quad y \le 0
\end{aligned}
$$

The random variable Y is said to have a *chi-square distribution with one degree of freedom*. This distribution is extremely important in many applications to statistics. △

4. Expectation of a Function of a Random Variable

Assume that X is a random variable and that ϕ is a real-valued function (say, a continuous function). We wish to discuss the expectation $E[\phi(X)]$ of the random variable $\phi(X)$.

Since E denotes a weighted average, it seems reasonable that the expectation of the random variable $\phi(X)$ should be given by

$$
E[\phi(X)] = \begin{cases} \sum_i \phi(x_i)p(x_i), & \text{in the discrete case} \\ \int_{-\infty}^{\infty} \phi(x)f(x)\,dx, & \text{in the continuous case} \end{cases} \tag{5.3}
$$

where $f(x)$ or $p(x_i)$ is the probability density function of X.

Equation (5.3) is true, but it does not follow directly from the definition (5.1) of expectation. It is, in fact, a theorem that requires proof.

To obtain $E[\phi(X)]$ by using the definition (5.1) directly, we proceed as follows: Denote the random variable $\phi(X)$ by Y, $Y = \phi(X)$. Let $f_Y(y)$ be the probability density function of this new random variable Y. Then, by (5.1),

$$
E[\phi(X)] = E[Y] = \begin{cases} \sum_j y_j f_Y(y_j), & \text{discrete case} \\ \int_{-\infty}^{\infty} y f_Y(y)\,dy, & \text{continuous case} \end{cases} \tag{5.4}
$$

The proof that (5.3) and (5.4) always give the same value for $E[\phi(X)]$ is beyond the level of this book and is omitted. Instead we illustrate the identity of (5.3) and (5.4) by some examples.

EXAMPLE 5.6. Suppose that you play a game in which you either win \$1 with probability $\frac{1}{4}$, lose \$1 with probability $\frac{1}{4}$, or tie with probability $\frac{1}{2}$. Let $X =$ your winnings. Then X is a random variable with range $\{-1, 0, 1\}$ and probability density function $p_{-1} = \frac{1}{4}$, $p_0 = \frac{1}{2}$, $p_1 = \frac{1}{4}$, $p_i = 0$ otherwise. (See Figure 5.3.)

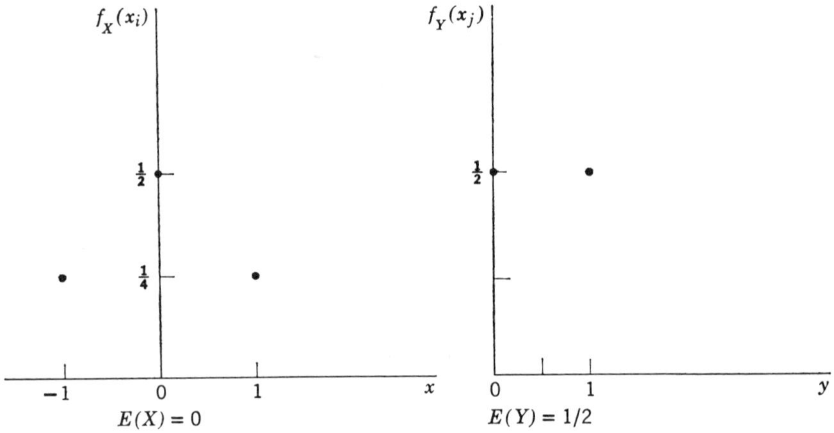

FIGURE 5.3 $\{p_i\}$ for X and Y and $Y = X^2$.

Let $Y = X^2$. Then Y has range $\{0, 1\}$, and represents the amount of money that changes hands in the game.
By (5.3),

$$E[Y] = E[X^2] = \sum_i x_i^2 f(x_i)$$

$$= (-1)^2 p_{-1} + 0^2 p_0 + 1^2 p_1$$

$$= \frac{1}{4} + 0 + \frac{1}{4} = \frac{1}{2}$$

On the other hand the alternative formula (5.4) can be applied directly if we note that Y has range $\{0, 1\}$, and probability density function q_j where $q_0 = \Pr[Y = 0] = \Pr[X = 0] = \frac{1}{2}$, $q_1 = \Pr[Y = 1] = \Pr[X^2 = 1] = \Pr[X = \pm 1] = \frac{1}{4} + \frac{1}{4} = \frac{1}{2}$, and $q_j = 0$ otherwise, (see Figure 5.3). Thus,

$$E[Y] = E[X^2] = \sum_j y_j q_j$$

$$= 0\left(\frac{1}{2}\right) + 1\left(\frac{1}{2}\right) = \frac{1}{2}$$

We notice that, in this case at least, (5.3) and (5.4) give the same result. △

5. Linear Properties of Expectation

The simplest type of function ϕ to which formula (5.3) might be applied is a linear function of the type that is discussed in Example 5.4. Suppose that X is a random variable with probability density function $f(x)$, and define

$$U = aX + b$$

where a and b are constants. Two random variables, X and U, that are related by this equation can differ only in origin and scale of measurement.

By (5.3),

$$E[U] = E[aX + b] = \int_{-\infty}^{\infty} (ax + b)f(x)\, dx$$

$$= a\int_{-\infty}^{\infty} xf(x)\, dx + b\int_{-\infty}^{\infty} f(x)\, dx$$

$$= aE[X] + b$$

in the continuous case. By replacing $\int$ with $\sum$, the same formula is seen to hold true in the discrete case. Thus

$$E[aX + b] = aE[X] + b \tag{5.5}$$

In other words, (5.5) says: "*The expectation of a linear function of a single random variable is the linear function obtained by replacing the random variable by its expectation.*"

Notice that, in particular, if $a = 0$, then $E[b] = b$; that is, the expectation of a constant is that constant.

6. Moments: Mean and Variance

One simple type of function ϕ, whose expectation might be of interest, is the power function, $\phi(X) = X^k$, for $k = 1, 2, 3, \ldots$. The expectation of this function is termed the kth *moment of the random variable* X, and is denoted by μ_k',

$$\mu_k' = E[X^k] = \begin{cases} \int_{-\infty}^{\infty} x^k f(x)\, dx, & \text{continuous case} \\ \sum_i x_i^k p(x_i), & \text{discrete case} \end{cases} \tag{5.6}$$

for $k = 1, 2, 3, \ldots$. The first moment, μ_1', is the ordinary expectation of X. This is used frequently, and is termed the *mean* of the random variable, usually denoted simply by μ or μ_X,

$$\mu = \mu_X = \mu_1' = E[X]$$

Of course, the integral or sum in (5.6) may not converge absolutely, hence, μ_k' may not exist for certain values of k.

EXAMPLE 5.7. Let X be a continuous random variable that has a constant probability density function over the range $0 \le x \le 2$ (a *uniform* distribution)

$$f(x) = \begin{cases} \dfrac{1}{2}, & \text{for } 0 \le x \le 2 \\ 0, & \text{otherwise} \end{cases}$$

In this case

$$\mu_k' = \int_{-\infty}^{\infty} x^k f(x)\, dx = \int_0^2 x^k \cdot \frac{1}{2}\, dx = \frac{2^k}{k+1}$$

In particular

$$\mu = \mu_1' = 1, \; \mu_2' = \frac{4}{3}$$

△

To eliminate from the discussion the effect of the position of the origin in the scale of measurement, it is more useful to work with powers of $X - \mu$, the *deviation from the mean*, rather than with powers of X. We define

$$\mu_k = E[(X - \mu)^k] \tag{5.7}$$

for $k = 1, 2, \ldots$, to be the kth *moment about the mean* of the random variable X.

Clearly

$$\mu_1 = E[X - \mu] = E[X] - \mu = \mu - \mu = 0, \qquad \text{by (5.5)}$$

The quantity

$$\mu_2 = E[(X - \mu)^2] \tag{5.8}$$

is of great importance in the field of statistics. It is called the *variance* of the distribution, and is symbolized by σ^2 or Var[X].

$$\operatorname{Var}[X] = \sigma^2 = \mu_2 = \begin{cases} \int_{-\infty}^{\infty} (x-\mu)^2 f(x)\,dx, & \text{continuous case} \\ \sum_i (x_i-\mu)^2 p(x_i), & \text{discrete case} \end{cases} \tag{5.9}$$

We notice that σ^2 is the weighted average of the values of $(X-\mu)^2$. Thus, if X has a "diffuse" or "spread-out" distribution, and takes values far away from μ with comparatively large probability, then σ^2 will be large. If X has a "concentrated" distribution, and takes only values near to μ with large probability, then σ^2 will be small. Thus, σ^2 is a measure of the *spread* of the distribution (Figure 5.4).

The computation of σ^2 frequently can be simplified as follows:

$$\begin{aligned} \sigma^2 &= E[(X-\mu)^2] \\ &= E[X^2 - 2\mu X + \mu^2] \\ &= E[X^2] - 2\mu E[X] + \mu^2, \qquad \text{by (5.5)} \\ &= E[X^2] - 2\mu^2 + \mu^2 \end{aligned}$$

or

$$\sigma^2 = \mu_2' - \mu^2 = E[X^2] - \{E[X]\}^2 \tag{5.10}$$

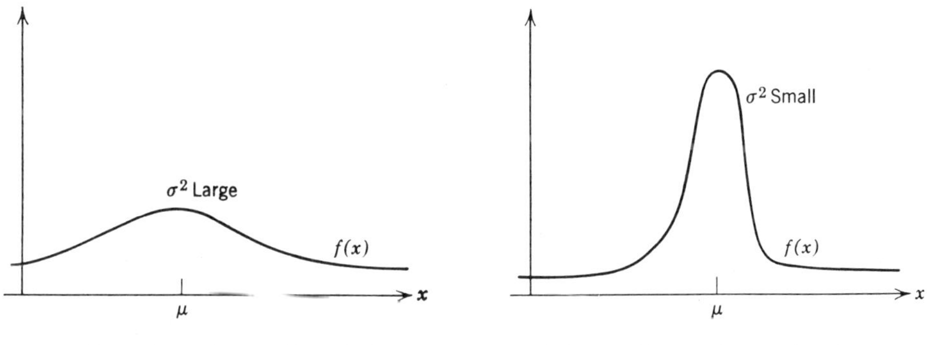

FIGURE 5.4 Illustration of σ^2 as a measure of "spread" of a density function.

EXAMPLE 5.8. Consider a gambling game in which a player wins \$1.00 with probability p or loses \$1.00 with probability q, $p+q=1$.

If X represents his winnings, then X has range $\{+1, -1\}$, and probability density function $p(1)=p$, $p(-1)=q$. X will have mean

$$\mu = E[X] = \sum x_i p(x_i) = 1 \cdot p(1) + (-1) \cdot p(-1) = p - q$$

and second moment

$$\mu_2' = E[X^2] = 1^2 \cdot p(1) + (-1)^2 \cdot p(-1) = p + q = 1$$

The variance σ^2 can be calculated directly using (5.9):

$$\begin{aligned}\sigma^2 = E[(X-\mu)^2] &= [1-(p-q)]^2 p(1) + [-1-(p-q)]^2 p(-1) \\ &= [(p+q)-(p-q)]^2 p + [-(p+q)-(p-q)]^2 q \\ &= 4q^2p + 4p^2q = 4pq(p+q) = 4pq\end{aligned}$$

The same result can be obtained using (5.10):

$$\begin{aligned}\sigma^2 = \mu_2' - \mu^2 &= 1 - (p-q)^2 \\ &= 1 - p^2 + 2pq - q^2 \\ &= 1 - (p^2 + 2pq + q^2) + 4pq \\ &= 1 - (p+q)^2 + 4pq \\ &= 1 - 1 + 4pq = 4pq\end{aligned}$$ △

EXAMPLE 5.9. Consider a random variable X that has the uniform distribution discussed in Example 5.7.

Using (5.10) we obtain

$$\sigma^2 = \mu_2' - \mu^2 = \left(\frac{4}{3}\right) - 1^2 = \frac{1}{3}$$ △

The quantity $\sigma = \sqrt{\sigma^2}$ is termed the *standard deviation* of the random variable X. The use of a single letter σ for this quantity, rather than for the variance, is primarily for dimensional reasons. σ has the same units as X, while σ^2 has the units of X^2. It is important to realize that the expectation $E[X]$ and the various moments of the random variable X are *numbers* determined by the distribution of the random variable, that is, by its probability density function.

It is possible for two random variables to have the same probability density function and, hence, the same moments, without being, in any sense, the same random variable. They may even be independent random variables. For example, if X is the number appearing on one roll of a die and Y is the number appearing on the second roll of the die, then X and Y clearly have the same ranges, probability density functions, and moments. However, we cannot say that $X = Y$ always. In fact, this is an unlikely event and has probability only $\frac{1}{6}$ of occurring in any given experiment.

An important theorem of analysis, which we shall not prove, states that if X and Y are random variables that have corresponding moments of all orders existing and equal, $E[X^k] = E[Y^k]$ for $k = 0, 1, 2, \ldots$, then X and Y have the same distribution (the same probability density function).

7. The Chebyshev Inequality

The statement in the preceding section that a large σ^2 means a diffuse distribution and a small σ^2 means a concentrated one can be made more precise by a means of the following theorem.

THEOREM. *Let the random variable X have mean μ and variance σ^2, and let t be any positive number, $0 < t < \infty$. Then*

$$\Pr[|X-\mu| \geq t] \leq \frac{\sigma^2}{t^2} \qquad (5.11)\,\triangle$$

This inequality, termed the *Chebyshev inequality*, places a bound on the probability that the random variable will differ from its mean by more than a fixed number t. This bound is directly proportional to σ^2 and inversely proportional to t^2.

Formula 5.11 can be proved as follows (in the continuous case):

PROOF. $\sigma^2 = E[(X-\mu)^2] = \int_{-\infty}^{\infty} (x-\mu)^2 f(x)\,dx$

$$= \int_{-\infty}^{\mu-t} (x-\mu)^2 f(x)\,dx + \int_{\mu-t}^{\mu+t} (x-\mu)^2 f(x)\,dx + \int_{\mu+t}^{\infty} (x-\mu)^2 f(x)\,dx$$

$$\geq \int_{-\infty}^{\mu-t} (x-\mu)^2 f(x)\,dx + \int_{\mu+t}^{\infty} (x-\mu)^2 f(x)\,dx$$

the middle integral being omitted

$$= \int_{|x-\mu|\geq t} (x-\mu)^2 f(x)\,dz > \int_{|x-\mu|\geq t} t^2 f(x)\,dx$$

$$= t^2 \int_{|x-\mu|\geq t} f(x)\,dx = t^2 \Pr[|X-\mu| \geq t]$$

Thus

$$\sigma^2 \geq t^2 \Pr[|X-\mu| \geq t]$$

and Chebyshev's inequality follows. $\triangle$

The proof in the discrete case is similar, with sums replacing integrals.

The Chebyshev inequality is very crude, and for most reasonable probability density functions much more precise bounds can be found for the probability of a given deviation from the mean. However, it has the prime advantage of universality. It applies to *all* random variables whose first and second moments exist. Thus it is very useful in proving general theorems.

8. Variance of a Linear Combination

Let X be a random variable having mean μ and variance σ^2, and let U be a random variable of the form $U = aX + b$, for some constants a and b. Thus, U differs from X only in origin and scale of measurement, and U has mean $\mu_U = E[U] = a\mu + b$, by (5.5). Thus

$$\begin{aligned}\mathrm{Var}[aX+b] &= \mathrm{Var}[U] \\ &= E[(U-\mu_U)^2] \\ &= E[(aX+b-a\mu-b)^2] \\ &= E[a^2(X-\mu)^2] \\ &= a^2E[(X-\mu)^2] \\ &= a^2\sigma^2\end{aligned}$$

$$\mathrm{Var}[aX+b] = a^2\,\mathrm{Var}[X] \tag{5.12}$$

or, equivalently, the standard deviation of $aX + b$ equals $|a|\,\sigma$. Thus the additive constant b has no effect on the variance or the standard deviation. This seems reasonable because movement of the origin does not affect the spread of the distribution. A change in scale by the factor a changes the standard deviation by the factor $|a|$.

If X and U are random variables that differ only in origin and scale of measurement (that is, if X and U are related by a linear equation of the form $U = aX + b$), then clearly there is no essential difference in the amount of information that is given by observing the value of X or of U.

To eliminate any effect of origin or scale, we frequently work with the *standardized random variable* X^*, instead of with X. This is defined as follows.

If the random variable X has mean μ and variance σ^2, then

$$X^* = \frac{X-\mu}{\sigma} \tag{5.13}$$

or, equivalently

$$X = \sigma X^* + \mu$$

Since $X^* = (1/\sigma)X - (\mu/\sigma)$, we observe that X^* always has mean 0 and variance 1 regardless of the values of μ and σ^2, using (5.5) and (5.12).

The use of standardized random variables frequently simplifies discussions. For example, the normal distribution, mentioned in Example 3.13 of Chapter 3, depends on two parameters, μ and σ^2, which are, in fact, the mean and the variance (see Chapter 6, Section 9). Thus it appears necessary that a table of its distribution function $F(x)$ be constructed for every pair μ, σ^2. The necessity for multiple tables is eliminated, however, by transforming all questions about a random variable X into questions about X^*. Thus the normal distribution function must only be tabulated for the single case $\mu = 0$, $\sigma^2 = 1$.

It is left to the reader, as an exercise, to show that if

$$U = aX + b$$

with $a \geq 0$, then

$$U^* = X^*$$

[Represent $E[U]$ and $\mathrm{Var}[U]$ by using (5.5) and (5.12) and find U^* by using (5.13).] This shows that standardization of random variables does, in fact, eliminate the effect of origin and scale.

9. Functions of Several Random Variables; Sums of Random Variables .

If X and Y are random variables and ϕ is a real-valued function of *two* variables, then the quantity U defined by

$$U = \phi(X, Y)$$

defines a new, single random variable whose value depends on the values of X and Y. For example, if X and Y represent the lifetimes of two randomly chosen pieces of electronic equipment, we may be interested in the *average* lifetime, $(X + Y)/2$, or the time at which the first failure occurs, $\min(X, Y)$, or the time at which the second failure occurs, $\max(X, Y)$. Each of these quantities is a random variable whose value depends on knowing the values of both X *and* Y. Each is a function of X and Y.

The distribution function, $F_U(u) = \Pr[U \leq u]$, of U can be found by integrating or summing the joint probability density function of X and Y over the region of (X, Y)-values for which $\phi(X, Y) \leq u$.

Instead of pursuing this question in general, we concentrate on the most important special case: the *sum* function.

Assume that

$$S = X + Y$$

where X and Y have joint probability density function $f(x, y)$ [or $p(x_i, y_j)$]. The distribution function of S is given by

$$\begin{aligned} F_S(s) &= \Pr[S \leq s] \\ &= \Pr[X + Y \leq s] \\ &= \iint_{x+y\leq s} f(x, y)\, dx\, dy \\ &= \int_{-\infty}^{\infty} \int_{-\infty}^{s-y} f(x, y)\, dx\, dy \end{aligned} \tag{5.14}$$

in the continuous case. The integration is over the region R in the (x, y)-plane sketched in Figure 5.5.

Now make the substitution $x = t - y$, (y fixed), in the inner integral of (5.14). This formula then becomes

$$\begin{aligned} F_S(s) &= \int_{-\infty}^{\infty} \int_{-\infty}^{s} f(t - y, y)\, dt\, dy \\ &= \int_{-\infty}^{s} \int_{-\infty}^{\infty} f(t - y, y)\, dy\, dt \end{aligned} \tag{5.15}$$

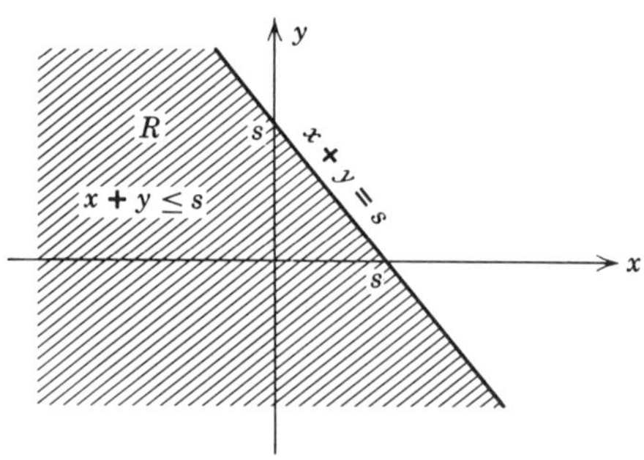

FIGURE 5.5 Region R defined by $x + y \leq s$.

The probability density function, $f_S(s)$, of S can now be found by differentiating the integral in (5.15):

$$f_S(s) = \frac{d}{ds} F_S(s) = \int_{-\infty}^{\infty} f(s-y, y)\, dy \tag{5.16}$$

In the special case where X and Y are *independent*, $f(x, y) = f_X(x) f_Y(y)$ and (5.16) becomes

$$f_S(s) = \int_{-\infty}^{\infty} f_X(s-y) f_Y(y)\, dy \tag{5.17}$$

If X and Y are *nonnegative* random variables, $X \geq 0$ and $Y \geq 0$, then $f_Y(y) = 0$ unless $y \geq 0$, and $F_X(s-y) = 0$ unless $s - y \geq 0$, $y \leq s$. Thus, in this case (5.17) becomes

$$f_S(s) = \begin{cases} \int_0^s f_X(s-y) f_Y(y)\, dy, & s \geq 0 \\ 0, & s \leq 0 \end{cases} \tag{5.18}$$

Formulas (5.17) and (5.18) carry over directly to the discrete case, with sums being replaced by integrals. Suppose that X and Y are *integer-valued* random variables with probability density functions p_i and q_j. Then (5.17) becomes

$$p_S(k) = \Pr[S = k] = \sum_j p_{k-j} q_j \tag{5.19}$$

in the independent case. Formula (5.18) becomes

$$p_S(k) = \Pr[S = k] = \sum_{j=0}^{k} p_{k-j} q_j \tag{5.20}$$

in the independent nonnegative case.

The distribution of sums of more than two random variables can be found by successively adding one term after another.

EXAMPLE 5.10. The lifetime of a certain type of electronic part is a random variable having a negative exponential distribution, with probability density function:

$$\lambda e^{-\lambda x}, \qquad x > 0$$

Assume the equipment that contains this part starts operating, and that as soon as the part breaks down a replacement of the same type is installed. Find the distribution of the time S until the replacement breaks down.

Here $S = X + Y$, where X and Y are the lifetimes of the original part and its replacement. Each will have the same probability density function, and they will be independent if the parts do not interact with each other. Thus, by (5.18), S will have probability density function

$$f_S(s) = \int_0^s \lambda\, e^{-\lambda(s-y)} \cdot \lambda e^{-\lambda y}\, dy$$

$$= \lambda^2\, e^{-\lambda s} \int_0^s dy$$

$$= \lambda^2\, e^{-\lambda s} s, \qquad s > 0 \qquad \triangle$$

10. Expectation of a Function of Several Random Variables

The problems involved in taking the expectation of a random variable that is a function of two or more other random variables, for example, $U = \phi(X, Y)$, are similar to the ones with functions of one random variable that were discussed in Section 4. Strictly speaking, the probability density function of U must be known to compute $E[U]$ directly from its definition. However, as in the single variable case, it can be proved that, instead, one can average $\phi(X, Y)$ over all pairs of values in the joint range of X and Y, using the joint probability density function of X and Y as weight function. That is, the following formula holds true

$$E[\phi(X, Y)] = \begin{cases} \displaystyle\int_{-\infty}^{\infty}\int_{-\infty}^{\infty} \phi(x, y) f(x, y)\, dx\, dy, & \text{continuous case} \\ \displaystyle\sum_i \sum_j \phi(x_i, y_j) p(x_i, y_j), & \text{discrete case} \end{cases} \tag{5.21}$$

where $f(x, y)$ [or $p(x_i, y_j)$] is the joint probability density function of X and Y. The proof of (5.21) is omitted. The formula generalizes in obvious fashion to functions of more than two random variables.

The simplest types of functions to consider are the sum function and the product function.

For a sum of random variables, (5.21) gives, in the continuous case,

$$E[X + Y] = \int_{-\infty}^{\infty}\int_{-\infty}^{\infty} (x + y) f(x, y)\, dx\, dy$$

$$= \int_{-\infty}^{\infty} x \int_{-\infty}^{\infty} f(x, y)\, dy\, dx + \int_{-\infty}^{\infty} y \int_{-\infty}^{\infty} f(x, y)\, dx\, dy$$

$$= \int_{-\infty}^{\infty} x\, f_X(x)\, dx + \int_{-\infty}^{\infty} y\, f_Y(y)\, dy$$

$$= E[X] + E[Y] \tag{5.22}$$

by using formulas (4.22) and (4.23) to obtain $f_X(x)$ and $f_Y(y)$ from $f(x, y)$. The proof of (5.22) in the discrete case is similar. Formula (5.22) can be extended to sums of more than two random variables and, when combined with (5.5), yields the following general *linear* property for expectation:

$$E\left[\sum_{i=1}^{n} a_i X_i\right] = \sum_{i=1}^{n} a_i E[X_i] \tag{5.23}$$

whenever $X_1, \ldots, X_n$ are random variables and $a_1, \ldots, a_n$ are constants.

A simple formula for the expectation of a product (corresponding to (5.22) for a sum) does not hold true in general. However, in the special case when X and Y are *independent*, we have

$$\begin{aligned} E[XY] &= \int_{-\infty}^{\infty}\int_{-\infty}^{\infty} xy\, f(x, y)\, dx\, dy \\ &= \int_{-\infty}^{\infty}\int_{-\infty}^{\infty} xy\, f_X(x) f_Y(y)\, dx\, dy \\ &= \int_{-\infty}^{\infty} x\, f_X(x)\, dx \int_{-\infty}^{\infty} y\, f_Y(y)\, dy \\ &= E[X]E[Y] \end{aligned}$$

the proof being similar in the discrete case. Thus, for independent random variables

$$E[XY] = E[X]E[Y] \tag{5.24}$$

Formula (5.24) generalizes without difficulty to

$$E[\phi_1(X)\phi_2(Y)] = E[\phi_1(x)]E[\phi_2(Y)] \tag{5.25}$$

when X and Y are independent and ϕ_1, ϕ_2 are two functions. Both (5.24) and (5.25) also hold true for products of more than two independent factors.

11. Variance of a Sum

Suppose X and Y are random variables that have expectations $\mu_X = E[X]$, $\mu_Y = E[Y]$, and variances $\sigma_X{}^2 = E[(X - \mu_X)^2]$, $\sigma_Y{}^2 = E[(Y - \mu_Y)^2]$. Let us attempt to compute the variance of their sum:

$$S = X + Y$$

By (5.22), S will have mean $\mu_S = \mu_X + \mu_Y$. Thus the variance of S is, by using (5.23),

$$\begin{aligned}\sigma_S{}^2 &= E[(S - \mu_S)^2]\\ &= E[(X + Y - \mu_X - \mu_Y)^2]\\ &= E[\{(X - \mu_X) + (Y - \mu_Y)\}^2]\\ &= E[(X - \mu_X)^2 + (Y - \mu_Y)^2 + 2(X - \mu_X)(Y - \mu_Y)]\\ &= E[(X - \mu_X)^2] + E[(Y - \mu_Y)^2] + 2E[(X - \mu_X)(Y - \mu_Y)]\end{aligned}$$

It follows that

$$\sigma_S{}^2 = \sigma_X{}^2 + \sigma_Y{}^2 + 2E[(X - \mu_X)(Y - \mu_Y)] \tag{5.26}$$

The last term on the right side of (5.26) is termed the *covariance* of X and Y and, in general, cannot be dispensed with. However, in the special case where X and Y are *independent*, by using (5.25), we find

$$\begin{aligned}E[(X - \mu_X)(Y - \mu_Y)] &= E[X - \mu_X]E[Y - \mu_Y]\\ &= (\mu_X - \mu_X)(\mu_Y - \mu_Y)\\ &= 0\end{aligned}$$

Thus, if X and Y are *independent*

$$\sigma_S{}^2 = \sigma_X{}^2 + \sigma_Y{}^2 \tag{5.27}$$

where $S = X + Y$.

Formula (5.27) readily extends to sums of more than two random variables and to linear combinations.

If

$$S = \sum_{i=1}^{n} a_i X_i$$

then

$$\sigma_S{}^2 = \sum_{i=1}^{n} a_i{}^2 \sigma_{X_i}{}^2 \tag{5.28}$$

if $X_1, X_2, \ldots, X_n$ are independent random variables, and $a_1, a_2, \ldots, a_n$ are constants.

EXAMPLE 5.11. The profit of a certain company in any given month is a random variable having mean \$10,000 and standard deviation \$1,000. Assuming that the profits in different months are independent, find the mean and standard deviation of the total profit S in an entire year.

Here $S = \sum_{i=1}^{12} X_i$, where $X_1, X_2, \ldots, X_n$ are independent random variables, each having mean 10,000 and variance $(1{,}000)^2$. By (5.23) and (5.28), S will have mean and variance:

$$\mu_S = \sum_{i=1}^{12} 10{,}000 = 120{,}000$$

$$\sigma_S{}^2 = \sum_{i=1}^{12} (1{,}000)^2 = 12{,}000{,}000$$

Thus the standard deviation of S is $\sqrt{12{,}000{,}000} = \$3{,}464$. Notice that the standard deviation of S is only about $3\frac{1}{2}$ times the standard deviation of each X, not 12 times as one might possibly guess. △

12. The Moment Generating Function

If X is a random variable, then the quantity $e^{X\theta}$ is a function of X for each fixed value of θ. Thus its expectation $E[e^{X\theta}]$will be a number depending on θ, provided that it exists. Hence, we are led to define the *function*:

$$M(\theta) = E[e^{X\theta}] = \begin{cases} \displaystyle\int_{-\infty}^{\infty} e^{x\theta} f(x)\, dx, & \text{continuous case} \\ \displaystyle\sum_i e^{x_i\theta} p(x_i), & \text{discrete case} \end{cases} \tag{5.29}$$

This function is termed the *moment-generating function* of the random variable X. If more than one random variable is involved in the same discussion, we use the notation $M_X(\theta)$ in place of $M(\theta)$.

Of course, the expectation defining $M(\theta)$ may not exist for *all* real numbers θ. (The integral or sum may not converge absolutely.) However, for most of the random variables that we are concerned with there is an interval of θ-values within which $M(\theta)$ does exist.

EXAMPLE 5.12. Let X have a negative exponential probability density function, $f(x) = \lambda e^{-\lambda x}$, $x > 0$. Then X has moment-generating function

$$\begin{aligned} M(\theta) = E[e^{X\theta}] &= \int_0^{\infty} e^{x\theta} \cdot \lambda e^{-\lambda x}\, dx \\ &= \lambda \int_0^{\infty} e^{-(\lambda-\theta)x}\, dx \\ &= \frac{\lambda}{\lambda - \theta} = \frac{1}{1 - (\theta/\lambda)} \end{aligned}$$

provided that $\theta < \lambda$. △

EXAMPLE 5.13. Let X represent the number of heads in n tosses of an honest coin. Then X has the probability density function given by (3.7) in Example 3.8:

$$p_i = \binom{n}{i} \frac{1}{2^n}, \qquad i = 0, 1, 2, \ldots, n.$$

The moment-generating function will be

$$\begin{aligned} M(\theta) &= E[e^{X\theta}] = \sum_{i=0}^{n} e^{i\theta} p_i \\ &= \left[\sum_{i=0}^{n} e^{i\theta} \binom{n}{i}\right] \frac{1}{2^n} = \frac{(1 + e^{\theta})^n}{2^n} \end{aligned}$$

using the binomial theorem. △

Clearly (5.29) represents a method of changing or transforming one function, the probability density function, into another function, the moment-generating function. We shall find that the moment-generating function of a random variable is often easier to determine than its actual probability density function, and that we can use $M(\theta)$ in place of the probability density function to answer many questions of interest. One reason for this is found in the following theorem, which we quote without proof.

THEOREM. *If two random variables have the same moment-generating functions, then they must have the same distribution and probability density functions.* △

If we can show that a random variable which is being studied has the same moment-generating function as that of another random variable with a *known* probability distribution, then this theorem assures us that the distribution of the original random variable must be the same.

The most important situation in which the use of the moment-generating function simplifies the discussion exists when independent random variables are added. If

$$S = X + Y$$

where X and Y are independent, then S has moment-generating function

$$\begin{aligned} M_S(\theta) &= E[e^{S\theta}] = E[e^{(X+Y)\theta}] \\ &= E[e^{X\theta}\, e^{Y\theta}] \\ &= E[e^{X\theta}]E[e^{Y\theta}] \\ &= M_X(\theta) M_Y(\theta) \end{aligned}$$

by using (5.25) with $\phi_1(X) = e^{X\theta}$, $\phi_2(Y) = e^{Y\theta}$, θ being held fixed. Thus, for independent random variables, the moment-generating function of a sum is simply the product of the moment-generating functions of the individual terms:

$$M_{X+Y}(\theta) = M_X(\theta) M_Y(\theta) \tag{5.30}$$

A similar formula holds true for sums of more than two independent terms. Clearly, (5.30) is much simpler than (5.17), which is the corresponding formula for probability density functions. The advantage becomes even more marked for sums of more than two terms.

EXAMPLE 5.14. In Example 5.12 the moment-generating function of a random variable that has a negative exponential probability density function was shown to be

$$\frac{1}{1-(\theta/\lambda)}.$$

If U has probability density function

$$f(u) = \lambda^2\, e^{-\lambda u}\, u, \qquad \text{for} \qquad u > 0$$

then U has moment-generating function

$$\begin{aligned} E[e^{U\theta}] &= \int_0^\infty e^{u\theta}\lambda^2\, e^{-\lambda u}\, u\, du \\ &= \lambda^2 \int_0^\infty e^{-(\lambda-\theta)u}\, u\, du \\ &= \frac{\lambda^2}{(\lambda-\theta)^2} = \frac{1}{[1-(\theta/\lambda)]^2}, \qquad \text{for} \qquad \theta < \lambda \end{aligned}$$

using integration by parts. If $S = X + Y$, where X and Y are independent negative exponentially distributed random variables, by (5.30) S will have moment-generating function

$$\left[\frac{1}{1-(\theta/\lambda)}\right]\left[\frac{1}{1-(\theta/\lambda)}\right] = \frac{1}{[1-(\theta/\lambda)]^2}$$

which is the same as the moment-generating function of U. Hence, by the preceding theorem, S and U must have the same probability density function. This provides an alternate solution to the problem studied in Example 5.10. △

The reason for the phrase *moment-generating* function is the following: $e^{X\theta}$ can be expanded in series

$$e^{X\theta} = 1 + X\theta + \frac{X^2\theta^2}{2!} + \cdots + \frac{X^n\theta^n}{n!} + \cdots$$

and, if the situation is sufficiently regular so that the linear property of expectation (5.23) can be applied to this *infinite* series, then the moment-generating function becomes

$$\begin{aligned} M(\theta) = E[e^{X\theta}] &= 1 + E[X]\theta + E[X^2]\frac{\theta^2}{2!} + \cdots + \frac{E[X^n]\theta^n}{n!} + \cdots \\ &= 1 + \mu\theta + \mu_2{}'\frac{\theta^2}{2!} + \cdots + \mu_n{}'\frac{\theta^n}{n!} + \cdots \end{aligned} \tag{5.31}$$

Thus the moments $\mu_n{}'$ can all be found by expanding $M(\theta)$ in powers of θ and by noting the coefficients of the terms $\theta^n/n!$. In particular, the mean and second moment, μ and $\mu_2{}'$, can be found by evaluating the first two derivatives at 0, as is usual in computing the coefficients of a Taylor series:

$$\mu = E[X] = M'(0), \qquad \mu_2{}' = E[X^2] = M''(0) \tag{5.32}$$

EXAMPLE 5.15. If X has a negative exponential distribution, we have found that

$$M(\theta) = \frac{1}{1 - (\theta/\lambda)}, \qquad \theta < \lambda$$

This can be expanded in series by using the formula for the sum of a geometric series (at least provided that $-\lambda < \theta < \lambda$)

$$\begin{aligned} M(\theta) &= 1 + \frac{\theta}{\lambda} + \frac{\theta^2}{\lambda^2} + \cdots + \frac{\theta^n}{\lambda^n} + \cdots \\ &= 1 + \frac{\theta}{\lambda} + \frac{2}{\lambda^2}\cdot\frac{\theta^2}{2!} + \cdots + \frac{n!}{\lambda^n}\cdot\frac{\theta^n}{n!} + \cdots \end{aligned}$$

It follows from (5.31) that

$$\mu = \frac{1}{\lambda}, \; \mu_2{}' = \frac{2}{\lambda^2}, \; \ldots, \; \mu^n = \frac{n!}{\lambda^n}, \ldots$$

Also

$$\sigma^2 = \mu_2{}' - \mu^2 = \frac{2}{\lambda^2} - \left(\frac{1}{\lambda}\right)^2 = \frac{1}{\lambda^2}$$

△

EXAMPLE 5.16. In Example 5.13 we showed that, if X is a discrete random variable representing the number of heads in n tosses of an honest coin, then X has moment-generating function $M(\theta) = (1 + e^\theta)^n/2^n$. By (5.32) the mean and second moment of X are given by

$$\mu = M'(0) = \left.\frac{n(1 + e^\theta)^{n-1} e^\theta}{2^n}\right|_{\theta=0} = \frac{n(1 + 1)^{n-1}}{2^n}$$

$$= \frac{n}{2}$$

$$\mu_2' = M''(0) = \left.\frac{[n(n-1)(1 + e^\theta)^{n-2}(e^\theta)^2 + n(1 + e^\theta)^{n-1}e^\theta]}{2^n}\right|_{\theta=0}$$

$$= \frac{[n(n-1)(1 + 1)^{n-2} + n(1 + 1)^{n-1}]}{2^n}$$

$$= \frac{n(n-1) + 2n}{4} = \frac{n^2 + n}{4}$$

Using (5.10) the variance of X is found to be

$$\sigma^2 = \mu_2' - \mu^2 = \frac{n^2 + n}{4} - \left(\frac{n}{2}\right)^2 = \frac{n}{4}$$

The student should try to compute μ and σ^2 directly from the definition of expectation by using the probability density function $p_i = \binom{n}{i}/2^n$, $i = 0, 1, \ldots, n$, in order to convince himself that the use of (5.32) offers some advantage in simplicity. △

13. Probability Generating Function, z-Transform, and Characteristic Function

If X is an *integer-valued*, discrete random variable with probability density function p_i, then the moment-generating function can be simplified by the substitution $z = e^\theta$. The resulting function of z is termed the *probability generating function* of X or, sometimes, the *z-transform* of X. It is defined by

$$\psi(z) = \psi_X(z) = E[z^X] = \sum_i z^i p_i$$

All the formulas of the previous section for moment-generating functions carry over to formulas that involve $\psi(z)$ if we change the variable by using the identity $z = e^\theta$. Thus, for example, (5.32) becomes

$$\mu = \psi'(1), \qquad \mu_2' = \psi''(1) + \psi'(1) \tag{5.33}$$

where one uses the fact that $M(\theta) = \psi(z)$ with $z = e^\theta$, and the differentiation formulas

$$M'(\theta) = \frac{d\psi}{dz} \cdot \frac{dz}{d\theta} = \psi'(z)e^\theta = \psi'(z)z$$

$$M''(\theta) = \frac{d}{d\theta} M'(\theta) = \frac{d}{dz} M'(\theta) \cdot \frac{dz}{d\theta} = [\psi''(z)z + \psi'(z)] \cdot z$$

as well as the fact that $z = 1$ when $\theta = 0$.

For instance, if X is the random variable that is discussed in Examples 5.13 and 5.16, X has probability generating function

$$\psi(z) = \sum_{i=0}^{n} \frac{z^i \binom{n}{i}}{2^n} = \frac{(1+z)^n}{2^n}$$

by using the binomial theorem. Thus

$$\mu = \psi'(1) = \left.\frac{n(1+z)^{n-1}}{2}\right|_{z=1} = \frac{n}{2}$$

$$\mu_2' = \psi''(1) + \psi'(1) = \left.\frac{n(n-1)(1+z)^{n-2}}{2^n}\right|_{z=1} + \frac{n}{2}$$

$$= \frac{n(n-1)}{4} + \frac{n}{2} = \frac{n^2+n}{4}$$

which is an even simpler computation than the one in Example 5.16.

We have ignored the possibility that the moment-generating function or probability-generating function might not exist. (The defining sum or integral might not converge absolutely.) There is no great loss in generality here, since most of what we have done could also be done by using still another generating function which always *does* exist. This is the *characteristic function* or *Fourier transform* of the random variable X. To define it, we must introduce complex numbers. The *characteristic function* is defined by

$$\phi(t) = E[e^{itX}] = E[\cos tX + i \sin tX]$$

for real t. (Here $i = \sqrt{-1}$.) Since

$$|e^{itX}| = \sqrt{\cos^2 tX + \sin^2 tX} = 1$$

the function whose expectation is being taken is bounded and, hence, the defining integral or sum must converge absolutely,

$$|E[e^{itX}]| = \left| \int_{-\infty}^{\infty} e^{itx} f(x)\, dx \right| \le \int_{-\infty}^{\infty} |e^{itx}| f(x)\, dx$$

$$= \int_{-\infty}^{\infty} 1 \cdot f(x)\, dx = 1.$$

These functions play a crucial role in the extensions of the theory that we have discussed. We do not develop them now because the tools that we have described are sufficient for our needs.

EXERCISES

1. A factory produces 25 percent defectives on the average. Choose 4 items at random and let $X =$ the number of defectives among them. Then X is a discrete random variable having range: 0, 1, 2, 3, 4. Compute the mean $\mu = E[X]$ and variance σ^2 of X directly from their definitions by completing the following table.

x	$p(x)$	$xp(x)$	$(x-\mu)^2 p(x)$
0			
1			
2			
3			
4			
Totals:	1	$\mu =$	$\sigma^2 =$

2. Let X be a continuous random variable having a uniform probability density function over the interval $-1 \le x \le 1$

$$f(x) = \begin{cases} \frac{1}{2}, & \text{for} \quad -1 < x \le 1 \\ 0, & \text{otherwise} \end{cases}$$

Find μ and σ.

3. Define the random variable Y by

$$Y = X^2$$

where X has the distribution in Exercise 2.

(a) Compute $E[Y]$ using formula (5.3).

(b) Compute $F_Y(y)$ using the fact that, for $y \geq 0$,

$$F_Y(y) = \Pr[Y \leq y] = \Pr[X^2 \leq y] = \Pr[-\sqrt{y} \leq X \leq \sqrt{y}]$$

(c) Compute $f_Y(y) = F_Y{}'(y)$.

(d) Compute $E[Y]$ using the definition (5.1) and compare it with the result obtained in (a).

4. Let X be a random variable that has a standard normal distribution ($\mu = 0$, $\sigma = 1$). Determine the probability density function of $Y = e^X$. This is termed a *log-normal* density function and is useful in many economic analyses.

5. An electrical device is termed a "full-wave" rectifier if the output of the device Y is related to the input of the device X by

$$Y = |X|$$

If the input has a standard normal distribution ($\mu = 0$, $\sigma = 1$), determine the probability density function of the output.

6. A simple electrical circuit contains a constant voltage source and a resistor. Suppose that the voltage E is known but that the resistor is randomly chosen from a stock whose resistance R has probability density function $f(r) = \lambda^2 r e^{-\lambda r}$, $r > 0$, $[f(r) = 0$ for $r \leq 0]$.

(a) Find the probability density function of the current I passing through the circuit which is determined by Ohm's law, $I = E/R$.

(b) Find the probability density function of the power W dissipated in the resistor, where $W = I^2 R$.

7. A folding 4-foot ruler is constructed by attaching together four 12 inch-pieces. The pieces are chosen independently from a population that has mean length $\mu = 12$ inches, and standard deviation σ. Thus, if S is the true length of the ruler,

$$S = X_1 + X_2 + X_3 + X_4$$

where X_1, X_2, X_3, X_4 are independent random variables that each have mean 12 inches and standard deviation σ.

(a) Find $E[S]$ and $\text{Var}[S]$.

(b) What should σ be if we want S to have standard deviation 0.1 inches?

8. If X has a uniform density function over the interval $-1 \leq x \leq 1$ (see Exercise 2), use the Chebyshev inequality to estimate $\Pr[|X| > \frac{3}{4}]$. Compare it with the true value of this probability.

9. Let X and U be random variables that differ only in origin and scale,

$$U = aX + b$$

for some constants a and b. Show that, for the standardized variables,

$$U^* = X^*$$

10. (a) If X is a discrete random variable that has the Poisson probability density function, discussed in Exercise 19 of Chapter 3,

$$p_i = \frac{\lambda^i e^{-\lambda}}{i!}, \qquad i = 0, 1, 2, \ldots$$

show that X has probability generating function:

$$\psi(z) = E[z^X] = e^{\lambda(z-1)}$$

(b) Using (5.33), show that

$$\mu = \lambda, \qquad \sigma^2 = \lambda$$

11. Show that if X is a random variable with range $[a, b]$

(a) $a \leq E[X] \leq b$

(b) $\mathrm{Var}[X] \leq \dfrac{(b-a)^2}{4}$

12. Show that

$$E[(X-a)^2]$$

is minimized when

$$a = E[X]$$

13. Find, in terms of μ and μ_2',

$$E[(X-\mu)^2]$$

14. Find in terms of μ_3', μ_2', μ,

$$E[(X-\mu)^3]$$

15. If X is a normally distributed random variable, find

$$\Pr[|X-\mu| > 3\sigma]$$

by using the normal tables. Estimate the same probability by using Chebyshev's inequality.

16. Prove Chebyshev's theorem (formula 5.11) if the random variable X is discrete.

17. Suppose that X is a random variable with mean 0 and variance σ^2. Suppose further that $Y = aX$, $a > 0$. Show that

$$\Pr[|Y| \geq t] \leq \frac{a^2\sigma^2}{t^2}$$

18. Suppose that the length of a manufactured part is a random variable X with mean 1 inch and variance 1×10^{-4}. The part is considered defective if the length is less than .98 inch or more than 1.02 inch. Find the Chebyshev bound on the probability of a defective part.

19. In electronic systems one adds redundant parts so that, if the primary part fails, the redundant parts can continue the operation of the system. Suppose that a system is designed with one primary part and one redundant part. The redundant part comes into operation only when the primary part fails. Let each part have a lifetime that is normally distributed random variable with mean life 500 hours and variance 50 hours. Find the mean life of the system, the variance of the system, and estimate the probability that the system will not live more than 950 hours, by using Chebyshev's inequality.

20. In designing systems with redundant parts one could design a parallel system in which the part and its redundant mate are placed in parallel. Delays in the changeover from a part to its redundant mate can be reduced if the redundant part is kept in a ready condition at all times. These designs are called "hot redundancy" systems. Thus, when a changeover occurs, there is no warmup period and the only delay is the one that results from switching, which can be considered negligible. However, since the mate is under load at all times, it may not be available when the changeover occurs. If the original part has lifetime X with distribution function $F_X(x)$, and if the lifetime under load Y of the redundant part is independent of X and has distribution function $F_Y(y)$, determine the distribution function of the lifetime Z of the system. (*Hint.* $Z = \max(X, Y)$, and $Z \leq z$ if, and only if, $X \leq z$ and $Y \leq z$.)

21. In Exercise 20, determine the probability density function $f_Z(z)$ of Z in terms of the probability density functions $f_X(x)$ and $f_Y(y)$.

22. In Exercise 19, a redundancy system was considered in which the redundant part was put under load only after the original part had failed. In this case, the lifetime of the system $Z = X + Y$, where X and Y are as in Exercise 20. This design is termed a "cold redundancy" system. If X and Y are independant with negative exponential densities with parameters λ_1 and λ_2, which type of redundancy hot or cold, would maximize the probability that the lifetime Z of the system will exceed a fixed number t? Is cold redundancy always best for every combination of λ_1, λ_2, and t?

23. Assume that the number of defects per day from a certain manufacturing process has a Poisson distribution (see Exercise 10), and that the numbers on different days are independent random variables. Show that the number of defects in two days also has a Poisson distribution.

24. If X has the triangular distribution discussed in Chapter 3, Exercise 11, find $M(\theta)$.

25. **(a)** If X has the gamma density discussed in Chapter 3, Exercise 17, find $M(\theta)$.

    (b) Show that the sum of independent random variables, each one of which has a gamma density with the same value of β, also has a gamma density.

26. Suppose that the number X of taxis arriving at a taxi stand at an airport during some time interval has a Poisson distribution. Suppose also that the number Y of taxi users who arrive during this time interval is an independent, identically distributed random variable. Then

$$Z = X - Y$$

represents the excess of taxis over users (positive or negative). Find an expression for the probability density function of Z.

6 • Special Distributions

1. Introduction

In many theoretical and applied problems, and under many seemingly diverse conditions, several probability density functions appear often enough that they are worth exploring in depth. In Chapters 3 to 5, and in the exercises to these chapters, we mentioned some of these density functions. In this chapter we discuss them in more detail.

The probability density functions that we shall explore in this chapter are the ones that appear most frequently in later chapters. We do not pretend that this chapter exhausts the list of useful distributions. Even among the ones that we do discuss, for a particular application, one of the distributions may be of paramount interest and the others of subsidiary interest. For example, in the important field of statistical hypothesis testing, the normal density function is of paramount interest. In the field of reliability theory, the negative exponential is of great importance. In the field of demand analysis, the Poisson density function is preeminent. The student should not assume that any one of the distributions in this chapter is a model of all reality.

In this chapter, in addition to using the results of Chapters 3 and 4 to illustrate the ideas of Chapter 5, we shall also give some threads that tie together many of these distributions. Again, the student is warned that the threads we choose by no means exhaust the possible interrelations of these distributions. These threads merely show that some interconnections do exist.

2. The Bernoulli Density Function

The *Bernoulli* density function is the density function of a discrete random variable X having 0 and 1 as its only possible values. It is given by

$$\begin{aligned} p_0 &= p(0) = \Pr[X = 0] = q \\ p_1 &= p(1) = \Pr[X = 1] = p \\ p_i &= p(i) = \Pr[X = i] = 0 \qquad \text{for} \qquad i \neq 0, 1 \end{aligned} \tag{6.1}$$

where $p + q = 1$.

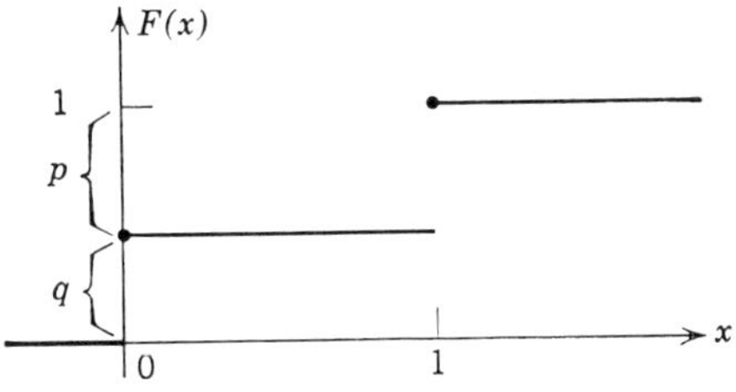

Figure 6.1 The Bernoulli distribution function.

If we consider the experiment of tossing a coin, and denote "tail" by 0, "head" by 1, that is, X (tail) $= 0$, X (head) $= 1$, then X will have such a Bernoulli density function. If the coin is "honest," then $p = q = \frac{1}{2}$.

Immediately, we have (Figure 6.1)

$$F(x) = \begin{cases} 0 & \text{for} \quad x < 0 \\ q & \text{for} \quad 0 \leq x < 1 \\ 1 & \text{for} \quad x \geq 1 \end{cases}$$

For this distribution

$$\mu = E[X] = 0 \cdot q + 1 \cdot p = p \tag{6.2}$$

$$\mu_2' = E[X^2] = 0^2 \cdot q + 1^2 \cdot p = p$$

$$\sigma^2 = \mu_2' - \mu^2 = p - p^2 = p(1 - p) = pq \tag{6.3}$$

The probability generating function is

$$\psi(z) = \psi_X(z) = E[z^X] = z^0 q + z^1 p = q + zp \tag{6.4}$$

3. The Binomial Density Function

Consider an experiment that has two possible outcomes, "success" and "failure" (or "heads" and "tails," "hit" and "miss," "defective" and "good," etc.). Suppose that these two outcomes have probabilities p and q, respectively, $p + q = 1$. Let us then consider the compound experiment that consists of a sequence of independent repetitions of this experiment. Equivalently, consider an independent sequence of events each one of which has the same probability p of occurring. This sequence is termed a *sequence of Bernoulli trials*.

Given a sequence consisting of a fixed number, for instance, n, of Bernoulli trials, we might naturally be interested in the random variable $X =$ the number of times the given event occurs in the sequence. Thus, if the trials involve the choosing of a part from a manufacturing process and the testing of it, X might represent the number of defective parts out of n parts tested. If we consider n repeated coin tosses, X might represent the number of heads.

Clearly, X is a discrete random variable with range 0, 1, 2, ..., n.

Consider the sequence $X_1, X_2, \ldots, X_n$ of random variables where X_i is 1 or 0 depending on whether the given event occurs on the ith trial or not (on whether the ith trial is a "success" or not). Then each X_i has the Bernoulli density function discussed in Section 2, and clearly

$$X = X_1 + X_2 + \cdots + X_n \tag{6.5}$$

since the sum on the right counts a 1 for every occurrence of the given event ("success") and, hence, simply counts the total number of occurrences. Since the events involved are independent, formulas (5.23), (5.28), and (5.30) can be applied to find the mean, the variance, and the probability-generating function of X from the mean, the variance, and the probability-generating function of the X_i given by (6.2), (6.3), and (6.4):

$$\begin{aligned}\mu = E[X] &= E[X_1] + E[X_2] + \cdots + E[X_n]\\ &= p + p + \cdots + p\\ &= np\end{aligned} \tag{6.6}$$

$$\begin{aligned}\sigma^2 = \operatorname{Var}(X) &= \operatorname{Var}(X_1) + \operatorname{Var}(X_2) + \cdots + \operatorname{Var}(X_n)\\ &= pq + pq + \cdots + pq\\ &= npq\end{aligned} \tag{6.7}$$

$$\begin{aligned}\psi(z) = E[z^X] &= E[z^{X_1+X_2+\cdots+X_n}]\\ &= E[z^{X_1}\cdot z^{X_2}\cdot \cdots \cdot z^{X_n}]\\ &= E[z^{X_1}]\cdot E[z^{X_2}]\cdot \cdots \cdot E[z^{X_n}]\\ &= \psi_{X_1}(z)\psi_{X_2}(z)\cdots\psi_{X_n}(z)\\ &= (q+zp)(q+zp)\cdots(q+zp)\\ &= (q+zp)^n\end{aligned} \tag{6.8}$$

If $p_0, p_1, \ldots, p_n$ represent the probability density function of X, $p_i = \Pr[X = i]$, then the probability-generating function of X is

$$\psi(z) = E[z^X] = \sum_{i=0}^{n} z^i p_i \tag{6.9}$$

However, $\psi(z)$ is also given by (6.8), which can be expanded in powers of z by using the binomial theorem:

$$\psi(z) = (q + zp)^n = \sum_{i=0}^{n} \binom{n}{i} q^{n-i}(zp)^i$$
$$= \sum_{i=0}^{n} \binom{n}{i} p^i q^{n-i} z^i \tag{6.10}$$

Since (6.9) and (6.10) represent the same function of z, the coefficients of the various powers of z can be compared. Thus we observe that the probability density function of X is given by

$$p_i = \binom{n}{i} p^i q^{n-i}, \qquad i = 0, 1, 2, \ldots, n \tag{6.11}$$

($p_i = 0$, otherwise).

That is, formula (6.11) gives the probability of i "successes" in n independent trials (Bernoulli trials) of an experiment that has probability p of "success" on each trial.

The fact that $\sum_{i=0}^{n} p_i = 1$ is clear, since

$$\sum_{i=0}^{n} p_i = \sum_{i=0}^{n} \binom{n}{i} p^i q^{n-i} = (q + p)^n = 1^n = 1$$

by using the binomial theorem.

Because of the presence of the binomial coefficients $\binom{n}{i}$ and also because of the extensive use of the binomial theorem in discussions, probabilities p_i, given by (6.11), are said to form a *binomial density function*, and any random variable X having this probability density function is said to be a *binomial* random variable.

Graphs of the binomial density function for various values of n and p are sketched in Figure 6.2.

4. The Geometric Density Function

Suppose that we are given a sequence of Bernoulli trials, that is, a sequence of independent events each of which has the same probability p of occurring. The binomial distribution enters when one counts the number of events that occur in a *fixed* number n of trials. Suppose instead that we count the number of trials (or events observed) until the first "failure" occurs. Let us denote this number by Y. Thus, Y is a random variable with range 1, 2, 3, ... To find its probability density function we note that the event $[Y = i]$ occurs if, and only if, one has a sequence of $i - 1$ "successes" followed by a "failure." Thus, we have

$$p_i = \Pr[Y = i] = p^{i-1}q \tag{6.12}$$

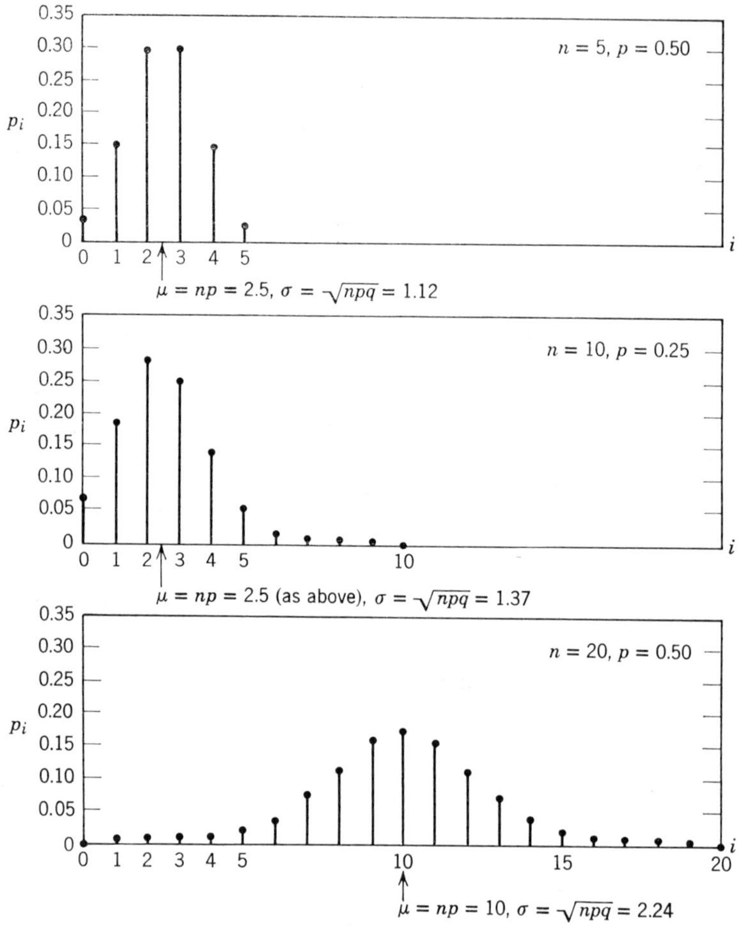

Figure 6.2 The binomial density function for several values of p, n.

$i = 1, 2, 3, \ldots$, where $q = 1 - p$ is assumed to be positive. Clearly,

$$\sum_{i=1}^{\infty} p_i = q + pq + p^2q + \cdots = \frac{q}{1-p} = \frac{q}{q} = 1$$

by using the formula for the sum of a geometric series.

Any random variable Y having range 1, 2, 3, ... and probability density function given by a formula of the form of (6.12) is said to have a *geometric* distribution, and (6.12) is termed a *geometric probability density function.*

Graphs of p_i for various values of p are sketched in Figure 6.3.

The probability generating function $\psi(z)$ of such a random variable Y is easily determined using the formula for the sum of a geometric series:

$$\begin{aligned}\psi(z) &= E[z^Y] = \sum_{i=1}^{\infty} z^i p_i \\ &= \sum_{i=1}^{\infty} z^i p^{i-1} q \\ &= zq(1 + zp + z^2p^2 + \cdots) \\ &= \frac{zq}{1 - zp} \qquad (6.13)\end{aligned}$$

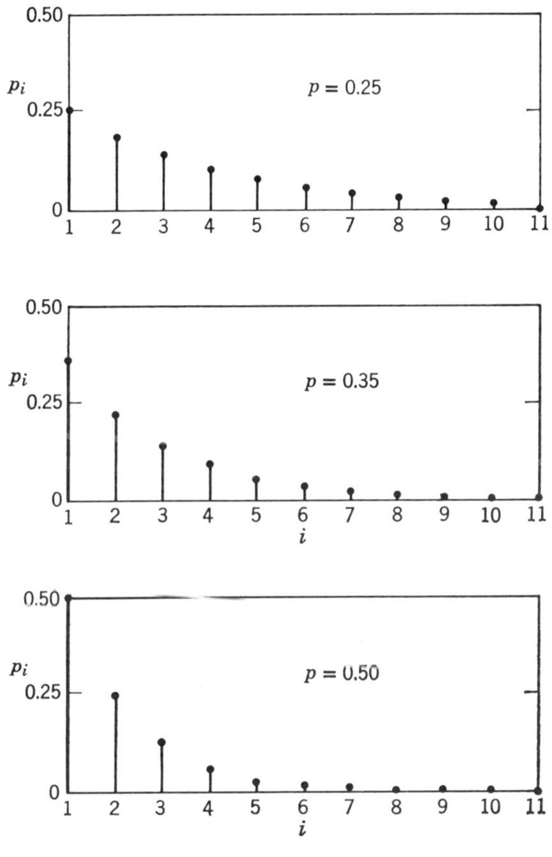

Figure 6.3 The geometric density function for several values of p.

The mean μ and the variance σ^2 can now be found by using formula (5.33.)

$$\psi'(z) = \frac{q}{(1 - zp)^2}$$

$$\psi''(z) = \frac{2pq}{(1 - zp)^3}$$

$$\mu = \psi'(1) = \frac{q}{(1 - p)^2} = \frac{q}{q^2} = \frac{1}{q} \tag{6.14}$$

$$\mu_2' = \psi''(1) + \psi'(1) = \frac{2pq}{(1 - p)^3} + \frac{1}{q} = \frac{2pq}{q^3} + \frac{1}{q}$$

$$= \frac{2p + q}{q^2}$$

$$\sigma^2 = \mu_2' - \mu^2 = \frac{2p + q}{q^2} - \frac{1}{q^2} = \frac{2p + q - 1}{q^2}$$

$$= \frac{2p + q - p - q}{q^2}$$

$$\sigma^2 = \frac{p}{q^2} \tag{6.15}$$

A random variable X is said to have a *modified geometric* probability density function if it has range 0, 1, 2, . . ., and probability density function

$$\Pr[X = i] = (1 - p)p^i, \qquad i = 0, 1, 2, \ldots \tag{6.16}$$

where p is some constant with $0 \le p < 1$. If we set

$$Y = X + 1$$

we notice that Y has range 1, 2, 3, . . ., and probability density function:

$$\Pr[Y = i] = \Pr[X = i - 1] = (1 - p)p^{i-1}$$

Thus, by comparison with (6.12), we observe that $Y = X + 1$ has a geometric distribution with $p = p$ and $q = 1 - p$. Since $X = Y - 1$, we have

$$E[X] = E[Y - 1] = E[Y] - 1$$

$$= \frac{1}{q} - 1 = \frac{1 - q}{q} = \frac{p}{q}$$

$$= \frac{p}{1 - p} \tag{6.17}$$

$$\mathrm{Var}[X] = \mathrm{Var}[Y] = \frac{p}{q^2} = \frac{p}{(1 - p)^2} \tag{6.18}$$

Random variables with modified geometric distribution will be important in our later work.

To discuss another important property, let us go back to the situation at the beginning of this section where Y represents the number of trials until the first "failure" in a sequence of Bernoulli trials. Now, suppose that we have observed a fixed number n of these trials and found them all to be "successes." We are concerned with the number Z of *additional* trials that must be observed until the first "failure." Thus $Z = Y - n$. We wish to study *conditional* probabilities involving Z, *given that* $Y > n$ (the first n trials observed were successes so that the number of trials until the first "failure" must be greater than n). Let us compute the following probabilities.

$$\begin{aligned} q_i = \Pr[Z = i \mid Y > n] &= \Pr[Y - n = i \mid Y > n] \\ &= \Pr[Y = n + i \mid Y > n] \\ &= \frac{\Pr[Y = n + i \text{ and } Y > n]}{\Pr[Y > n]} \end{aligned}$$

by using the definition of conditional probability. For $i = 1, 2, 3, \ldots,$ $Y = n + i$ implies that $Y > n$. Thus the event $[Y = n + i \text{ and } Y > n]$ is the same as the event $[Y = n + i]$. It follows that

$$\begin{aligned} q_i &= \Pr[z = i \mid Y > n] \\ &= \frac{\Pr[Y = n + i]}{\Pr[Y > n]} \\ &= \frac{p_{n+i}}{\sum_{j=n+1}^{\infty} p_j} \\ &= \frac{p^{n+i-1}q}{\sum_{j=n+1}^{\infty} p^{j-1}q} \\ &= \frac{p^{n+i-1}q}{p^n q/(1-p)} \\ &= p^{i-1}q = p_i \end{aligned} \tag{6.19}$$

Thus we observe that, conditional on $Y > n$, the number of trials remaining until the first failure, $Z = Y - n$, has the *same* probability density function as Y had originally. This is termed the *forgetfulness property* of the geometric probability density function. It is related to the Markov property of certain sequences of random variables, which was discussed in Chapter 4. If a sequence of "successes" is observed in a sequence of Bernoulli trials, one need not "remember" how long the sequence is to determine probabilities for the number of additional trials needed until the first "failure."

It is important to note that the converse is also true here. If Y is any discrete random variable with range: 1, 2, 3, ..., and having this forgetfulness property, then Y must have a geometric probability density function. To see this we observe that the forgetfulness property implies that $p_i = \Pr[Y=i]$ satisfies

$$p_i = \Pr[Y=i] = \Pr[Y-n=i \mid Y>n] \tag{6.20}$$

for each integer n and i in the range 1, 2, 3, ... By the same reasoning that we used in deriving (6.19) we notice that the right side of (6.20) is given by

$$\frac{p_{i+n}}{\sum_{j=n+1}^{\infty} p_j}$$

Thus

$$p_i = \frac{p_{i+n}}{\sum_{j=n+1}^{\infty} p_j}$$

Now set $n=1$ and rearrange.

$$\frac{p_{i+1}}{p_i} = \sum_{j=2}^{\infty} p_j = 1 - p_1 = p$$

where the constant $1-p_1$ is denoted by p. Thus $p_{i+1} = p_i p$. Now p_i can be expressed as $p_{i-1}p$, etc., and eventually we obtain

$$p_i = p_1 p^{i-1} = (1-p)p^{i-1} = p^{i-1}q, \qquad \text{for } i = 1, 2, 3, \ldots$$

where $q = 1-p$. Thus we observe that p_i satisfies formula (6.12) and that Y has a geometric probability density function.

5. The Poisson Density Function

If X is a discrete random variable with range 0, 1, 2, ..., and

$$p_i = \Pr[X=i] = \frac{\lambda^i e^{-\lambda}}{i!} \qquad i = 0, 1, 2, \ldots \tag{6.21}$$

$$p_i = 0, \text{ otherwise}$$

then the set $\{p_i\}$ is said to be a *Poisson* probability density function. Here λ represents some given positive constant (parameter). A graph of the Poisson density function for several values of λ is given in Figure 6.4.

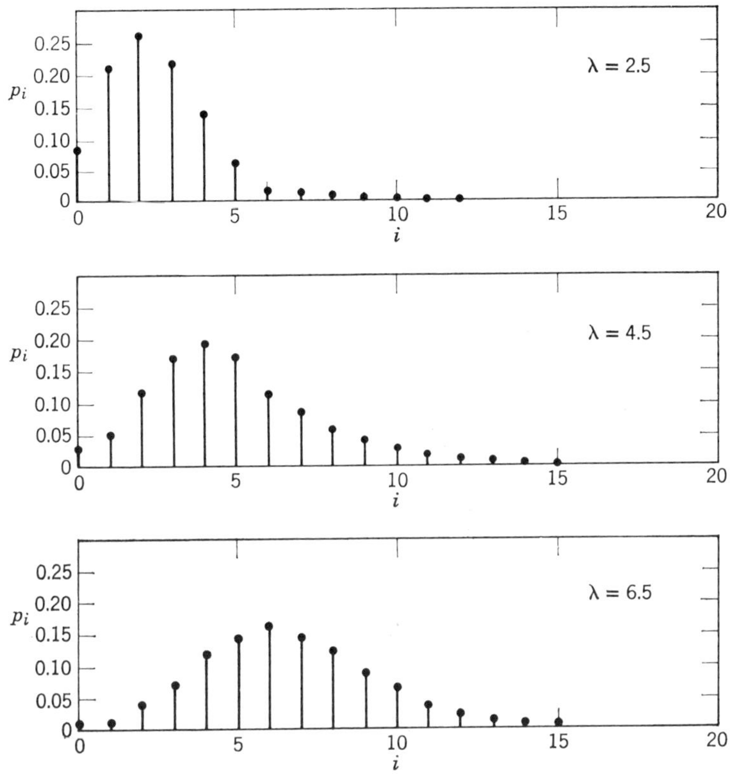

Figure 6.4 The Poisson density function for several values of λ.

This function occurs in many guises. First, notice that the binomial probability density function, $\binom{n}{i} p^i q^{n-i}$, can be written as

$$\frac{n(n-1)(n-2)\cdots(n-i+1)}{i!}\,p^i q^{n-i}$$

If we now multiply and divide by n^i to get

$$\frac{1\left(1-\frac{1}{n}\right)\left(1-\frac{2}{n}\right)\cdots\left(1-\frac{i-1}{n}\right)}{i!}\,(np)^i\left(1-\frac{np}{n}\right)^{n-i}$$

and let $n \to \infty$ in such a way that $np = \lambda$ remains constant, each term in the product $1(1-1/n)\cdots[1-(i-1)/n]$ approaches 1, while $(np)^i$ is λ^i. Also

$$\left(1-\frac{np}{n}\right)^{n-i} = \left(1-\frac{\lambda}{n}\right)^n\left(1-\frac{\lambda}{n}\right)^{-i} \to e^{-\lambda}(1) = e^{-\lambda}$$

Hence, in the limit as $n \to \infty$ with $np = \lambda$ (and as $p = \lambda/n \to 0$)

$$\binom{n}{i} p^i q^{n-i} \to \frac{\lambda^i e^{-\lambda}}{i!} \tag{6.22}$$

Thus the Poisson probability density function can be used as a convenient approximation to the binomial probability density function in the case of large n and small p.

$$\binom{n}{i} p^i q^{n-1} \approx \frac{e^{-\lambda}\lambda^i}{i!}, \qquad \text{where} \qquad \lambda = np$$

In Table 6.1 the probability density functions for two binomial random variables, one with $n = 5$, $p = 0.20$ and the other with $n = 20$, $p = 0.05$, are compared with one another and with that of a Poisson random variable with $\lambda = 1$. Since $np = 1$ in each of these cases, the same Poisson should provide an approximation to each of these binomials. We observe that the approximation is better in the case of the larger n and the smaller p. The two binomial random variables could be interpreted as the number of defective items in a sample of size 5 (or 20) from a production process that has probability 0.20 (or 0.05) of any given item being defective.

TABLE 6.1[a]

	Binomial		Poisson
i	$n = 5$, $p = 0.20$	$n = 20$, $p = 0.05$	$\lambda = 1$
0	0.328	0.359	0.368
1	0.410	0.377	0.368
2	0.205	0.189	0.184
3	0.051	0.060	0.061
⋮	⋮	⋮	⋮

[a] The Poisson as an Approximation to the Binomial p_i.

In reliability theory we can show that, under reasonable assumptions, if X represents the number of radio tubes in a large system that will burn out during a period of time of length t, then

$$\Pr[X = i] = \frac{(\lambda t)^i e^{-\lambda t}}{i!}, \qquad i = 0, 1, 2, \ldots \tag{6.23}$$

Thus, X has a Poisson distribution with the constant λ being replaced by a new constant λt which is proportional to the time of observation. (See Section 7 of Chapter 7).

In certain waiting-line type problems which we discuss later, we find that the number of parts produced, the number of customers served, or the number of messages passed through a communication center in some fixed interval of time will have a probability density function that has approximately the Poisson form.

We must now show that the $\{p_i\}$ given by the Poisson probability density function is, in fact, a valid probability density function. That is,

$$0 \le p_i \le 1$$

$$\sum_{i=0}^{\infty} p_i = 1$$

Easily, we determine that $p_i \ge 0$, since each term in the formula for p_i is positive. It suffices, then, to show that

$$\sum_{i=0}^{\infty} p_i = \sum_{i=0}^{\infty} \lambda^i \frac{e^{-\lambda}}{i!} = 1$$

But this is true since the sum can be written as

$$\sum_{i=0}^{\infty} p_i = e^{-\lambda} \sum_{i=0}^{\infty} \frac{\lambda^i}{i!}$$

$$= e^{-\lambda}\left[1 + \frac{\lambda}{1} + \frac{\lambda^2}{2!} + \frac{\lambda^3}{3!} + \cdots\right]$$

The sum in the brackets is the infinite series expansion of e^{λ}. Hence

$$\sum_{i=0}^{\infty} p_i = e^{-\lambda} \sum_{i=0}^{\infty} \frac{\lambda^i}{i!} = e^{-\lambda}\, e^{\lambda} = 1$$

The probability generating function can also be found readily:

$$\psi(z) = E[z^X] = \sum_{i=0}^{\infty} z^i p_i$$

$$= \sum_{i=0}^{\infty} z^i \frac{\lambda^i e^{-\lambda}}{i!} = e^{-\lambda} \sum_{i=0}^{\infty} \frac{(z\lambda)^i}{i!}$$

$$= e^{-\lambda} \cdot e^{z\lambda}$$

by the series expansion of $e^{z\lambda}$. Thus

$$\psi(z) = e^{-\lambda(1-z)} \tag{6.24}$$

Again (5.33) can be used to compute μ, μ_2', and σ^2:

$$\psi'(z) = \lambda e^{-\lambda(1-z)}, \qquad \psi''(z) = \lambda^2 e^{-\lambda(1-z)}$$

$$\mu = \psi'(1) = \lambda \tag{6.25}$$

$$\mu_2' = \psi''(1) + \psi'(1) = \lambda^2 + \lambda$$

$$\sigma^2 = \mu_2' - \mu^2 = (\lambda^2 + \lambda) - \lambda^2 = \lambda \tag{6.26}$$

Thus a random variable having a Poisson density function has equal mean and variance.

6. The Uniform Density Function

If X is a discrete random variable with a *finite* range $x_1, x_2, \ldots, x_N$, one of the simplest probability densities to consider is one for which each point in the range has equal probability. In this case, the probability density function is given by

$$p(x_i) = \frac{1}{N}, \qquad \text{for } x_i \text{ in the range}$$

$$p(x) = 0, \qquad \text{otherwise} \tag{6.27}$$

where N represents the number of points in the range. A random variable that has this distribution is said to have a *uniform* probability density. These distributions occur frequently in applications. For example if, in a sequence of Bernoulli trials, the second "success" is known to occur on the $N + 1$st trial, a simple computation shows that the number of the trial on which the first "success" occurs will have this uniform density.

If we attempt to extend this concept to the case of a discrete random variable that has an infinite range $x_1, x_2, x_3, \ldots$, we soon arrive at an impasse. Since $\sum p(x_i) = 1$, it is impossible that all the terms $p(x_i)$ be the same. An infinite series of equal *positive* numbers would diverge (and hence could not equal 1), while if each $p(x_i) = 0$, the sum would always be 0, never 1. Thus a discrete random variable with infinite range cannot have a uniform density function.

Paradoxically, the definition *does* extend to the case of a continuous random variable X having range some finite interval $a \leq x \leq b$. Such a random variable is said to have a *uniform* density if its probability density function is constant within the range:

$$f(x) = \begin{cases} c, & a \leq x \leq b \\ 0, & \text{otherwise} \end{cases} \tag{6.28}$$

Since $\int_{-\infty}^{\infty} f(x)\,dx = \int_a^b f(x)\,dx = \int_a^b c\,dx = 1$, it follows that $c = 1/(b-a)$ and that

$$f(x) = \begin{cases} \dfrac{1}{b-a}, & a \leq x \leq b \\ 0, & \text{otherwise} \end{cases} \tag{6.29}$$

In this case, the distribution function $F(x) = \int_{-\infty}^{x} f(t)\,dt$ will be given by

$$F(x) = \begin{cases} 0, & x < a \\ \dfrac{x-a}{b-a}, & a \leq x \leq b \\ 1, & x > b \end{cases} \tag{6.30}$$

The functions $f(x)$ and $F(x)$ are sketched in Figure 6.5.

The term "at random" is frequently used to refer to such a random variable. If we say that a value is chosen *at random* in the interval $a \leq x \leq b$, it

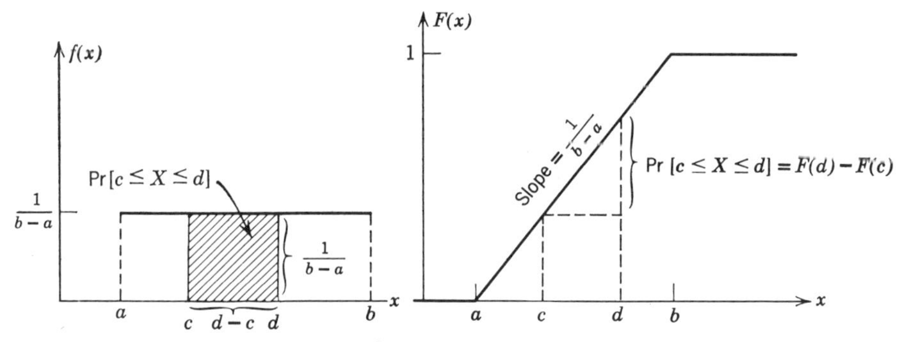

Figure 6.5 The uniform density and distribution function in the continuous case.

means that a value is observed of a random variable that has a uniform density over the interval $a \le x \le b$.

The mean, the variance, and the moment-generating functions of a random variable that has a uniform probability density function are easily computed:

$$\mu = E[X] = \int_a^b \frac{x}{b-a}\,dx = \frac{a+b}{2} \tag{6.31}$$

$$\sigma^2 = E[X^2] - \mu^2 = \int_a^b \frac{x^2}{b-a}\,dx - \left(\frac{a+b}{2}\right)^2 = \frac{(b-a)^2}{12} \tag{6.32}$$

$$M(\theta) = E[e^{X\theta}] = \int_a^b \frac{e^{x\theta}}{b-a}\,dx = \frac{e^{b\theta} - e^{a\theta}}{\theta(b-a)} \tag{6.33}$$

The student should check the calculations involved.

EXAMPLE 6.1 In a particular computer storage system, information is retrieved by entering the system at a random point and systematically checking through all the items in the system at a uniform speed. If it takes b seconds to run through the entire system, the time taken to reach any given item of information is a random variable that has a uniform distribution over the interval from time 0 to time b. If it takes a seconds to start the search, the time required will have a uniform distribution from time a to time $a + b$. △

7. The Negative Exponential Density Function

A continuous random variable having range $0 \le x < \infty$ is said to have a *negative exponential distribution* if it has a probability density function of the form

$$f(x) = \begin{cases} a\,e^{-ax}, & 0 \le x < \infty \\ 0, & x < 0 \end{cases} \tag{6.34}$$

where a is some positive constant. The corresponding distribution function is

$$F(x) = \int_{-\infty}^{x} f(t)\,dt = \begin{cases} 1 - e^{-ax}, & 0 \le x < \infty \\ 0, & x < 0 \end{cases} \tag{6.35}$$

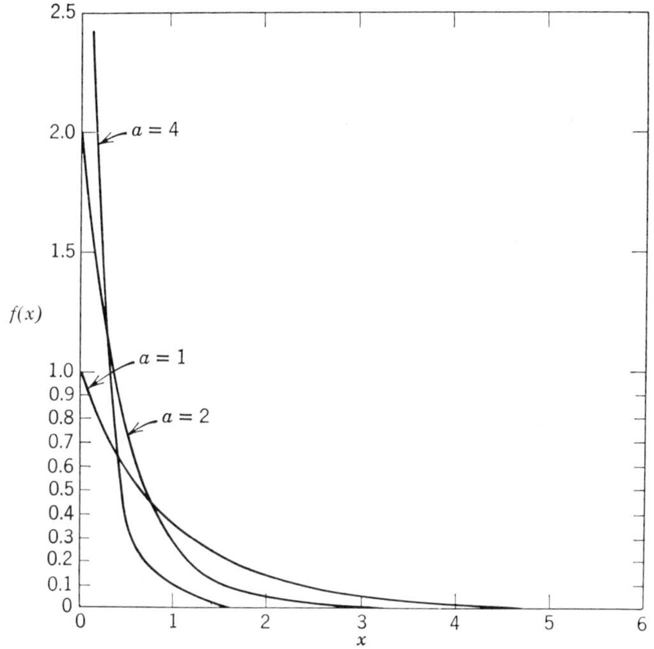

Figure 6.6 The negative exponential density function for several values of a.

The graph of $f(x)$ is sketched for several different values of a in Figure 6.6.

The mean, the variance, and the moment-generating function of a random variable X that has this distribution are easily calculated:

$$\mu = E[X] = \int_0^\infty x\, a\, e^{-ax}\, dx$$

$$= \frac{1}{a} \int_0^\infty u\, e^{-u}\, du, \qquad u = ax$$

$$= \frac{1}{a} \left[-u\, e^{-u} \Big|_0^\infty + \int_0^\infty e^{-u}\, du \right]$$

$$= \frac{1}{a} \tag{6.36}$$

using integration by parts,

$$
\begin{aligned}
E[X^2] &= \int_0^\infty x^2\, ae^{-ax}\, dx \\
&= \frac{1}{a^2}\int_0^\infty u^2 e^{-u}\, du \\
&= \frac{1}{a^2}\left[-u^2e^{-u}\Big|_0^\infty + 2\int_0^\infty ue^{-u}\, du\right] \\
&= \frac{1}{a^2}[0+2] = \frac{2}{a^2}
\end{aligned}
$$

$$
\operatorname{Var}[X] = \frac{2}{a^2} - \left(\frac{1}{a}\right)^2 = \frac{1}{a^2} \tag{6.37}
$$

$$
\begin{aligned}
M(\theta) = E[e^{X\theta}] &= \int_0^\infty e^{x\theta} ae^{-ax}\, dx \\
&= a\int_0^\infty e^{-(a-\theta)x}\, dx \\
&= \frac{a}{a-\theta}\int_0^\infty e^{-u}\, du, \qquad u = (a-\theta)x \\
&= \frac{a}{a-\theta} = \left(1 - \frac{\theta}{a}\right)^{-1}
\end{aligned} \tag{6.38}
$$

provided that $\theta < a$.

We notice that

$$
M(\theta) = \frac{a}{a-\theta} = \frac{1}{1-\left(\dfrac{\theta}{a}\right)} = 1 + \frac{\theta}{a} + \frac{\theta^2}{a^2} + \cdots + \frac{\theta^n}{a^n} + \cdots \tag{6.39}
$$

provided that $|\theta/a| < 1$ or $-a < \theta < a$. Formula (5.31) for moment generating functions can now be applied, and we observe that the nth moment of X, $E[X^n]$, is the coefficient of $\theta^n/n!$ in the series expansion of $M(\theta)$. If we write the nth term of (6.39) as

$$
\frac{n!}{a^n}\cdot\frac{\theta^n}{n!}
$$

we find that

$$
E[X^n] = \int_0^\infty x^n a e^{-ax}\, dx = \frac{n!}{a^n}
$$

or

$$
\int_0^\infty x^n e^{-ax}\, dx = \frac{n!}{a^{n+1}} \tag{6.40}
$$

an integration formula that will be useful later.

The negative exponential probability density occurs frequently in applications. It frequently enters as the density of random variables that represent the lifetimes of certain types of equipment, or as the time interval between events that occur in random sequence (orders to a wholesaler, telephone calls at an exchange, demands for computer service, successive arrivals of cosmic rays, or clicks on a Geiger counter, etc.). The primary reason for the great importance of this distribution is that a random variable with this probability density has a certain "forgetfulness" property analogous to the one of the discrete geometric density considered in Section 4.

Suppose that X has this negative exponential density and suppose we know that X exceeds some given value s, $X > s$. For example, if we interpret X as the lifetime of a certain piece of equipment, suppose we have observed that this equipment has already been operating for s hours. We may then be interested in the distribution of the *remaining* lifetime, $X' = X - s$. Suppose that we denote by $G(x)$ the *conditional* distribution function of X', given that $X > s$. Thus, for $x \geq 0$, we have

$$\begin{aligned} G(x) &= \Pr[X' \leq x \mid X > s] \\ &= \Pr[X - s \leq x \mid X > s] \\ &= \Pr[X \leq x + s \mid X > s] \\ &= \frac{\Pr[X \leq x + s \text{ and } X > s]}{\Pr[X > s]} \\ &= \frac{\Pr[s < X \leq x + s]}{\Pr[X > s]} \end{aligned}$$

by using the definition (2.9) of conditional probability. Thus (see Figure 6.7):

$$G(x) = \frac{\int_s^{x+s} f(t)\, dt}{\int_s^{\infty} f(t)\, dt} \tag{6.41}$$

$$= \frac{\int_s^{x+s} ae^{-at}\, dt}{\int_s^{\infty} ae^{-at}\, dt}$$

$$= \frac{e^{-as} - e^{-a(x+s)}}{e^{-as}}$$

$$= 1 - e^{-ax} \tag{6.42}$$

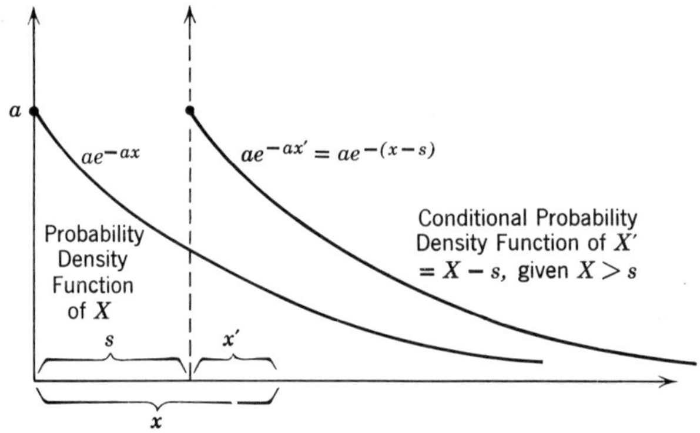

Figure 6.7 The comparison between densities of a negative exponentially distributed lifetime X and the remaining lifetime $X' = X - s$, given that the lifetime exceeds s.

If we compare (6.42) with (6.35), we see that $G(x)$ is identical with the distribution function $F(x)$ of the original random variable X without any conditional term. In words this says that *the distribution of the remaining life of the equipment does not depend on how long the equipment has been operating*. The equipment "forgets" how long it has been running, and its eventual breakdown is the result of some suddenly-appearing failure, not of gradual deterioration.

The negative exponential density is the only continuous probability density that has this property. To show this, assume that X is a continuous nonnegative random variable for which the conditional distribution function of $X' = X - s$, given that $X > s$, equals the initial distribution function, $F(x)$.

Thus by (6.41)

$$G(x) = \frac{\int_s^{s+x} f(t)\, dt}{\int_s^{\infty} f(t)\, dt} = F(x)$$

or

$$\frac{F(x+s) - F(s)}{1 - F(s)} = F(x) \tag{6.43}$$

Since $F(0) = 0$ (X being nonnegative), we can rearrange the terms of (6.43) as follows, dividing each side by x:

$$\frac{F(x+s) - F(s)}{x} = [1 - F(s)] \left[\frac{F(x) - F(0)}{x}\right]$$

Now let $x \to 0+$, and use the definition of a derivative:

$$F'(s) = [1 - F(s)]F'(0)$$

Let $U(s) = 1 - F(s)$, $F'(0) = a =$ constant. Thus

$$U'(s) = -F'(s) = -U(s)a$$

This simple differential equation can be solved to give

$$U(s) = Ce^{-as}$$

Since $U(0) = 1 - F(0) = 1$, it follows that $C = 1$. Thus

$$U(s) = e^{-as} = 1 - F(s)$$
$$F(s) = 1 - e^{-as}$$

for $s > 0$. Thus we observe that X must have the distribution function of a negative exponentially distributed random variable.

The preceding discussion can be summed up in this theorem:

THEOREM. *Let X be a continuous random variable with range $0 \leq x < \infty$. For each fixed $s > 0$, assume that, conditional on X being greater than s, the probability density function of $X - s$ is the same as the unconditional probability density function of X. Then X has a negative exponential probability density function. Conversely, any random variable having a negative exponential probability density function has this property.* △

EXAMPLE 6.2. Experience indicates that the remaining life of a certain type of light bulb is unaffected by previous use. A sample of 100 of these bulbs is found to have an average life of 1000 hours. If two new bulbs are selected and plugged in, estimate the probability that both will be burning 1200 hours later.

In view of the previous theorem, the first sentence of this example can be interpreted as requiring that the lifetime X of any light bulb be a random variable that has a probability density function of the form ae^{-ax}, $x > 0$. By (6.36), X has mean $E[X] = 1/a$. The average of the lives of a large number of light bulbs would provide an approximation to the mean life of a bulb, $1000 \approx 1/a$. Thus, $a \approx 1/1000$. If X and Y are the lives of the two new bulbs, then they will be independent random variables (why?), each having probability density function:

$$ae^{-ax} \approx \frac{1}{1000} e^{-x/1000}, \qquad x > 0$$

Thus the probability that both will be burning after 1200 hours is given by

$$\Pr[X > 1200 \text{ and } Y > 1200] = \Pr[X > 1200] \Pr[Y > 1200]$$

using the independence assumption

$$= \{\Pr[X > 1200]\}^2$$

since X and Y have the same densities

$$= \left\{\int_{1200}^{\infty} ae^{-ax}\, dx\right\}^2$$

$$= (e^{-1200a})^2$$

$$\approx \{e^{-1200/1000}\}^2 = e^{-2.4} = .0907 \qquad \triangle$$

8. The Gamma Density Function

A useful generalization of the negative exponential probability density function is the gamma density. A continuous random variable X with range $0 < x < \infty$ is said to have a *gamma density* if its probability density function has the form

$$f(x) = Cx^{b-1}e^{-ax}, \qquad 0 < x < \infty \tag{6.44}$$

where a and b are positive constants. We readily check that the integral $\int_0^\infty x^{b-1}e^{-ax}\, dx$ converges for $a > 0,\ b > 0$. Thus, if the constant C is chosen so that

$$C \int_0^\infty x^{b-1}e^{-ax}\, dx = 1 \tag{6.45}$$

formula (6.44) does represent a probability density function. The precise value of the constant C is not particularly important. By substituting $u = ax$ in (6.45), we obtain, after a simple manipulation, $C = a^b/\Gamma(b)$, where $\Gamma(b) = \int_0^\infty u^{b-1}e^{-u}\, du$ is termed the *gamma function* (hence, the name of this density). If b is an integer, then by using formula (6.40), $\Gamma(b) = (b - 1)!$. In this case

$$C = \frac{a^b}{(b-1)!}$$

Notice that the negative exponential density is a special case of the gamma density with $b = 1$.

Another important special case is $a = \frac{1}{2}$ and $b = n/2$ where n is a positive integer. A random variable that has this density is said to have a *chi-square*

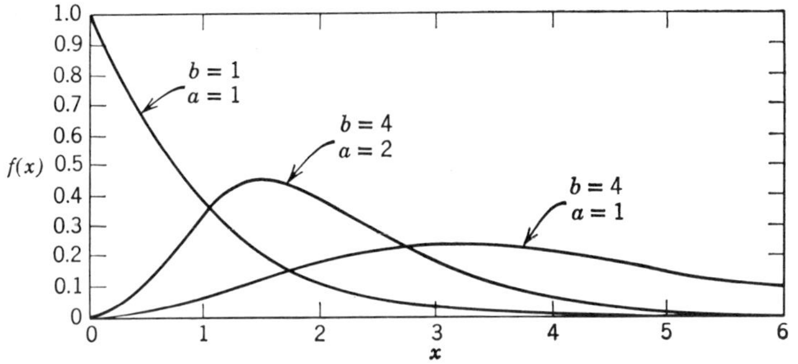

Figure 6.8 The gamma density function for several values of a and b.

distribution with n degrees of freedom. These distributions have important applications in statistics.

Some gamma density functions for various values of a and b are sketched in Figure 6.8. The moment generating function of a random variable having a gamma density is easily computed:

$$\begin{aligned} M(\theta) &= E[e^{X\theta}] \\ &= C\int_0^\infty e^{x\theta}x^{b-1}e^{-ax}\,dx \\ &= C\int_0^\infty x^{b-1}e^{-(a-\theta)x}\,dx \end{aligned}$$

This integral will converge provided that $a-\theta>0$, $\theta<a$. If we substitute

$$x=\frac{a}{a-\theta}\cdot y, \qquad dx=\frac{a}{a-\theta}\,dy$$

we obtain

$$\begin{aligned} M(\theta) &= C\int_0^\infty \left(\frac{a}{a-\theta}\right)^{b-1} y^{b-1}\cdot e^{-ay}\cdot\frac{a}{a-\theta}\,dy \\ &= \left(\frac{a}{a-\theta}\right)^{b} C\int_0^\infty y^{b-1}e^{-ay}\,dy \\ &= \left(\frac{a}{a-\theta}\right)^{b} \end{aligned} \tag{6.46}$$

since $C\int_0^\infty y^{b-1}e^{-ay}\,dy=1$ by (6.45). Thus

$$M(\theta)=\left(1-\frac{\theta}{a}\right)^{-b}=1+\frac{b}{a}\cdot\theta+\frac{b(b+1)}{a^2}\cdot\frac{\theta^2}{2}+\cdots \tag{6.47}$$

using the binomial theorem. From (5.31) we observe that

$$\mu = E[X] = \frac{b}{a} \qquad \text{and} \qquad \mu_2' = E[X^2] = \frac{b(b+1)}{a^2}$$

X has mean and variance

$$\mu = \frac{b}{a}, \qquad \sigma^2 = \mu_2' - \mu^2 = \frac{b}{a^2} \tag{6.48}$$

EXAMPLE 6.3. The length of time required to repair an article at a certain appliance repair shop is known to be a random variable X that has approximately a gamma distribution. By observing the repair times for several objects, the mean and the variance of X are estimated to be 2 and 1, respectively, if X is measured in hours. (The details of how this sample can be used to estimate a population mean and variance belong properly in a course in *statistics* and are omitted here.) Estimate the probability that the next defective appliance to arrive will be repaired in less than one hour.

We use the *method of moments* to estimate a and b. If μ and σ^2 are estimated to be 2 and 1, then by (6.48)

$$\frac{b}{a} \approx 2, \qquad \frac{b}{a^2} \approx 1$$

These equations are easily solved to give $a \approx 2$, $b \approx 4$. Thus the required probability is

$$\begin{aligned}
\Pr[X < 1] &= C \int_0^1 x^{b-1} e^{-ax}\,dx \\
&= \frac{a^b}{(b-1)!} \int_0^1 x^{b-1} e^{-ax}\,dx \\
&\approx \frac{2^4}{3!} \int_0^1 x^3 e^{-2x}\,dx \\
&= \frac{1}{6} \int_0^2 u^3 e^{-u}\,du, \qquad \text{where } u = 2x
\end{aligned}$$

This integral can be evaluated, using integration by parts, or it can be found in a *table of incomplete gamma* functions. $\Pr[X < 1] = 0.143$. △

By comparing formulas (6.47) and (6.38), we observe that the moment generating function of a gamma distributed random variable is the bth power of the moment generating function of a negative exponentially distributed random variable. We recall that, when independent random variables are added, their moment-generating functions multiply (formula 5.30). Thus, if b is an integer, we have this important theorem:

THEOREM. *If $X_1, X_2, \ldots, X_b$ are independent random variables, each having a negative exponential density with mean $1/a$, then $X = X_1 + \cdots + X_b$ will have a gamma density of the form* (6.44). △

EXAMPLE 6.4. The lifetime of a certain electronic tube is known to be a negative exponentially distributed random variable with mean 1,000 hours. Three of these tubes are available (the original and two spares) and, when one burns out, it is replaced by a spare. Determine the distribution of the time X until the last spare burns out.

By (6.36), $\mu = 1/a = 1000$, $a = 1/1000$. Thus the lifetimes X_1, X_2, X_3 of the 3 tubes will be random variables, independent, since they do not interact with each other, and each will have density function $(1/1000)e^{-x/1000}$, $x > 0$. The time X until the last burn-out is given by $X = X_1 + X_2 + X_3$. By the property above, X must have a gamma density with $a = 1/1000$, $b = 3$. △

9. The Normal Density Function

The normal density function was introduced in Chapter 3, Example 3.13. A continuous random variable is said to have a normal distribution if its range is all real numbers and its probability density function has the form:

$$f(x) = \frac{1}{\sqrt{2\pi}\sigma} e^{-(x-\mu)^2/2\sigma^2}, \qquad -\infty < x < \infty \tag{6.49}$$

Here μ and σ are arbitrary parameters, $\sigma > 0$, which we shall show represent the mean and standard deviation of the random variable. A sketch of the graph of $f(x)$ is given in Figure 3.18. In the case $\mu = 0$, $\sigma = 1$, its distribution function was tabulated in Example 3.13, and the use of this table to compute general normal probabilities was illustrated there.

To show that $f(x)$ given by (6.49) is a true probability density function, we need the following identity:

$$\int_{-\infty}^{\infty} e^{-z^2/2}\, du = \sqrt{2\pi} \tag{6.50}$$

This identity is easily derived (see Exercise 12).

Thus

$$\int_{-\infty}^{\infty} f(x)\,dx = \frac{1}{\sqrt{2\pi}\sigma}\int_{-\infty}^{\infty} e^{-(x-\mu)^2/2\sigma^2}\,dx$$

$$= \frac{1}{\sqrt{2\pi}}\int_{-\infty}^{\infty} e^{-z^2/2}\,dz, \qquad \text{where } z = \frac{x-\mu}{\sigma}$$

$$dz = \frac{dx}{\sigma}$$

$$= \frac{1}{\sqrt{2\pi}}\cdot\sqrt{2\pi} = 1$$

The moment generating function of a random variable X that has this distribution is also easily obtained:

$$M(\theta) = E[e^{X\theta}] = \frac{1}{\sqrt{2\pi}\,\sigma}\int_{-\infty}^{\infty} e^{x\theta}\cdot e^{-(x-\mu)^2/2\sigma^2}\,dx$$

We again make the substitution $z = (x-\mu)/\sigma$, $dz = dx/\sigma$, $x = \sigma z + \mu$.

$$M(\theta) = \frac{1}{\sqrt{2\pi}}\int_{-\infty}^{\infty} e^{(\sigma z+\mu)\theta}\, e^{-z^2/2}\,dz$$

$$= e^{\mu\theta}\cdot\frac{1}{\sqrt{2\pi}}\int_{-\infty}^{\infty} e^{-(z^2-2\sigma\theta z)/2}\,dz$$

$$= e^{\mu\theta}\cdot e^{\sigma^2\theta^2/2}\cdot\frac{1}{\sqrt{2\pi}}\int_{-\infty}^{\infty} e^{-(z^2-2\sigma\theta z+\sigma^2\theta^2)/2}\,dz$$

$$= e^{\mu\theta+\sigma^2\theta^2/2}\cdot\frac{1}{\sqrt{2\pi}}\int_{-\infty}^{\infty} e^{-(z-\sigma\theta)^2/2}\,dz$$

$$= e^{\mu\theta+\sigma^2\theta^2/2}\cdot\frac{1}{\sqrt{2\pi}}\int_{-\infty}^{\infty} e^{-z'^2/2}\,dz', \qquad z' = z-\sigma\theta$$

$$= e^{\mu\theta+\sigma^2\theta^2/2}\cdot\frac{1}{\sqrt{2\pi}}\cdot\sqrt{2\pi}$$

Thus

$$M(\theta) = e^{\mu\theta+\sigma^2\theta^2/2} \tag{6.51}$$

The mean and variance can be computed by expanding (6.51) in powers of θ:

$$\begin{aligned} M(\theta) &= e^{\mu\theta} \cdot e^{\sigma^2\theta^2/2} \\ &= \left(1 + \mu\theta + \frac{\mu^2\theta^2}{2} + \cdots\right)\left(1 + \frac{\sigma^2\theta^2}{2} + \cdots\right) \\ &= 1 + \mu\theta + (\mu^2 + \sigma^2)\frac{\theta^2}{2} + \cdots \end{aligned}$$

Since the coefficients of the powers of θ give the moments, we have

$$\begin{aligned} E[X] &= \mu \\ E[X^2] &= \mu^2 + \sigma^2 \\ \mathrm{Var}[X] &= E[X^2] - (E[X])^2 = \mu^2 + \sigma^2 - \mu^2 = \sigma^2 \end{aligned}$$

and we notice that the parameters μ and σ^2 do indeed represent the mean and variance here (see Chapter 5, Section 8).

The normal distribution is the most important probability distribution for use in statistics. In this book, however, it is used, for the most part, only as a convenient approximation to certain other distributions. Thus we do not give an extended discussion here.

The two most important approximation properties from our point of view are the following.

I. *If X has the binomial distribution of Section* 3, *then X has approximately a normal distribution with $\mu = np$, $\sigma = \sqrt{npq}$. More precisely, the distribution function of $X^* = (X - np)/\sqrt{npq}$ approaches the distribution function of the standard normal* ($\mu = 0$, $\sigma = 1$) as $n \to \infty$.

II. *If X has the Poisson distribution discussed in Section* 5, *then it has approximately a normal distribution with $\mu = \lambda$, $\sigma = \sqrt{\lambda}$. More precisely, the distribution function of $X^* = (X - \lambda)/\sqrt{\lambda}$ approaches the standard normal distribution function as $\lambda \to \infty$.*

The proofs of I and II are omitted here. In each case the proof involves showing that the moment-generating function of X^* approaches that of the standard normal.

The approximations I and II are illustrated by Tables 6.2 and 6.3.

We observe from Table 6.2 that the normal distribution function is a good approximation for the binomial distribution function for values of p

TABLE 6.2[a]

x	Binomial $\Pr[X \leq x]$	Normal $\Pr[X \leq x]$
	$n = 10,\ p = 0.5$	
0	0.0010	0.0008
2	0.0547	0.0287
4	0.3770	0.2643
6	0.8282	0.7357
8	0.9894	0.9713
10	1.0000	0.9992
	$n = 100,\ p = 0.5$	
40	0.0284	0.0228
45	0.1841	0.1587
50	0.5398	0.5000
55	0.8644	0.8413
60	0.9824	0.9772
	$n = 10,\ p = 0.05$	
0	0.5987	0.2327
1	0.9138	0.7673
2	0.9884	0.9854
3	0.9909	0.9999
4	0.9919	0.9999+
	$n = 100,\ p = 0.05$	
0	0.0059	0.0110
2	0.1183	0.0838
4	0.4360	0.3228
6	0.7660	0.6772
8	0.9369	0.9162
10	0.9885	0.9890
12	0.9985	0.9993

[a] The Normal Approximation to the Binomial.

about $\frac{1}{2}$ and n large. Even for p small the approximation is fair for $n = 100$. For small values of n (for instance, 10), the approximation is reasonably poor and, for small values of p, it is very poor. We conclude that, if p is near $\frac{1}{2}$ and n is large, the normal serves as a good approximation to the binomial. If p is very different from $\frac{1}{2}$, the normal is a poorer approximation even for moderate

n values. However, we observed in Section 5 that for p small, the Poisson distribution is a good approximation to the binomial. Thus we conclude that the binomial can be approximated either by the normal (if p is approximately $\frac{1}{2}$) or by the Poisson (if p is small). In either case the approximation improves as n increases. (See Exercise 14 for a method of improving the approximation of the binomial by the normal.)

TABLE 6.3[a]

x	$\lambda = 1$ $\Pr[X \leq x]$ Poisson	$\Pr[X \leq x]$ Normal
0	0.3679	0.1587
1	0.7358	0.5000
2	0.9197	0.8413
3	0.9810	0.9772
4	0.9963	0.9987
5	0.9994	0.9999
	$\lambda = 4$	
x		
0	0.0183	0.0228
2	0.2381	0.1587
4	0.6289	0.5000
6	0.8894	0.8413
8	0.9787	0.9772
10	0.9972	0.9987
	$\lambda = 9$	
x		
0	0.0001	0.0013
3	0.0212	0.0228
6	0.2067	0.1587
9	0.5874	0.5000
12	0.8758	0.8413
15	0.9780	0.9772
18	0.9976	0.9987

[a] The normal as an Approximation to the Poisson.

We observe here that the approximation is better for larger λ.

10. The Cauchy Density Function

A mathematically interesting density function can arise in the following way.

Suppose that a pointer is supported 1 unit above the x-axis (shown in Figure 6.9).

When the pointer is spun, the angle Y takes on values in the interval $-\pi/2 \leq Y \leq \pi/2$. Assume that the random variable Y has a uniform distribution over this interval. If we denote by X the point on the x-axis toward which the pointer is pointing, then X is a continuous random variable that has range $-\infty < X < \infty$. Clearly, X and Y are related by the identity

$$X = \tan Y$$

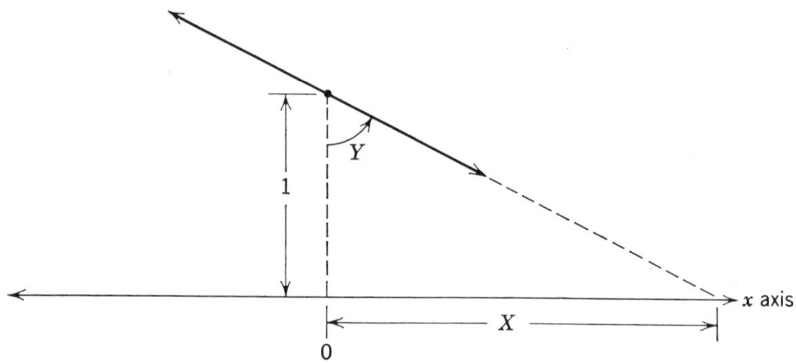

Figure 6.9 A physical apparatus to generate a Cauchy distributed random variable (X).

and, from this, the distribution of X can be determined as follows.

$$\begin{aligned} F(x) &= \Pr[X \leq x] = \Pr[\tan Y \leq x] \\ &= \Pr[Y \leq \tan^{-1} x] \\ &= \int_{-\pi/2}^{\tan^{-1} x} \frac{1}{\pi}\, dy \\ &= \frac{1}{\pi}\left[\tan^{-1} x + \frac{\pi}{2}\right] \end{aligned}$$

Hence, X has probability density function:

$$f(x) = F'(x) = \frac{1}{\pi(1 + x^2)}, \qquad -\infty < x < \infty$$

This density function was discussed in Example 5.3. A random variable that has this density function is said to have a *Cauchy distribution*. A sketch of the graph of $f(x)$ is given in Figure 5.2.

EXERCISES

1. A communication system has probability 0.05 of transmitting any given symbol incorrectly. What is the probability that 3 symbols are incorrect in a message of 20 symbols?

2. Determine μ and σ^2 for a random variable that has a binomial distribution directly from their definitions and the probability density function (6.11).

$$\left[\textit{Hint.}\ i\binom{n}{i} = i \cdot \frac{n!}{i!\,(n-i)!} = n\binom{n-1}{i-1};\ i^2\binom{n}{i} = [i(i-1)+i]\binom{n}{i}.\right]$$

3. If X has a binomial distribution, show that

$$\mu_3 = E[(X-\mu)^3] = npq(q-p)$$

4. Two honest coins are tossed until the first double heads occurs.
   (a) Compute the probability that 4 tosses are required.
   (b) What is the mean number of tosses required?

5. A lot of manufactured goods are sampled by testing items sequentially until the first defective is found. If k or more items are tested before the first defective is found, the lot is accepted as meeting specifications.
   (a) If $k = 4$ and the lot contains 10% defective items, what is the probability that it will be accepted?
   (b) What value should be given to k if one wishes that a lot having 40% defectives be rejected with probability at least .80?

6. Determine the moments about zero for the Weibull density function given in Chapter 3, Exercise 25, for the special case $x_0 = 0$.

7. Determine the moments about zero for the beta density function defined in Chapter 3, Exercise 21.

8. Determine the third moment about the mean, $\mu_3 = E[(X-\mu)^3]$, for the Poisson density function. From the results of Section 5 and this exercise, what do you assume is true for μ_k?

9. Determine the probability generating function for the modified geometric density function given in Section 4.

10. Prove formulas (6.31), (6.32), (6.33).

11. The density function

$$f(x) = \tfrac{1}{2}e^{-|x|}, \qquad -\infty < x < \infty$$

is called the Laplace distribution. It arises as the density of the difference of two independent negative exponentially distributed random variables. Determine the mean and variance of a random variable that has this density.

12. If $I = \int_{-\infty}^{\infty} e^{-x^2/2}\,dx$, show that $I = \sqrt{2\pi}$. $\Big[$*Hint.* $I = \int_{-\infty}^{\infty} e^{-y^2/2}\,dy$ as well, and therefore

$$I^2 = \int_{-\infty}^{\infty} e^{-x^2/2}\,dx \int_{-\infty}^{\infty} e^{-y^2/2}\,dy = \int_{-\infty}^{\infty}\int_{-\infty}^{\infty} e^{-(x^2+y^2)/2}\,dx\,dy$$

Evaluate this double integral using polar coordinates.]

13. If X has a standard normal distribution, show that X^2 has a gamma distribution with $a = b = \tfrac{1}{2}$.

14. In the tables that compare the distribution function of the binomial and the Poisson with the normal, part of the discrepancy results from the fact that the binomial and Poisson are discrete and the normal is continuous. For the binomial or Poisson, $\Pr[X \le k]$ and $\Pr[X < k+1]$ are the same, since X must be an integer. The corresponding normal probabilities are different. The approximation is improved by taking a value midway between k and $k+1$ and by comparing $\Pr[X \le k]$ for the binomial or Poisson with $\Pr[X \le k + \tfrac{1}{2}]$ for the normal. Check this statement by constructing the corresponding tables in the case where $n = 100$, $p = \tfrac{1}{2}$, and $\lambda = 9$.

7 • Stochastic Processes: Examples

1. Introduction

We often consider situations in which observations are made over a period of time and that are influenced by chance or random effects, not just at a single instant but throughout the entire interval of time or the sequence of times that we are considering. This situation is termed a *stochastic process*. Examples are very diverse. The variation of traffic at a certain intersection that involves the growth and dissipation of traffic jams forms a stochastic process. Another example would be the daily variation in the size of a certain company's inventory. Other instances include the behavior of particles subject to random impacts, shot noise in vacuum tubes, molecular diffusion, radioactive decay, variations in the quality of the products of a factory, genetic mutations, variation in stock market activity, etc.

In a rough sense, a stochastic process is a phenomenon that varies to some degree unpredictably as time goes on. The unpredictability here implies that if we observed an entire time-sequence of the process on several different occasions, under presumably "identical" conditions, the resulting observation-sequences, in general, would be different. Thus, probability enters, but not in the sense that each result of a random experiment determines only a single number. Instead, the random experiment determines the behavior of some system for an entire sequence or interval of time. That is, the result of the random experiment is a sequence or series of values, *a function*, not just a single number.

Thus, for example, if we observe the passage of cars at a certain street intersection for a 24-hour period, we are observing a realization of a random process, a single sample point from a certain random experiment. It is easy to conceive of events connected with this observation for which probabilities would be defined. The event "72 cars pass between 2 p.m. and 3 p.m." would have a certain probability, and clearly it would be different from the probability of the event "72 cars pass between 2 a.m. and 3 a.m."

2. Stochastic Processes

Since a stochastic process involves the behavior of a system over time, in defining such a process, we must start by specifying the *time set* T involved. This might be some time *interval* in a situation in which measurements are made continuously, such as keeping track of the cars that pass a certain intersection. In this case, we speak of a *continuous parameter* process. In other situations, T might be a *sequence* of times. This would happen, for example, if a company's inventory is only recorded once a month, or if a person's total winnings in a game of chance are recorded after each play. In this case we speak of a *discrete parameter* process.

In this book we only consider one type of discrete T, consisting of a sequence of consecutive integers

$$T = \{0, 1, 2, 3, \ldots\} \text{ or } T = \{1, 2, 3, \ldots\}$$

and one type of continuous T, consisting of all times subsequent to some given origin

$$T = \{t: 0 \leq t < \infty\}$$

Let us suppose that at each point t of the time set T we can observe a measurement or random variable X_t. Thus we assume that some random experiment is given, and that for each sample point or experimental outcome there corresponds not a single number (as in the case of a random variable) but an entire *function* X_t. If the sample point or experimental outcome is denoted by s, then we can properly denote the function by

$$X_t(s) \qquad \text{for} \qquad t \in T$$

Such a *random function of t* is termed a *stochastic process* or *random process*. A single sequence of observations (or function) X_t that corresponds to a single sample point s is termed a *realization* of the stochastic process. Several possible realizations of stochastic processes are sketched in Figures 7.1 to 7.3.

The *range* of X_t (the collection of numerical values that the random variables X_t can take on) is termed the *state space* of the process, and the values of X_t are termed *states*. If X_t represents some *measurement* (temperature, voltage, etc.), then the state space might be an interval of real numbers. If X_t represents a *count*, or a whole number of objects (items in inventory, dollars won in a sequence of games, etc.), then the state space might consist of a sequence

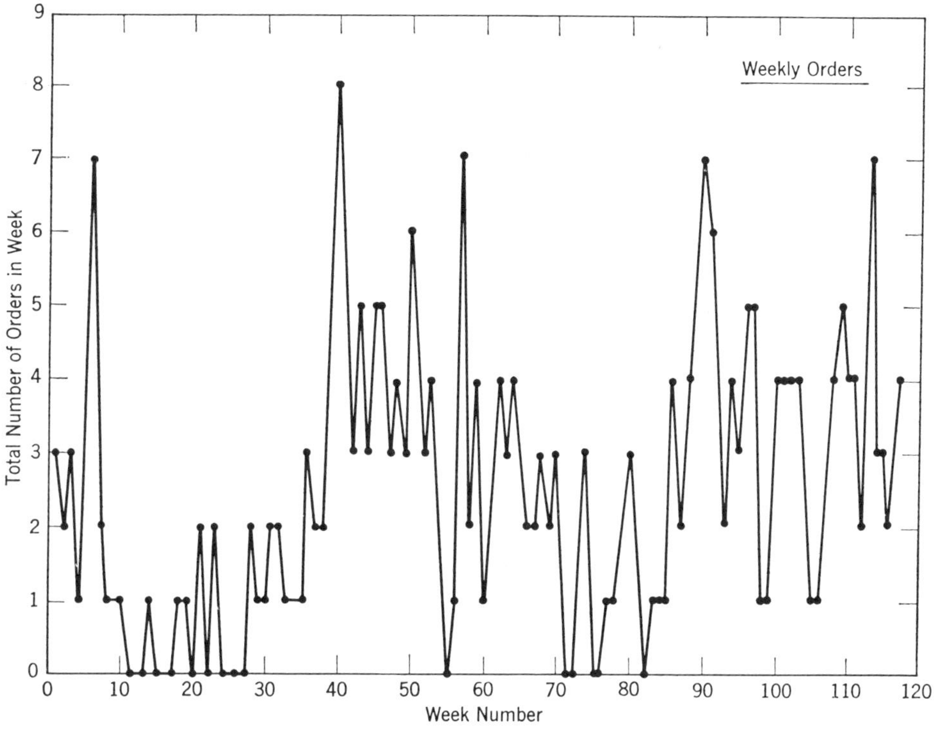

Figure 7.1 The demand per week for spare parts. (This graph has been prepared from data supplied by Dr. H. P. Galliher).

(finite or infinite) of integers. In the first case, we speak of a *continuous state* process, in the second, we speak of a *discrete state* process. In our subsequent discussion we shall assume that all state spaces are discrete unless it is specifically stated otherwise. These processes are frequently termed *random chains.*

Notice that for a fixed value of t, X_t is a *random variable* that describes the state of the process at time t. Given any finite collection $t_1, t_2, \ldots, t_n$ of times, then $X_{t_1}, X_{t_2}, \ldots, X_{t_n}$ is a set of n jointly distributed random variables. We say that the probability structure of the process X_t is completely determined provided that the joint distribution or density function of each such set of random variables is determined. Basically, analysis of a stochastic process involves determining these joint distributions and using them to predict the behavior of the process in the future, given certain behavior in the past.

In Sections 10, 11, and 12 of Chapter 4, we discussed sequences of random variables, $X_1, X_2, \ldots, X_n, \ldots$. Clearly, we can consider this sequence a

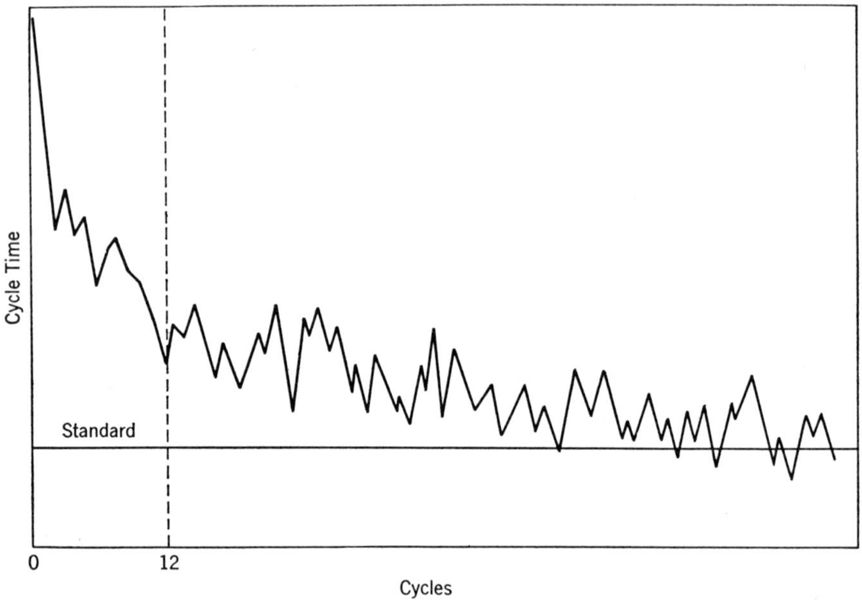

Figure 7.2 A typical learning curve for a repetitive short-cycle operation. (From Hancock, W. M. et al *Learning Curve Research on Manual Operations*, MTM Assoc. Ann Arbor, Michigan, 1965.)

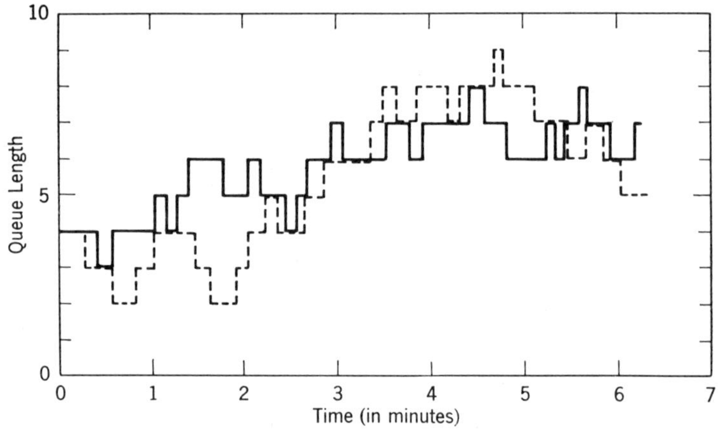

Figure 7.3 Two independent realizations of a waiting-line process, each beginning with four individuals.

discrete parameter stochastic process X_t in which the time set T consists of the integers $\{1, 2, 3, \ldots\}$.

Before we embark on a discussion of classification and general properties of stochastic processes, it is important that the student acquire an intuitive feel for the preceding ideas and an acquaintance with certain special processes.

The balance of this chapter is devoted to the discussion of certain important special examples of stochastic processes. The ideas illustrated in these examples will be expanded in subsequent chapters, with greater generality and applicability.

3. Independent Sequences

In Section 10 of Chapter 4 the concept of a random sequence, $X_1, X_2, X_3, \ldots,$ was discussed. Clearly such a sequence forms a discrete parameter stochastic process with time set $T = \{1, 2, 3, \ldots\}$. In Section 11 of Chapter 4 the important special case of *independent, identically distributed* random variables was introduced. A sequence of this kind can be considered as resulting from independent repetitions of the same random experiment, to each repetition of which is applied an associated measurement or random variable X.

For example, if a piece of electronic equipment contains a tube that is replaced whenever it fails, the sequence $X_1, X_2, X_3, \ldots$ might represent the sequence of lifetimes of the successive tubes. In this case, each random variable must be *positive* since it is the lifetime of a tube. Such a sequence of independent, identically distributed random variables, each of which has a positive range, is termed a *renewal process*. These processes are of importance in reliability theory.

Given a sequence $X_1, X_2, X_3, \ldots$ of random variables, another sequence or stochastic process can be constructed by considering the sequence of *partial sums*

$$S_1, S_2, S_3, \ldots$$

where $S_n = X_1 + X_2 + \cdots + X_n = \sum_{i=1}^{n} X_i$. If the original X_i's form an independent sequence, the sequence of partial sums will not be independent, but will form a Markov dependent sequence in the sense of Section 12, Chapter 4. This situation recurs frequently (see Section 5 of this chapter).

4. Two-State Markov Sequences

In Section 12 of Chapter 4 the concept of a *Markov dependent* sequence was introduced. We stated that it was a random sequence or discrete random process $\{X_1, X_2, \ldots\}$ which had the property that, conditional on knowing the value of X_{n-1} ("the present state"), the probability that X_n will take on some given value ("a future state") will be unaffected by knowledge of the values of $X_1, X_2, \ldots, X_{n-2}$ ("the past states"). That is,

$$\Pr[X_n = i_n \mid X_1 = i_1, \ldots, X_{n-1} = i_{n-1}] \\ = \Pr[X_n = i_n \mid X_{n-1} = i_{n-1}]$$

In the case of a *stationary* Markov dependent sequence, those probabilities were assumed not to depend on n, and we defined the quantities

$$p_{ij} = \Pr[X_n = j \mid X_{n-1} = i] \tag{7.1}$$

as the *one-step transition probabilities* of the process. Clearly they are conditional probabilities and satisfy

$$\sum_j p_{ij} = 1 \qquad \text{for each } i \tag{7.2}$$

where the sum is extended over the entire state space of the process.

The simplest nontrivial example of this process would be one in which the state space consists of only two states. They could be labeled "success" and "failure," or 0 and 1. In this case, there would be four one-step transition probabilities, $p_{00}, p_{01}, p_{10}, p_{11}$, that satisfy $p_{00} + p_{01} = 1, p_{10} + p_{11} = 1$.

Many real-world processes behave in this way, at least, to a first approximation. For instance, the two states could refer to a symbol either correctly transmitted or garbled in a communications system, to a machine operating or under repair in a factory, to a storage position empty or filled in a computer storage system, or to many other situations. To fix our ideas, let us consider a military application in which a Markovian model has been successfully used.

In firing the main gun on a tank we find that each shell fired hits or misses its target with a probability that depends on the success of the previously-fired round. If the previous shot was a hit, then the tank commander has the target zeroed in, and has a relatively high probability of repeating the hit with his next shot. If the previous shot was a miss, then the aim must be adjusted, and the probability is lower of making a hit on the next shot. The tank and target may both be moving. Thus, once the result of a certain shot is known, the success or failure of previous shots is of negligible importance in predicting the success of the next shot. Thus, to a reasonable approximation, we can assume that we are observing a stationary Markov process with state space $\{0, 1\}$, where 0 denotes a "hit" and 1 denotes a "miss."

Let us assume that battlefield observations enable us to estimate the one-step transition probabilities as follows:

$$p_{00} = \frac{3}{4}, \qquad p_{01} = \frac{1}{4}$$

$$p_{10} = \frac{1}{2}, \qquad p_{11} = \frac{1}{2}$$

Thus, for example, if a given shot hits, the probability is $\frac{3}{4}$ that the next shot will hit, etc.

Denote the process by $X_1, X_2, X_3, \ldots$. The random variable X_n takes values 0 or 1 depending on whether the nth shot is a hit or a miss.

To study the behavior of the process, we wish to evaluate all probabilities of the form

$$p_{ij}^{(n)} = \Pr[X_{k+n} = j \mid X_k = i], \qquad i, j = 0, 1 \tag{7.3}$$

for $n = 1, 2, 3, \ldots$, where these probabilities do not depend on k, since the process is assumed stationary. In this notation $p_{ij}^{(1)} = p_{ij}$.

Note that $p_{i0}^{(n)} + p_{i1}^{(n)} = 1$ for each i, since X_{k+n} must equal 0 or 1.

Now set $i = 0$, and assume that $X_k = 0$ for some fixed k. Thus, for any n, the event $X_{k+n+1} = 0$ has conditional probability $p_{00}^{(n+1)}$. Now let us consider what must happen at the preceding step [$(k+n)$th step] in order that the event $X_{k+n+1} = 0$ holds true. Either $X_{k+n} = 0$ and a one-step transition from state 0 to state 0 takes place in passing from the $(k+n)$th to the $(k+n+1)$st stage, or $X_{k+n} = 1$ and the transition is from state 1 to state 0. These two events have probabilities $p_{00}^{(n)}\, p_{00}$ and $p_{01}^{(n)} p_{10}$ (conditional on $X_k = 0$).*

Since these are the only two mutually exclusive ways in which the event $X_{k+n+1} = 0$ can occur, we have

$$p_{00}^{(n+1)} = p_{00}^{(n)}\, p_{00} + p_{01}^{(n)}\, p_{10} \tag{7.4}$$

for $n = 1, 2, 3, \ldots$.

* Intuitively this seems obvious. To prove it rigorously one must reason as follows:

$$\Pr[X_{k+n+1} = 0 \quad \text{and} \quad X_{k+n} = 0 \mid X_k = 0]$$

$$= \frac{\Pr[X_{k+n+1} = 0 \text{ and } X_{k+n} = 0 \text{ and } X_k = 0]}{\Pr[X_k = 0]}, \qquad \text{by (2.9)}$$

$$= \frac{\Pr[X_{k+n} = 0 \text{ and } X_k = 0]\Pr[X_{k+n+1} = 0 \mid X_{k+n} = 0 \text{ and } X_k = 0]}{\Pr[X_k = 0]}, \qquad \text{by (2.10)}$$

$$= \Pr[X_{k+n} = 0 \mid X_k = 0]\, \Pr[X_{k+n+1} = 0 \mid X_{k+n} = 0 \text{ and } X_k = 0]$$

$$= p_{00}^{(n)} \Pr[X_{k+n+1} = 0 \mid X_{k+n} = 0], \qquad \text{by (4.56)}$$

$$= p_{00}^{(n)}\, p_{00}$$

with a similar argument for $\Pr[X_{n+k+1} = 0 \text{ and } X_{n+k} = 1 \mid X_k = 0]$.

Since $p_{00}^{(n)} + p_{01}^{(n)} = 1$, equation (7.4) can be rewritten as

$$p_{00}^{(n+1)} = p_{00}^{(n)}\, p_{00} + [1 - p_{00}^{(n)}]p_{10}$$

or

$$p_{00}^{(n+1)} = p_{00}^{(n)}\,(p_{00} - p_{10}) + p_{10} \tag{7.5}$$

for $n = 1, 2, 3, \ldots$.

Formula (7.5) is a *recursion formula* or *difference equation* of the form

$$p_{00}^{(n+1)} = p_{00}^{(n)}\,\alpha + \beta \tag{7.6}$$

with $\alpha = p_{00} - p_{10}$ and $\beta = p_{10}$. By applying this formula repeatedly we can solve for $p_{00}^{(n)}$ in terms of $p_{00}^{(1)}$:

$$\begin{aligned} p_{00}^{(n+1)} &= p_{00}^{(n)}\,\alpha + \beta \\ &= [p_{00}^{(n-1)}\,\alpha + \beta]\,\alpha + \beta \\ &= p_{00}^{(n-1)}\,\alpha^2 + \beta(1 + \alpha) \\ &\text{etc.} \\ &= p_{00}^{(1)}\,\alpha^n + \beta(1 + \alpha + \cdots + \alpha^{n-1}) \end{aligned} \tag{7.7}$$

If $\alpha \neq 1$, the formula for the sum of a geometric series yields

$$p_{00}^{(n+1)} = p_{00}^{(1)}\,\alpha^n + \beta\left(\frac{1-\alpha^n}{1-\alpha}\right) \tag{7.8}$$

Setting $p_{00}^{(1)} = p_{00}$, $\alpha = p_{00} - p_{10}$, $\beta = p_{10}$ and changing from n to $n - 1$, we have

$$\begin{aligned} p_{00}^{(n)} &= p_{00}(p_{00} - p_{10})^{n-1} + p_{10}\left[\frac{1 - (p_{00} - p_{10})^{n-1}}{1 - p_{00} + p_{10}}\right] \\ &= \left[p_{00} - \frac{p_{10}}{1 - p_{00} + p_{10}}\right](p_{00} - p_{10})^{n-1} + \frac{p_{10}}{1 - p_{00} + p_{10}} \end{aligned} \tag{7.9}$$

provided that $p_{00} - p_{10} \neq 1$. Similarly, we obtain

$$p_{11}^{(n)} = \left[p_{11} - \frac{p_{01}}{1 - p_{11} + p_{01}}\right](p_{11} - p_{01})^{n-1} + \frac{p_{01}}{1 - p_{11} + p_{01}} \tag{7.10}$$

Formulas (7.9) and (7.10) determine all of the $p_{ij}^{(n)}$, since

$$p_{01}^{(n)} = 1 - p_{00}^{(n)}, \qquad p_{10}^{(n)} = 1 - p_{11}^{(n)}$$

In the tank-firing example considered, $p_{00} = \frac{3}{4}$, $p_{10} = \frac{1}{2}$. Thus, $p_{00} - p_{10} = \frac{1}{4}$, $1 - p_{00} + p_{10} = \frac{3}{4}$, and (7.9) yields

$$p_{00}^{(n)} = \left[\frac{3}{4} - \frac{1/2}{3/4}\right]\left(\frac{1}{4}\right)^{n-1} + \frac{1/2}{3/4} = \left(\frac{1}{12}\right)\left(\frac{1}{4}\right)^{n-1} + \frac{2}{3} \tag{7.11}$$

Similarly

$$p_{11}^{(n)} = \left(\frac{1}{6}\right)\left(\frac{1}{4}\right)^{n-1} + \frac{1}{3} \tag{7.12}$$

$$p_{01}^{(n)} = 1 - p_{00}^{(n)} = -\left(\frac{1}{12}\right)\left(\frac{1}{4}\right)^{n-1} + \frac{1}{3} \tag{7.13}$$

$$p_{10}^{(n)} = 1 - p_{11}^{(n)} = -\left(\frac{1}{6}\right)\left(\frac{1}{4}\right)^{n-1} + \frac{2}{3} \tag{7.14}$$

We notice that

$$\lim_{n\to\infty} p_{00}^{(n)} = \lim_{n\to\infty} p_{10}^{(n)} = \frac{2}{3} \tag{7.15}$$

$$\lim_{n\to\infty} p_{01}^{(n)} = \lim_{n\to\infty} p_{11}^{(n)} = \frac{1}{3} \tag{7.16}$$

Formulas (7.15) and (7.16) give the long-run probabilities of a "hit" and a "miss" on a given shot. We observe that as n increases, the probability of a "hit" tends to be the same regardless of whether the sequence of shots started

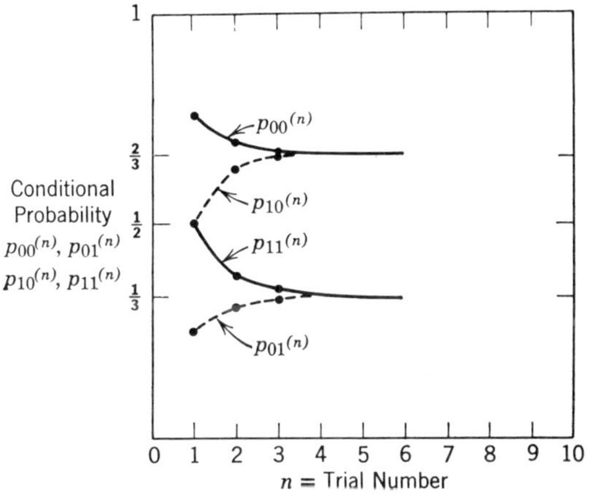

Figure 7.4 Conditional probabilities for a tank firing problem as a 2-state Markov chain.

with a "hit" or with a "miss." The probabilities $p_{ij}^{(n)}$ and the ways in which they approach their limiting values are sketched in Figure 7.4.

This simple example illustrates many ideas that we shall discuss in detail in subsequent chapters. We shall be concerned with

1. general methods of computing probabilities like $p_{ij}^{(n)}$

2. conditions under which $p_{ij}^{(n)}$ approaches a limiting value p_j as $n \to \infty$
3. conditions under which this limiting value is the same regardless of the initial state i
4. conditions under which the limits p_j themselves form a probability density function (that is, add up to 1)
5. ways of computing the limits p_j

5. Random Walks

In Example 4.22 the concept of a random walk was introduced. This is a sequence $Y_0, Y_1, Y_2, \ldots$ of random variables that forms a stationary Markov process. The state space consists of all integers, positive or negative, and the one-step transition probabilities are given by

$$p_{i,i+1} = p, \qquad p_{i,i-1} = q$$

for i any integer, $p + q = 1$. All other terms p_{ij} are taken to be 0. Thus, one-step transitions of $+1$ and -1 are the only possibilities, and they have probabilities p and q, respectively. Clearly, we can consider this process as a special case of the *partial sum* process $S_n = X_1 + \cdots + X_n$, which we defined in Section 3 of this chapter, with the summand random variables X_i taking ± 1 as their only possible values.

We can easily calculate the n-step transition probabilities:

$$p_{ij}^{(n)} = \Pr[Y_{k+n} = j \mid Y_k = i]$$

To pass from the state i to state j in n steps of ± 1, some number, l, of these steps must be $+1$, while the remainder, $n - l$, must be -1. Since a net change of $j - i$ is needed, one has

$$l - (n - l) = j - i$$

$$l = \frac{n + j - i}{2}$$

Thus, $p_{ij}^{(n)}$ will be 0 unless $l = (n + j - i)/2$ is an integer between 0 and n, $0 \le l \le n$. If l is such an integer, then the number of different ways in which such a sequence of transitions can be performed will equal the number of ways in which l of the n steps can be chosen to be $+1$, $\binom{n}{l}$. Each such sequence of l transitions of $+1$ and $n - l$ of -1 will have probability $p^l q^{n-l}$. Thus we have

$$p_{ij}^{(n)} = \binom{n}{l} p^l q^{n-l}$$

or

$$p_{ij}^{(n)} = \binom{n}{\frac{n+j-i}{2}} p^{(n+j-i)/2} q^{(n-j+i)/2} \tag{7.17}$$

provided that $(n+j-i)/2$ is an integer

$$0 \leq \frac{n+j-i}{2} \leq n$$

Otherwise $p_{ij}{}^{(n)} = 0$

Random walks occur in many applied problems. Much research has been carried out by studying the properties of these processes, and this research has suggested many of the key generalizations that have led to advances in the field of random processes as a whole.

We mention here only a few important results; each is stated for the special case of a simple random walk with $p = q = \frac{1}{2}$. The proofs and analyses involved in each one of these theorems have initiated a whole field of mathematical research.

(a) CENTRAL LIMIT THEOREM

For large n, $X_n/\sqrt{n}$ has approximately a normal distribution. More precisely

$$\lim_{n\to\infty} \Pr[X_n/\sqrt{n} \leq t] = \frac{1}{\sqrt{2\pi}} \int_{-\infty}^{t} e^{-x^2/2}\, dx \tag{7.18}$$

for every t. △

(b) STRONG LAW OF LARGE NUMBERS

As n increases, X_n/n approaches 0 with probability 1. More precisely

$$\Pr[\lim_{n\to\infty} X_n/n = 0] = 1 \qquad \triangle \tag{7.19}$$

(c) LAW OF THE ITERATED LOGARITHM

For large n, $X_n/\sqrt{2n\, \ln\ln n}$ takes on values close to 1 many times, but few values over 1. More precisely

$$\Pr\left[\frac{X_n}{\sqrt{2n\, \ln\ln n}} > 1 - \varepsilon \text{ infinitely often}\right] = 1$$

$$\Pr\left[\frac{X_n}{\sqrt{2n\, \ln\ln n}} > 1 + \varepsilon \text{ infinitely often}\right] = 0 \tag{7.20}$$

for every $\varepsilon > 0$. △

6. Gambler's Ruin

A player plays a sequence of independent games against an opponent. At each game he wins \$1 with probability p or loses \$1 with probability $q = 1 - p$. Let the sequence $X_0, X_1, X_2, \ldots$ represent his capital after the 0th, 1st, 2nd, …, etc., trials. Assume that his initial capital is \$ k and his opponent's initial capital is \$ $N - k$, where \$ N represents their total joint resources. If either his capital or his opponent's capital drops to 0 (if he or his opponent is *ruined*), then the game ceases, and all subsequent X_i's can be taken to be constant (0 or N). Thus the sequence $X_0, X_1, \ldots$ acts somewhat similarly to the random walk discussed in the preceding section, with the difference that, instead of all integer values being possible states, each X_i must lie between 0 and N, $0 \leq X_i \leq N$. It will be a stationary Markov process with state space $\{0, 1, \ldots, N\}$.

We are interested in computing the probability t_k that the player will be ruined if he starts with initial capital of \$ k:

$$t_k = \Pr[X_n = 0 \text{ for some } n \mid X_0 = k]$$

Assume that the total joint resources available, N, remains fixed.

In order that the player eventually be ruined, starting with \$ k, we observe that he must either lose the first trial (probability q) and go on to be ruined starting with \$ $k - 1$, or that he must win the first trial (probability p) and go on to be ruined starting with \$ $k + 1$. This line of thought leads to the following difference equation.

$$t_k = qt_{k-1} + pt_{k+1}, \qquad 0 < k < N \tag{7.21}$$

Since $1 = p + q$, formula (7.21) can be written in the form

$$(p + q)t_k = qt_{k-1} + pt_{k+1}$$

which can be rewritten as

$$q(t_k - t_{k-1}) = p(t_{k+1} - t_k)$$

or

$$t_{k+1} - t_k = \left(\frac{q}{p}\right)(t_k - t_{k-1}) \tag{7.22}$$

$0 < k < N$.

By the repeated application of formula (7.22), we obtain

$$t_{k+1} - t_k = \left(\frac{q}{p}\right)^k (t_1 - t_0)$$
$$= \left(\frac{q}{p}\right)^k (t_1 - 1) \tag{7.23}$$

since clearly $t_0 = 1$ (ruin is certain starting with zero capital). Formula (7.23) is valid for $0 \leq k < N$. Since

$$t_i - t_0 = (t_i - t_{i-1}) + (t_{i-1} - t_{i-2}) + \cdots + (t_1 - t_0)$$

it follows from (7.23) that

$$t_i - t_0 = \left\{\left(\frac{q}{p}\right)^{i-1} + \left(\frac{q}{p}\right)^{i-2} + \cdots + 1\right\}(t_1 - 1) \tag{7.24}$$

or

$$t_i - 1 = \frac{1 - (q/p)^i}{1 - (q/p)} (t_1 - 1), \qquad 0 < i \leq N \tag{7.25}$$

by using the formula for the sum of a geometric series (which is valid here provided $p \neq q$). To find $t_1 - 1$, set $i = N$ in (7.25). Clearly $t_N = 0$ (ruin is impossible if one starts with all the money). Therefore

$$-1 = \frac{1 - (q/p)^N}{1 - (q/p)} (t_1 - 1)$$

or

$$t_1 - 1 = -\frac{1 - q/p}{1 - (q/p)^N}$$

Now substitute this into (7.25) and obtain

$$t_i - 1 = -\frac{1 - (q/p)^i}{1 - (q/p)^N}$$

or, by replacing i by k,

$$t_k = 1 - \frac{1 - (q/p)^k}{1 - (q/p)^N} = \frac{(q/p)^k - (q/p)^N}{1 - (q/p)^N}, \qquad p \neq q \tag{7.26}$$

This gives the probability of a player who starts with \$ k eventually being ruined, provided that $p \neq q$.

If $p = q$, then t_k can be found even more easily from (7.24). The result is

$$t_k = 1 - \frac{k}{N}, \qquad \text{if } p = q \tag{7.27}$$

To study the situation of play against an infinitely rich adversary, let $N \to \infty$. Since $(q/p)^N \to 0$ if $p > q$, formula (7.26) yields:

$$\lim_{N\to\infty} t_k = \left(\frac{q}{p}\right)^k, \qquad \text{if } p > q$$

$$\lim_{N\to\infty} t_k = 1, \qquad \text{if } p < q$$

By using formula (7.27), we notice that the second case also holds true if $p = q$. thus

$$\lim_{N\to\infty} t_k = \begin{cases} \left(\dfrac{q}{p}\right)^k, & \text{if } p > q \\ 1 \quad , & \text{if } p \leq q \end{cases} \tag{7.28}$$

We observe that, if the player has a better than even chance of winning any given trial, $p > q$, then he has a positive probability $1 - (q/p)^k$ of *never* being ruined, even if his opponent is infinitely rich. On the other hand, if $p \leq q$, then his ruin is sure.

This process with finite N differs from the random walk that we considered in the previous section in that the states 0 and N are *absorbing* states. If the system ever enters either of these states, it never can leave. The n-step transition probabilities $p_{ij}{}^{(n)}$ can be calculated. However, unlike the 2-state Markov process considered in Section 4, the limiting probabilities $\lim_{n\to\infty} p_{ij}{}^{(n)}$ will depend on i as well as on j. For example

$$\lim_{n\to\infty} p_{00}{}^{(n)} = 1, \qquad \text{while} \qquad \lim_{n\to\infty} p_{N0}{}^{(n)} = 0$$

Although we have discussed this process by using gambling game terminology, similar situations arise in many physical and real-world processes that concern random walks or random motions between absorbing barriers.

7. The Poisson Process

The examples considered thus far in this chapter have all been discrete parameter processes. We now discuss an important continuous parameter process. Our discussion here is from the point of view of a very special application—

long-distance telephone calls between two cities. However, the general situation applies to a vast number of important applications in many fields.

Let us suppose that a telephone engineer takes a survey of the demands for service on the long-distance lines between two cities during the hour from 9 to 10 a.m. of each business day. His survey establishes the following observations about the calls made over these lines.

(a) The distribution of the number of calls coming in during any given time interval appears to depend only on the length of the time interval. The longer the interval, the more calls tend to come in.

(b) The calls appear to come in independently. That is, an accidental excess or deficiency of calls in some given time interval has no effect on the number of calls during any other time interval.

(c) The chance of two or more calls coming in during a small time interval is very small compared to the chance of one call coming in.

The engineer proceeds to construct a mathematical probabilistic model to describe the situation. He hypothesizes that he has been observing the outcomes of a certain random experiment, which has, connected with it, an appropriate sample space and probability distribution. His observations, (a) to (c), lead him to make the following assumptions about this random experiment:

(a′) The probability of n calls coming in during a time interval of length t depends only on the *length* t of the time interval, and not on the initial and final points of the particular interval. Let this probability be denoted by $p_n(t)$, for $n = 0, 1, 2, \ldots$. Thus

$$p_n(t) \geq 0$$

$$\sum_{n=0}^{\infty} p_n(t) = 1 \tag{7.29}$$

and

$$p_0(0) = 1 \qquad \text{and} \qquad p_n(0) = 0 \qquad \text{for } n > 0 \tag{7.30}$$

since the probability is 1 that zero calls arrive in zero time. To simplify this model, let us also assume that each $p_n(t)$ is continuous and differentiable for $0 \leq t < \infty$.

(b′) Suppose two successive nonoverlapping time intervals of lengths t and τ are given. We assume in this case that the events "k calls in the first interval" and "$n - k$ calls in the second interval" are independent and, hence, that the probability of *both* happening is $p_k(t)p_{n-k}(\tau)$. Thus the probability of n

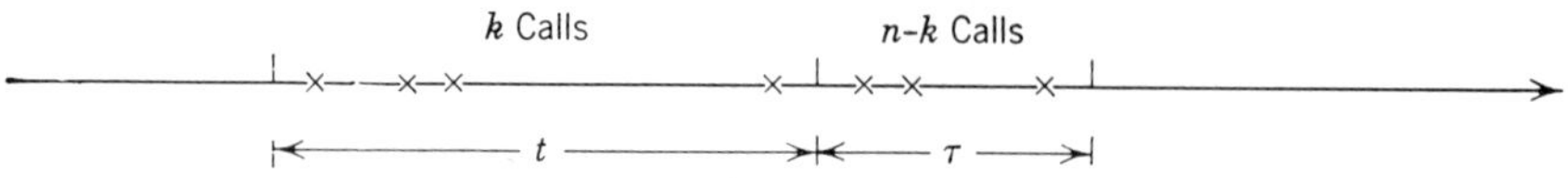

Figure 7.5 An illustration of the events "n calls in the interval of length $t+\tau$" and "k calls in the interval of length t and $n-k$ calls in the interval of length τ."

calls coming in during the overall time interval of length $t+\tau$ is obtained by adding up these terms for all $k \leq n$ (see Figure 7.5)

$$p_n(t+\tau) = \sum_{k=0}^{n} p_k(t)p_{n-k}(\tau) \tag{7.31}$$

(c′) Both the probability $p_1(t)$ of one call and the probability of more than one call

$$\sum_{n=2}^{\infty} p_n(t) = 1 - p_0(t) - p_1(t)$$

have the value 0 when $t = 0$, by (7.30). Hence, observation (c) can be modeled by assuming:

at $t = 0$, $p_1(t)$ has a positive derivative, while the probability of more than one call has derivative 0 (see Figure 7.6):

$$p_1'(0) = \lim_{\Delta t \to 0} \frac{p_1(\Delta t)}{\Delta t} = \lambda > 0 \tag{7.32}$$

$$\frac{d}{dt}\sum_{n=2}^{\infty} p_n(t)\bigg|_{t=0} = \lim_{\Delta t \to 0} \frac{1}{\Delta t}\sum_{n=2}^{\infty} p_n(\Delta t) = 0 \tag{7.33}$$

It follows that

$$p_n'(0) = \lim_{\Delta t \to 0} \frac{p_n(\Delta t)}{\Delta t} = 0 \qquad \text{for} \qquad n > 1 \tag{7.34}$$

while

$$\begin{aligned} p_0'(0) &= \lim_{\Delta t \to 0} \frac{p_0(\Delta t) - 1}{\Delta t} \\ &= -\lim_{\Delta t \to 0} \frac{p_1(\Delta t) + \sum_2^\infty p_n(\Delta t)}{\Delta t} \\ &= -\lambda \end{aligned} \tag{7.35}$$

By using this model, the engineer wants to answer questions and wants to compute probabilities for the process. For example, the following questions

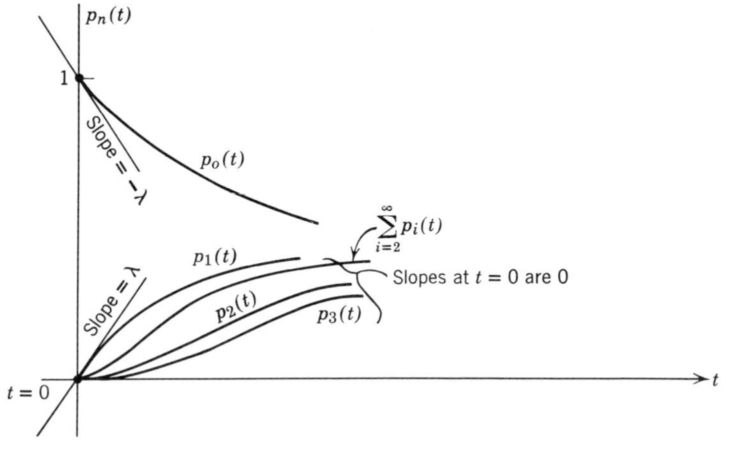

Figure 7.6 An illustration of the slope of the probabilities $p_i(t)$, *at* $t = 0$.

might be asked: What is the mean number of calls per minute? What is the probability of more than 10 calls coming in during a given 3-minute period? What is the chance of more than 10 calls coming in during *any* 3-minute period in a given hour?, etc. These questions would be important in deciding whether more long-distance lines were needed.

In particular, the engineer would like to compute the probabilities $p_n(t)$ for every n and t. To do this, the key formula is (7.31), which allows him to break up his probabilities into smaller pieces. Then these pieces can be analyzed by using (7.32) to (7.35), and can be reassembled. This can be accomplished most simply by differentiating each side of (7.31) with respect to τ, and then by setting $\tau = 0$. The result is

$$\frac{\partial}{\partial \tau} p_n(t+\tau)\bigg|_{\tau=0} = p_n'(t) = \sum_{k=0}^{n} p_k(t) p_{n-k}'(0)$$

By (7.32) to (7.35), $p_{n-k}'(0) = 0$ except when $k = n$ or $n - 1$. Thus the preceding formula can be written as

$$p_n'(t) = -\lambda p_n(t) + \lambda p_{n-1}(t) \qquad \text{for} \qquad n \geq 1 \tag{7.36}$$

$$p_0'(t) = -\lambda p_0(t) \tag{7.37}$$

This system of differential equations for the unknown functions $p_n(t)$must be solved, subject to the initial conditions (7.30). This system is particularly simple, since (7.37) can be solved directly to give $p_0(t) = e^{-\lambda t}$. If this is substituted into (7.36) with $n = 1$, the equation

$$p_1'(t) = -\lambda p_1(t) + \lambda e^{-\lambda t}$$

is obtained, which has solution $p_1(t) = e^{-\lambda t}\,\lambda t$. This can be substituted again into (7.36) with $n = 2$, so that the system can be solved recursively. The general solution is

$$p_n(t) = e^{-\lambda t}\frac{(\lambda t)^n}{n!} \tag{7.38}$$

We observe that the number of calls in time t has a Poisson distribution with mean λt and variance λt (Chapter 6, Section 5). If the time is measured in minutes, then the mean number of calls per minute is λ.

Any probabilistic system in which the observations consist of a sequence of events that occur at random points on a time scale is termed a *counting process*. A counting process that satisfies assumptions (a′) to (c′) is termed a *Poisson process*, since the resulting distribution is Poisson.

Actually, assumptions (a′) to (c′) are much too strong. They can be weakened considerably (particularly the differentiability assumption), and the same Poisson process will result.

If X_t is a random variable representing the number of events (calls) that occur up to time t, then X_t is a *continuous parameter* random process, $0 \le t < \infty$, as opposed to the *discrete parameter* random processes, X_n, $n = 0, 1, 2, \ldots$, discussed in the previous sections. With this notation $p_n(t) = \Pr[X_t = n]$, and assumption (b′) implies that X_t is a Markov process.

The Poisson process provides an approximation (in many cases, a very good approximation) to many real-world situations. A list of cases where this process has proved useful would be very long. Telephone traffic, radioactive disintegration, orders to a factory, breakdown of machinery, arrivals on a queue or conveyor belt, insurance claims, genetic mutations, occurrence of epidemics, etc., are but a few examples.

Comments. (i) The event "no calls in the first t minutes," which has probability $p_0(t) = e^{-\lambda t}$, can also be stated as "the first call occurs after time t." Thus, if T is the random variable giving the time from 0 to the first call, we have

$$\Pr[T > t] = e^{-\lambda t}$$

Thus, T has distribution function

$$\Pr[T \le t] = 1 - e^{-\lambda t} \tag{7.39}$$

and probability density function

$$f(t) = \lambda e^{-\lambda t} \tag{7.40}$$

Thus, T has a negative exponential distribution (Chapter 6, Section 7). By use of the independence property (b′) we can show that the time interval between *any* two successive calls is a continuous random variable having the same exponential density (7.40). The sequence

$$T_1, T_2, T_3, \ldots$$

of time intervals between successive calls is a sequence of *independent, identically distributed, exponentially distributed* random variables. This fact provides an alternative characterization of the Poisson process.

(ii) Many probabilities connected with the process can be computed by using (7.38) and the defining axioms (a′) to (c′). For example, the event "the third call comes in during the second minute" can be decomposed into the union of the following events: "no calls in the first minute and three or more calls in the second minute," "one call in the first minute and two or more calls in the second minute," and "two calls in the first minute and one or more calls in the second minute." By using (7.38) and the independence axiom (b′), the probability of this will be

$$\begin{aligned}
p_0(1)\left[\sum_{n=3}^{\infty} p_n(1)\right] &+ p_1(1)\left[\sum_{n=2}^{\infty} p_n(1)\right] + p_2(1)\left[\sum_{n=1}^{\infty} p_n(1)\right] \\
&= p_0(1)[1 - p_0(1) - p_1(1) - p_2(1)] + p_1(1)[1 - p_0(1) - p_1(1)] \\
&\qquad + p_2(1)[1 - p_0(1)] \\
&= e^{-\lambda}\left[1 - e^{-\lambda} - e^{-\lambda}\lambda - e^{-\lambda}\frac{\lambda^2}{2}\right] + e^{-\lambda}\lambda[1 - e^{-\lambda} - e^{-\lambda}\lambda] + e^{-\lambda}\frac{\lambda^2}{2}[1 - e^{-\lambda}] \\
&= e^{-\lambda}\left[\left(1 + \lambda + \frac{\lambda^2}{2}\right) - e^{-\lambda}(1 + 2\lambda + 2\lambda^2)\right]
\end{aligned}$$

Other probabilities are more complicated. For example, an insurance company that is studying its risk might be interested in the probability that the number of claims *ever* will exceed a linear function of time of the form $\alpha t + \beta$. The study of such an event goes beyond the level of this treatment.

EXERCISES

1. Flip a coin 50 times and let $X_n = -1$ if a head appears and $X_n = +1$ if a tail appears. Plot the results of your flips as X_n versus n.
2. Repeat Exercise 1 twice more and plot the three sets of flips on the same graph.

3. Identify the stochastic process for which you have three realizations in Exercises 1 and 2. Is it an independent sequence? A Markov sequence? A Gambler's Ruin problem?

4. Define a new sequence for Exercise 1 by

$$Y_1 = X_1$$

$$Y_2 = \frac{X_1 + X_2}{2}$$

$$\vdots$$

$$Y_n = \frac{X_1 + X_2 + \cdots + X_n}{n}$$

Plot Y_n versus n.

5. Repeat Exercise 4 for each sequence in Exercise 2.

6. Identify the $\{Y_n\}$ sequence. Is it an independent sequence? A Markov sequence? A Gambler's Ruin sequence?

7. Define new sequences for the results obtained in Exercises 1 and 2 by letting

$$Z_1 = X_1$$

$$Z_2 = \frac{X_1 + X_2}{2}$$

$$\vdots$$

$$Z_n = \frac{X_n + X_{n+1}}{2}, \qquad n = 1, 2, \ldots, 49$$

Plot graphs of the $\{Z_n\}$ sequence for each set of data in Exercises 1 and 2.

8. Give reasons why $\{Z_n\}$ of Exercise 7 is not Markovian.

9. For a "symmetric" random walk, $p = q$, use the results of Exercise 15 in Chapter 2 to show that $p_{00}{}^{(n)} \to 0$ as $n \to \infty$. Interpret the meaning of this result in the context of the random walk.

10. Using the results given for the Poisson process in Section 7, determine the probability that (a) the second call comes in the third minute, and (b) the third call has not come in after five minutes.

8 • Discrete Parameter Markov Processes: The Finite, Irreducible Case

1. Introduction

The concept of a *Markov dependent random sequence* was introduced in Chapter 4, Section 12, and examples were discussed in Chapter 7. In view of their many important applications these sequences are discussed in detail in this chapter and the two following chapters.

2. Some Definitions

Let us consider a *random sequence* or *discrete parameter* random process

$$\{X_n\} = \{X_0, X_1, X_2, \ldots\}$$

and let us confine our attention to the case where the *state space* (the set of possible *values* of the random variables X_n) is *discrete*. Since any discrete set can be indexed by a set of integers, we need only consider the case where the state space consists of a set of integers of the form: 0, 1, 2, ... This set may be finite or infinite; however, for most of this chapter, we concentrate on the finite case. The statement "$X_n = i$" can be read, "the system is in state i after n steps." The state i might refer to an actual count (the capital of a gambler after n plays, the size of an inventory after n weeks, the number of garbled symbols in a communication system after n transmissions, the number of individuals in a waiting line after the nth individual completes service, etc.), or i may simply be an index symbol (0 for a hit, 1 for a miss, in the tank-firing example).

Let us review the concepts and the notation that were introduced in Chapter 4.

The probability structure of a random sequence (or discrete parameter random process) is determined by the joint probabilities:

$$p(j_0, j_1, \ldots, j_k) = \Pr[X_0 = j_0, X_1 = j_1, \ldots, X_k = j_k] \tag{8.1}$$

for every finite k and for every sequence $j_0, j_1, \ldots, j_k$ of states. Such a process is termed a *Markov process* or *Markov chain* if, for each k, the conditional probability that the system is in a given state after k steps, knowing the states of the system at all preceding steps, is the same as the conditional probability knowing only the state at the immediately preceding step. Intuitively speaking, knowledge of $X_0, X_1, \ldots, X_{k-1}$ gives no more information for predicting the value of X_k than does knowledge of X_{k-1} alone. In symbols, if we use the somewhat ambiguous notation

$$p(j_k \mid j_0, j_1, \ldots, j_{k-1}) = \Pr[X_k = j_k \mid X_0 = j_0, X_1 = j_1, \ldots, X_{k-1} = j_{k-1}] \quad (8.2)$$

then a discrete parameter Markov process is a random sequence satisfying the identity

$$\begin{aligned} p(j_k \mid j_0, j_1, \ldots, j_{k-1}) &= \Pr[X_k = j_k \mid X_{k-1} = j_{k-1}] \\ &= p(j_k \mid j_{k-1}) \end{aligned} \quad (8.3)$$

for each k and for each sequence $j_0, j_1, \ldots, j_k$ of states.

If (8.3) holds true for every k, then we can make repeated use of the identity $\Pr[A \cap B] = \Pr[A] \Pr[B \mid A]$ and obtain

$$\begin{aligned} p(j_0, j_1, \ldots, j_k) &= p(j_0, j_1, \ldots, j_{k-1}) p(j_k \mid j_0, j_1, \ldots, j_{k-1}) \\ &= p(j_0, j_1, \ldots, j_{k-1}) p(j_k \mid j_{k-1}) \\ &\;\;\vdots \\ &= p(j_0) p(j_1 \mid j_0) \ldots p(j_k \mid j_{k-1}) \end{aligned} \quad (8.4)$$

Thus, all joint probabilities that are needed to describe the process are determined in terms of the simple conditional probabilities $p(j_k \mid j_{k-1})$, termed the *one-step transition probabilities*, and the set of *initial probabilities* $p(j_0)$.

In all the examples we consider here, the one-step transition probabilities do not depend on *which* step is being considered

$$\Pr[X_k = j \mid X_{k-1} = i] = \Pr[X_r = j \mid X_{r-1} = i] \quad (8.5)$$

for each k and r (provided that X_{k-1} and X_{r-1} have nonzero probabilities of equaling i). This process is said to have a *stationary transition mechanism*, and only discrete Markov processes with this property will be considered. In this case, our notation can be simplified by denoting the one-step transition probabilities by

$$p_{ij} = \Pr[X_k = j \mid X_{k-1} = i] \quad (8.6)$$

which does not depend on k, and the initial probabilities by

$$p_i = p_i^{(0)} = \Pr[X_0 = i] \tag{8.7}$$

for $i, j = 0, 1, 2, \ldots$.

Thus (8.4) can be written as

$$p(j_0, j_1, \ldots, j_k) = p_{j_0} p_{j_0 j_1} p_{j_1 j_2} \cdots p_{j_{k-1} j_k} \tag{8.8}$$

Clearly, we have

$$\sum_j p_{ij} = 1 \tag{8.9}$$

for each i, since by starting in state i after $k - 1$ steps, the process will have to be in *some* state j after the next step.

The *matrix* **P** whose entries are the one-step transition probabilities p_{ij},

$$\mathbf{P} = (p_{ij}) = \begin{pmatrix} p_{00} & p_{01} & p_{02} \cdots \\ p_{10} & p_{11} & p_{12} \cdots \\ \vdots & \vdots & \vdots \end{pmatrix}$$

is termed the *transition matrix* of the process, and condition (8.9) simply asserts that the sum of the entries in each row of **P** is 1. Any matrix that has non-negative entries with row-sums all 1 is termed a *stochastic matrix*.*

The vector

$$\mathbf{p}^0 = (p_1, p_1, p_2, \ldots)$$

is termed the *initial probability vector*.

We define the *n-step transition probabilities* $p_{ij}^{(n)}$ by

$$p_{ij}^{(n)} = \Pr[X_{k+n} = j \mid X_k = i]. \tag{8.10}$$

for $i, j = 0, 1, 2, \ldots$, and for each $n \geq 0$. $p_{ij}^{(n)}$ is the probability of the system ending up in state j after n transitions, starting from state i. That is, $p_{ij}^{(n)}$ is the probability that the process goes from state i to state j in n steps. Clearly, $p_{ij}^{(1)} = p_{ij}$.

Similarly

$$p_i^{(n)} = \Pr[X_n = i] \tag{8.11}$$

is the unconditional probability that the process is in state i after n steps.

* The use of matrix and vector notation is extremely useful in many probability discussions. Only the most elementary concepts of matrices and simple matrix multiplication will be assumed in this book.

3. Examples and Preliminary Remarks

Some examples of finite Markov chains have been given in Chapters 4 and 7. Thus the tank-firing example in Section 4 of Chapter 7 is an example of a finite Markov chain. There were two states of interest. The gambler's ruin problem in Section 6 of Chapter 7 is an example of a finite Markov chain. The random walk problem in Section 5 is not a *finite* Markov chain, although we observed that it *is* a Markov chain. Other examples of Markov chains are given below.

EXAMPLE 8.1. The Gambler's Ruin example can be extended as follows. Two players, A and B, play a game that consists of a sequence of independent trials for stakes of \$1 per trial. Let the probability of A winning any given trial be p, and the probability of B winning be q. If tie games are possibly allowed, then $p+q \leq 1$, and the probability of a tie is $r = 1 - p - q$. This is a somewhat more general assumption than the one made in Chapter 7 where ties were ruled out. On a tie, no money changes hands.

Suppose that A starts with k dollars, B starts with $N - k$ dollars, and thus a total of N dollars is available for play. We are interested in the behavior of A's fortunes.

Let $X_n =$ the amount of money that A has after n trials ($X_0 = k$). Clearly, if $X_n = x$, then X_{n+1} must equal $x+1$, $x-1$, or x, with probabilities p, q, r, regardless of the values of X_i for $i < n$.

Thus $\{X_n\}$ forms a Markov chain with one-step transition probabilities given by

$$
\begin{aligned}
p_{x,x+1} &= \Pr[X_{n+1} = x+1 \mid X_n = x] = p \\
p_{x,\,x-1} &= \Pr[X_{n+1} = x-1 \mid X_n = x] = q \\
p_{x,x} &= \Pr[X_{n+1} = x \mid X_n = x] = r
\end{aligned}
$$

for $0 < x < N$, $p_{00} = p_{NN} = 1$, and all other $p_{ij} = 0$.
Clearly, the game terminates if A's fortune ever reaches 0 or N.

The initial probabilities are given by

$$
\begin{aligned}
p_k &= \Pr[X_0 = k] = 1 \\
p_i &= \Pr[X_0 = i] = 0 \qquad \text{for} \qquad i \neq k
\end{aligned}
$$

The matrix $\mathbf{P}$ of one-step transition probabilities has the form:

$$
\begin{array}{c|l}
 & \text{(state next play)} \\
\text{(state now) } X_n & X_{n+1}\colon \; 0 \quad 1 \quad 2 \quad 3 \quad \cdots \quad N-1 \quad N \\
\hline
\end{array}
$$

$$
\mathbf{P} = \begin{array}{c}
0 \\ 1 \\ 2 \\ \cdot \\ \cdot \\ \cdot \\ N-1 \\ N
\end{array}
\begin{pmatrix}
1 & 0 & 0 & 0 & \cdots & 0 & 0 \\
q & r & p & 0 & \cdots & 0 & 0 \\
0 & q & r & p & \cdots & 0 & 0 \\
\cdot & \cdot & \cdot & \cdot & & \cdot & \cdot \\
\cdot & \cdot & \cdot & \cdot & & \cdot & \cdot \\
\cdot & \cdot & \cdot & \cdot & & \cdot & \cdot \\
0 & 0 & 0 & 0 & \cdots & r & p \\
0 & 0 & 0 & 0 & \cdots & 0 & 1
\end{pmatrix}
$$

The initial probability vector is

$$\mathbf{p}^0 = (0, 0, \ldots, 0, 1, 0, \ldots, 0)$$

where the 1 occupies the kth position (counting the initial term as the zeroth position).

EXAMPLE 8.2. In a certain production process items pass through 3 manufacturing stages. At the end of each stage the items are inspected, and are either junked, sent through the stage again for reworking, or passed on to the next stage. Suppose that at each stage the probability is p that the item will be junked, q that it will be reworked, and r that it will be passed on, $p + q + r = 1$. The process has five states that can be indexed as follows.

0: item junked
1: item completed
2: item in first manufacturing stage
3: item in second manufacturing stage
4: item in third manufacturing stage

Let us assume that the process can be modeled by a Markov chain $X_0, X_1, X_2, \ldots$, in which X_n gives the state of the item after n inspection points. Thus, $X_4 = 3$ means that the item is in the second manufacturing stage after 4 inspections. Clearly, once an item is junked or is completed, it cannot shift into another state. The transition matrix is

$$
\mathbf{P} = \begin{pmatrix}
1 & 0 & 0 & 0 & 0 \\
0 & 1 & 0 & 0 & 0 \\
p & 0 & q & r & 0 \\
p & 0 & 0 & q & r \\
p & r & 0 & 0 & q
\end{pmatrix}
$$

If we assume that the item initially starts in the first manufacturing stage, the initial probability vector will be

$$\mathbf{p}^0 = (0, 0, 1, 0, 0)$$

EXAMPLE 8.3. A two-state process has been used as a simple model for weather changes. If the weather is classified as State 0: *dry*, or State 1: *wet*, then the transition matrix might, for example, have the form:

$$\mathbf{P} = \begin{pmatrix} \frac{3}{4} & \frac{1}{4} \\ \\ \frac{1}{3} & \frac{2}{3} \end{pmatrix}$$

The actual entries would have to be estimated by using meteorological records. Thus, if an examination of these records showed that, out of 1,000 days, 600 were dry, and that 450 of the 600 dry days were followed by other dry days, we would estimate p_{00} by $450/600 = \frac{3}{4}$. △

EXAMPLE 8.4. One theoretical model for tank warfare (see Section 4, Chapter 7) expresses the firing mechanism as a two-state Markov process in which the states are 0: a *hit*, 1: a *miss*. Thus X_n is 1 or 0 depending on whether the nth shot is a miss or a hit. For example, p_{10} would be the probability of the tank's hitting on a certain shot after it had missed on the previous shot.

This model can be generalized to study the interaction of two opposing tanks shooting at each other.

EXAMPLE 8.5. A Markov model has been used to study the behavior of satellite vehicles. Here we suppose that at a sequence of points in time a vehicle reentering from space is given corrective signals. These signals are successful or not. The transition probability matrix can be structured as in the tank example where the states are now 0 for a "Success" and 1 for a "Failure." △

EXAMPLE 8.6. In the study of waiting lines, one tries to structure the process by defining the number of individuals who are waiting as a Markov process. Hence, one considers the length of the waiting line at certain points in time (perhaps at those time points when someone completes his service). These points are called "epochs." $X_n = j$ implies that the number in the waiting line at the nth epoch is j. The state space is taken to be 0, 1, 2, ... A one-step transition matrix will have the following form (see Chapters 10 and 13 for a discussion):

$$\mathbf{P} = \begin{pmatrix} p_{00} & p_{01} & p_{02} & \cdots \\ p_{10} & p_{11} & p_{12} & \cdots \\ 0 & p_{21} & p_{22} & \cdots \\ 0 & 0 & p_{32} & \cdots \\ \cdot & \cdot & \cdot & \cdot \\ \cdot & \cdot & \cdot & \cdot \\ \cdot & \cdot & \cdot & \cdot \end{pmatrix}$$

We are interested in knowing $\Pr[X_n = j]$, the probability that there are j individuals in the waiting line at the nth epoch.

EXAMPLE 8.7. An interesting two-state Markov process arises in learning theory. Here we suppose that at the nth trial either a correct response is given or an incorrect one. Which response occurs depends only on the previous response. Then learning theory attempts to study the pattern of (correct, incorrect) responses. △

We notice that Example 8.6 is not a *finite* Markov chain in general, but that all of the others are. If we restricted the number of people who could wait in Example 8.6, it too would be a finite Markov chain and would form part of the area studied in this chapter.

There is yet another property of interest to us in this chapter for which we can see examples by comparing Examples 8.1 or 8.2 and 8.3. Examples 8.1 and 8.2 contain at least one state, say i, with

$$\begin{aligned} p_{ii} &= 1 \\ p_{ij} &= 0 \qquad \text{for} \qquad i \neq j \end{aligned}$$

In the terminology of Chapter 7, both Examples 8.1 and 8.2 contain *absorbing states*, which one can make a transition *into* but not *out of*. Examples 8.3, 8.4, 8.5, 8.6, and 8.7 do not have absorbing states, and for these chains eventual transitions from any state to any other state are possible, although they may not be possible in one step. Chains such as the ones in Examples 8.3, 8.4, 8.5, 8.6, and 8.7 with this property are said to be *irreducible* Markov chains. Chains such as the ones in Examples 8.1 and 8.2 are said to be *nonirreducible*. We shall give precision to these definitions in Section 7. For now, we merely note the existence of these types of chains, and remark that in this chapter our attention will be concentrated on irreducible chains of the type exemplified by Examples 8.3, 8.4, 8.5, and 8.7. Thus, in this chapter we are concerned with *finite, irreducuble* Markov chains that have *stationary transition mechanisms.*

4. n-step Transition Probabilities

Recall that the n-step transition probabilities are given by (8.10)

$$p_{ij}^{(n)} = \Pr[X_n = j \mid X_0 = i]$$

For example

$$p_{ij}^{(2)} = \Pr[X_2 = j \mid X_0 = i]$$

That is, $p_{ij}^{(2)}$ is the probability of being in state j after the second step, given that the process was in state i initially. We are interested in determining these probabilities from the known 1-step transition probabilities p_{ij}. To this end, consider a path that the process might take to get from state i to state j in two steps (Figure 8.1):

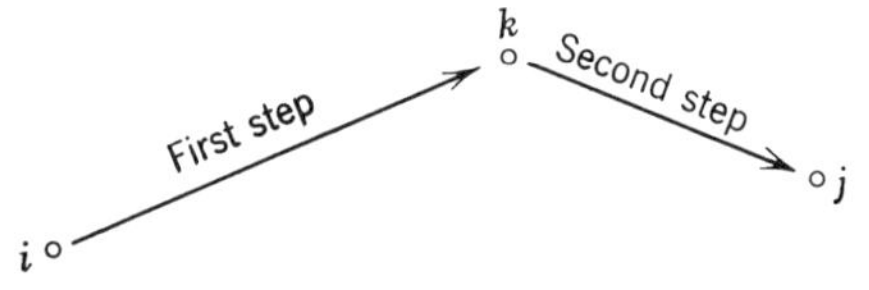

Figure 8.1 A possible path from state i to state j in two steps.

Suppose that k is the state (possibly even i or j) that the process visits on the first step. Then for this path the probability of going from i to k to j in exactly two steps is

$$p_{ik}p_{kj} = \Pr[X_1 = k \mid X_0 = i]\Pr[X_2 = j \mid X_1 = k] = \Pr[X_2 = j;\, X_1 = k \mid X_0 = i]$$

More generally, since the chain has a stationary transition mechanism, this is the probability of going from i to k to j in any two consecutive steps (not necessarily the first two).

Since in the first step the process must visit *some* state k, we observe that (see Figure 8.1)

$$p_{ij}^{(2)} = \sum_k p_{ik}p_{kj} \tag{8.12}$$

Let us denote the matrix whose entries are $p_{ij}^{(2)}$ by $\mathbf{P}^{(2)}$ and, more generally, let us denote the matrix whose entries are $p_{ij}^{(n)}$ by $\mathbf{P}^{(n)}$. By using the well-known formula for computing the elements of a product of two given matrices, we observe that (8.12) is the formula for determining the elements of the product matrix. Formula (8.12) can be written in matrix form as

$$\mathbf{P}^{(2)} = \mathbf{P} \cdot \mathbf{P} = \mathbf{P}^2 \tag{8.13}$$

where $\mathbf{P}$ is the matrix with entries p_{ij}. That is, the matrix of two-step transition probabilities is formed by squaring the matrix of one-step transition probabilities. It easily follows that this result continues and, in general,

$$\mathbf{P}^{(n)} = \mathbf{P}^n \tag{8.14}$$

that is, the matrix of n-step transition probabilites can be formed by multiplying the matrix of one-step transition probabilities by itself n times. Thus the problem of finding n-step transition probabilities reduces to the determination of powers of a given matrix.

From the matrix relation $\mathbf{P}^{n+m} = \mathbf{P}^n\mathbf{P}^m$, we can also deduce an important relation about the elements of $\mathbf{P}^{(n)}$:

$$\mathbf{P}^{(n+m)} = \mathbf{P}^{(n)}\mathbf{P}^{(m)} \tag{8.15}$$

or, in terms of the transition probabilities,

$$p_{ij}{}^{(n+m)} = \sum_k p_{ik}{}^{(n)} p_{kj}{}^{(m)} \tag{8.16}$$

If we take $p_{ik}{}^{(0)} = 0$ for $i \neq k$ and $p_{kk}{}^{(0)} = 1$; that is, if we define $\mathbf{P}^{(0)}$ to be the identity matrix $\mathbf{I}$, (8.15) holds true for all integers $m, n \geq 0$.

Equation (8.15), or its equivalent form (8.16), is termed the *Chapman-Kolmogorov Equation*, and is one of the basic relations used in Markov process theory.

One interpretation of (8.16) is as follows. To go from state i to state j in $(n + m)$ steps, one must go from i to some intermediate state k in n steps, and then proceed to j from k in m steps. Because of the Markov property, the process of going from i to k in n steps is simply $p_{ik}{}^{(n)}$. Likewise, because of the Markov property, the process of going from k to j does not depend on which state i formed the "first" state. Hence the probability of going from k to j in m steps is $p_{kj}{}^{(m)}$. Furthermore, the events "go from i to k in n steps" and "go from k to j in m steps" are independent. Hence the probability of going from i to j in $(n + m)$ steps by following the path i, k, j is $p_{ik}{}^{(n)}p_{kj}{}^{(m)}$. The probability of going from i to j no matter what path is followed is then $\sum_k p_{ik}{}^{(n)}p_{kj}{}^{(m)}$ and, of course, this is $p_{ij}{}^{(n+m)}$.

5. Examples

EXAMPLE 8.4 (continued). For this example, given in Section 3, suppose that

$$\mathbf{P} = \begin{pmatrix} \frac{3}{4} & \frac{1}{4} \\ \frac{1}{2} & \frac{1}{2} \end{pmatrix}$$

Then by direct multiplication we have

$$\mathbf{P}^{(3)} = \mathbf{P}^3 = \begin{pmatrix} \frac{43}{64} & \frac{21}{64} \\ \frac{21}{32} & \frac{11}{32} \end{pmatrix}$$

$$\mathbf{P}^{(5)} = \mathbf{P}^2\mathbf{P}^3 = \begin{pmatrix} \frac{683}{1024} & \frac{341}{1024} \\ \frac{341}{512} & \frac{171}{512} \end{pmatrix} = \begin{pmatrix} .6689 & .3330 \\ .6660 & .3339 \end{pmatrix}$$

$$\mathbf{P}^{(10)} = \mathbf{P}^5\mathbf{P}^5 = \begin{pmatrix} .666 & .333 \\ .666 & .333 \end{pmatrix}$$

Thus, for example, the probability that the 11th shot fired from the tank (the 10th shot after the first) hits its target, given that the initial shot hit is .666. The probability that the 4th shot hits the target, given that the initial shot missed the target is 21/32.

EXAMPLE 8.8. Suppose that we have a piece of equipment that can be in one of three conditions: (a) running, (b) under repair, or (c) idle—awaiting more work to do. We observe this equipment only when it changes states. Let $X_n = 0$ if the nth change of state puts it into a running condition; $X_n = 1$ if the nth change of state puts it into a repair condition; and $X_n = 2$ if the nth change of state puts it into an idle condition. One possible assumption is that, if the equipment is down for repair, the next change of state must be to a running condition. (If no work is waiting, the repair crew is taken off the job, so that no transition from under repair to idle is possible.) If the equipment is idle, the next change of state has it in a running condition. But, if it is running, it can change to an idle condition or it can break down and be placed under repair. For simplicity, suppose that the last two possibilities are equally likely. It is clear that $\{X_n\}$ is a Markov Chain and that

$$\mathbf{P} = \begin{pmatrix} 0 & \frac{1}{2} & \frac{1}{2} \\ 1 & 0 & 0 \\ 1 & 0 & 0 \end{pmatrix}$$

From this we find that

$$\mathbf{P}^{(2)} = \begin{pmatrix} 1 & 0 & 0 \\ 0 & \frac{1}{2} & \frac{1}{2} \\ 0 & \frac{1}{2} & \frac{1}{2} \end{pmatrix}$$

$$\mathbf{P}^{(3)} = \begin{pmatrix} 0 & \frac{1}{2} & \frac{1}{2} \\ 1 & 0 & 0 \\ 1 & 0 & 0 \end{pmatrix}$$

And we observe, in general, that

$$\mathbf{P}^{(2n)} = \begin{pmatrix} 1 & 0 & 0 \\ 0 & \frac{1}{2} & \frac{1}{2} \\ 0 & \frac{1}{2} & \frac{1}{2} \end{pmatrix}, \quad n = 1, 2, \ldots$$

$$\mathbf{P}^{(2n-1)} = \begin{pmatrix} 0 & \frac{1}{2} & \frac{1}{2} \\ 1 & 0 & 0 \\ 1 & 0 & 0 \end{pmatrix}, \quad n = 1, 2, \ldots$$

△

We notice that this chain exhibits rather peculiar behavior. If we start in state 0, we can return to state 0 only after an even number of steps. Thus the time between visits to state 0 exhibits a periodicity. It must be a multiple of 2.

The same type of behavior is true for states 1 and 2. The time between visits to either of these states must always be a multiple of 2. Any Markov Chain that exhibits this behavior is called a *periodic Markov Chain with period* 2. Clearly, this definition can be extended to define periodic chains with period other than 2. If some state (but not necessarily all states, as in our example) has this property of periodicity, we say the *state* is *periodic*. Periodic chains and states are troublesome in the theory of Markov Chains. To study their behavior, one often must develop theorems that exclude the periodic case, and then reanalyze the theorems for the periodic case. In practice, periodic chains occur seldom enough so that it is desirable to avoid the troubles they create in our elementary exposition of the theory. As a general rule, we shall dismiss periodic Markov Chains from our future discussion. We shall make this exclusion explicit by using the word *aperiodic* (not periodic) in the hypotheses of a given theorem, or in the general discussion preceding certain groups of theorems. Many of the theorems are true even in the periodic case, but we shall not be concerned with this.

6. nth-step State Probabilities

Let us define $\mathbf{p}^{(n)} = (p_0{}^{(n)}, p_1{}^{(n)}, \ldots)$ as the *vector of state probabilities after n transitions*. The element $p_k{}^{(1)}$ of the state vector $\mathbf{p}^{(1)}$ is the *probability of being in state k after* 1 *step*. It follows that

$$p_k{}^{(1)} = \Pr[X_1 = k] = \sum_j \Pr[X_0 = j]\Pr[X_1 = k \mid X_0 = j] = \sum_j p_j p_{jk} \quad (8.17)$$

In words, the left-hand side is the probability of being in state k after 1 transition. The right-hand side can be interpreted simply as the probability of being in some state j to start, and in one step moving from j to k. Since we can start in any state j (with probability $p_j = \Pr[X_0 = j]$), we must add all the ways that we can be in state k, regardless of the initial state, to determine $p_k{}^{(1)}$.

Another interpretation of (8.17) is that

$$\Pr[X_1 = k; X_0 = j] = \Pr[X_0 = j]\Pr[X_1 = k \mid X_0 = j]$$

by the definition of conditional probabilities. The left-hand side is the joint probability of being in state k at time 1 (after 1 step) and being in state j at time 0 (to start). The marginal probability $\Pr[X_1 = k]$ can then be obtained from this joint probability as

$$\Pr[X_1 = k] = \sum_j \Pr[X_1 = k, X_0 = j] = \sum_j \Pr[X_0 = j]\Pr[X_1 = k \mid X_0 = j]$$

Formula (8.17) can be written in vector and matrix form

$$\mathbf{p}^{(1)} = \mathbf{p}^{(0)}\mathbf{P} \tag{8.18}$$

where the elements of the vector $\mathbf{p}^{(1)}$ are the $p_k^{(1)}$ in (8.17).

If we continue this process for $n = 2$, it follows that

$$p_k^{(2)} = \Pr[X_2 = k] = \sum_j p_j^{(1)} p_{jk}$$

or, in vector and matrix form, that

$$\mathbf{p}^{(2)} = \mathbf{p}^{(1)}\mathbf{P} \tag{8.19}$$

By induction it follows easily that

$$\mathbf{p}^{(n)} = \mathbf{p}^{(n-1)}\mathbf{P} \tag{8.20}$$

Thus the components of the vector $\mathbf{p}^{(n)}$ satisfy the relation

$$p_j^{(n)} = \sum_k p_k^{(n-1)} p_{kj} \tag{8.21}$$

We also observe that, if (8.17) is introduced into (8.19) for $\mathbf{p}^{(1)}$, we then have

$$\mathbf{p}^{(2)} = \mathbf{p}^{(0)}\mathbf{P}^2$$

Here $\mathbf{P}^2 = \mathbf{P} \cdot \mathbf{P}$ is the square of the matrix $\mathbf{P}$. In general

$$\mathbf{p}^{(n)} = \mathbf{p}^{(0)}\mathbf{P}^n, \qquad n = 1, 2, 3, \ldots \tag{8.22}$$

Thus we notice that $\mathbf{p}^{(n)}$, the vector of state probabilities after n steps, is easily found in terms of the vector of initial conditions $\mathbf{p}^{(0)}$ and the nth power of the matrix of one-step transition probabilities.

Observe the distinction between $p_{ij}^{(n)}$ and $p_j^{(n)}$. The former is a conditional probability. The latter is a marginal probability. That is, $p_{ij}^{(n)}$ is the probability of being in state j *given* that the process was in state i, n steps previously. On the other hand, $p_j^{(n)}$ is just the probability of being in state j after n steps, regardless of the initial state. Formula (8.22), which can be written in component form as

$$p_j^{(n)} = \sum_k p_k p_{kj}^{(n)} \tag{8.23}$$

can be interpreted as follows. The probability of being in state j after n steps is the weighted average of the probabilities of going to j in n steps, given that the process was in k at the start, weighted by the probability of being in state k to start.

EXAMPLE 8.4 (continued). Let us return to the tank example, (Example 8.4) that was discussed in Sections 3 and 5. Suppose that a tank commander on first encountering an enemy fires his first round for ranging purposes, and suppose that it is fairly unlikely that the first round hits. More precisely, let us suppose that the probability of a hit on the initial shot is $\frac{1}{4}$, and that the probability of a miss on the initial shot is $\frac{3}{4}$. One question then might be: What is the probability that the fourth shot hits? By (8.22) we observe that

$$\mathbf{p}^{(3)} = \mathbf{p}^{(0)}\mathbf{P}^3$$

or that

$$\mathbf{p}^{(3)} = \left(\frac{1}{4}, \quad \frac{3}{4}\right)\begin{pmatrix} \frac{43}{64} & \frac{21}{64} \\ \frac{21}{32} & \frac{11}{32} \end{pmatrix}$$

$$= \left(\frac{43}{256} + \frac{63}{128}, \qquad \frac{21}{256} + \frac{33}{128}\right)$$

or, approximately

$$\mathbf{p}^{(3)} = (.66, \quad .34)$$

Thus the probability that the fourth shot (the third after the initial ranging shot) hits is .66, and the probability that it misses is .34. Notice that if we know that the first round missed, the probability that this fourth round hits is .656 (or 21/32); while, if we know that the initial shot hit, the probability of this fourth round hitting is .672 (or 43/64). Thus we argue that, if the tank commander can observe his first shot and can know that it hit the target, he has gained information that increases his chance of hitting with his fourth shot.

Notice also that the probability vector for his 11th shot is approximately

$$\mathbf{p}^{(10)} = \left(\frac{1}{4}, \frac{3}{4}\right)\begin{pmatrix} .666 & .333 \\ .666 & .333 \end{pmatrix}$$

$$= (.666, .333)$$

Thus, after 11 shots, the probability of a hit is not much different than it was after 4 shots (.66 versus .666).

For future reference we note that, if the tank commander has initial hitting probabilities given by the vector

$$\mathbf{p}^{(0)} = \left(\frac{2}{3}, \frac{1}{3}\right)$$

it follows by direct multiplication that

$$\mathbf{p}^{(n)} = \mathbf{p}^{(0)}\mathbf{P}^n = \mathbf{p}^{(0)}$$

That is, for this initial probability vector, the vector $\mathbf{p}^{(n)}$ is *invariant* to changes in the time variable n. Any vector with this property is called a *stationary vector*. We shall return to the discussion of stationary vectors later.

7. Classification of States

In the study of discrete Markov chains, it is convenient to classify the states so that we can identify certain types of chains. Let us make these definitions.

(a) Suppose that the process starts in some state j. If k is a state so that $p_{jk}{}^{(n)} > 0$ for some n, then we say that state k can be *reached* from state j. If, in addition, $p_{kj}{}^{(m)} > 0$ for some m, the state j can be reached from k, and we say that states j and k *communicate*.

(b) If C is a set of states so that no state outside of C can be reached from any state in C, the set C is said to be *closed*. C has the property that once the process enters it, it cannot leave. If, in addition, each pair of states in C communicate, C is termed a *closed communicating class*.

(c) If a closed set contains only one state, that state is termed an *absorbing state*. (Recall our discussion of Examples 8.1 and 8.2, Section 3.)

(d) If a discrete Markov chain contains no closed sets with the exception of the set of all states, then the chain is said to be *irreducible*. A chain is irreducible if every state can be reached from every state. In this chapter, we are concerned primarily with Markov chains that are *irreducible*.

8. Examples

EXAMPLE 8.1 (continued). In the Gambler's Ruin (Example 8.1), the process clearly has two absorbing states: the "ruin" state 0 in which the player has lost all his money, and the state N in which the player has won all the money. Once either of these states are entered, no exit is possible. △

EXAMPLE 8.4 (continued). Recall the tank duel Example 8.4. There,

$$\mathbf{P} = \begin{pmatrix} a & 1-a \\ b & 1-b \end{pmatrix}$$

If $a \neq 0, 1$, and if $b \neq 0, 1$, then the Markov Chain is irreducible. Each state can be reached from the other. On the other hand, if $a = 1$, the state 1 ("miss") cannot be reached from the state 0 ("hit"). Hence, 0 is a closed set. Furthermore, since the closed set contains only one state, this state is an absorbing state. It has the property that once a "hit" is made, hits continue thereafter. We might think of this as the behavior of a perfect tracking-firing mechanism. Once the gun system locks on target, it hits with every firing.

EXAMPLE 8.2 (continued). The transition matrix for the production item's progress through manufacturing stages had the form:

$$\mathbf{P} = \left(\begin{array}{c:c:ccc} 1 & 0 & 0 & 0 & 0 \\ \hdashline 0 & 1 & 0 & 0 & 0 \\ \hdashline p & 0 & q & r & 0 \\ p & 0 & 0 & q & r \\ p & r & 0 & 0 & q \end{array}\right)$$

This matrix can be decomposed into blocks, as indicated by the dotted lines, to take the form

$$\mathbf{P} = \begin{pmatrix} \mathbf{A}_1 & \mathbf{0} & \mathbf{0} \\ \mathbf{0} & \mathbf{A}_2 & \mathbf{0} \\ \mathbf{B}_1 & \mathbf{B}_2 & \mathbf{B}_3 \end{pmatrix} \tag{8.24}$$

where

$$\mathbf{A}_1 = (1),\ \mathbf{A}_2 = (1),\ \mathbf{B}_1 = \begin{pmatrix} p \\ p \\ p \end{pmatrix},\ \mathbf{B}_2 = \begin{pmatrix} 0 \\ 0 \\ r \end{pmatrix},\ \mathbf{B}_3 = \begin{pmatrix} q & r & 0 \\ 0 & q & r \\ 0 & 0 & q \end{pmatrix}$$

and each $\mathbf{0}$-symbol means a matrix of 0's. The sets $\{0\}$ (item junked) and $\{1\}$ (item completed) are each closed sets. Since each contains only one state, 0 and 1 are absorbing states. Clearly, the chain is not irreducible, and the set of states $\{2, 3, 4\}$ is not closed. △

A decomposition of $\mathbf{P}$ into subblocks similar to (8.24) is possible for any nonirreducible Markov chain, although the states may have to be reordered first.

EXAMPLE 8.6 (continued). This waiting-line example is an irreducible chain with an infinite number of states. △

9. Limiting Distributions

Frequently, we wish to study a Markov chain that has been operating for some time. Thus we focus attention on the behavior of the state probabilities $p_j^{(n)}$ for large n. More precisely, let us consider the quantities:

$$v_j = \lim_{n \to \infty} p_j^{(n)} \tag{8.25}$$

The quantities v_j are sometimes referred to as *steady state probabilities.* In designing a physical system, there are often "start up" effects that are different from what can be expected in the "long run." A designer might be mainly interested in the "long run" behavior that is described by the limiting probabilities v_j.

From another point of view, v_j might be interpreted as the *long-run proportion* of time, the process spends in state j.

Reasonably, we can expect that, over a long period of time, the influence of the initial state in which the process started may die out and, thus, that the limiting probabilities v_j may not depend on the initial state. If this is the case, then v_j can also be interpreted as the limit of the n-step transition probabilities $p_{ij}^{(n)}$, $v_j = \lim_{n \to \infty} p_{ij}^{(n)}$, since $p_{ij}^{(n)}$ is simply the probability of the process being in state j after n steps, given that it was initially in state i. Thus, if each v_j does not depend on the initial state, the matrix $\mathbf{P}^{(n)} = (p_{ij}^{(n)})$ will converge to a matrix $\mathbf{V}$ as $n \to \infty$, each row of which is identical to the vector $\mathbf{v}$ with components v_j,

$$\mathbf{P}^{(n)} \to \mathbf{V} = \begin{pmatrix} \mathbf{v} \\ \mathbf{v} \\ \cdot \\ \cdot \\ \cdot \end{pmatrix} \qquad \text{as } n \to \infty$$

where $\mathbf{v} = (v_0, v_1, \ldots, v_j, \ldots)$

Clearly, we must be concerned with the questions: Do the limits defining the v_j's exist? If they exist, do they form a probability distribution; that is, do they add to 1, $\sum v_j = 1$? How can the v_j's be computed?

If the limits $v_j = \lim_{n \to \infty} p_j^{(n)} = \lim_{n \to \infty} p_{ij}^{(n)}$ exist and do not depend on the initial state, then by letting $n \to \infty$ in the identity (8.21)

$$p_j^{(n)} = \sum_k p_k^{(n-1)} p_{kj}$$

we obtain

$$v_j = \sum_k v_k p_{kj}, \qquad j = 0, 1, 2, \ldots \tag{8.26}$$

or, equivalently

$$\mathbf{v} = \mathbf{vP} \tag{8.27}$$

Formula (8.26) (or formula (8.27)) gives a system of linear equations that must be satisfied by the quantities v_j.

Any vector $\mathbf{x} = (x_0, x_1, x_2, \ldots)$ with $x_j \geq 0, \sum x_j = 1$, that satisfies the system (8.26) or (8.27)

$$x_j = \sum_k x_k p_{kj} \qquad \text{or} \qquad \mathbf{x} = \mathbf{xP} \tag{8.28}$$

is termed a *stationary probability vector* of the process.

The following theorems are stated without proof.

THEOREM 1. *In any aperiodic Markov chain the limits* $v_j = \lim_{n\to\infty} p_j^{(n)}$ *all exist.* △

THEOREM 2. *In any irreducible, aperiodic Markov chain the limits* $v_j = \lim_{n\to\infty} p_j^{(n)} = \lim_{n\to\infty} p_{ij}^{(n)}$ *do not depend on the initial distribution.* △

THEOREM 3. *In any finite, irreducible, aperiodic Markov chain the limit vector* $\mathbf{v} = (v_0, v_1, \ldots, v_j, \ldots)$ *is the unique stationary probability vector of the process.* △

These theorems generally are sufficient to allow us to discuss and to determine the limiting probabilities in finite Markov chains encountered in practice.

Theorem 3 implies that, for a finite, aperiodic, irreducible Markov chain, the matrix $\mathbf{P}^{(n)} = (p_{ij}^{(n)})$ approaches a matrix that has every row equal, each row being identical with the stationary vector $\mathbf{v} = (v_0, v_1, \ldots)$

$$\lim_{n\to\infty} \mathbf{P}^{(n)} = \begin{pmatrix} \mathbf{v} \\ \mathbf{v} \\ \cdot \\ \cdot \\ \cdot \end{pmatrix} = \begin{pmatrix} v_0, v_1, v_2, \ldots \\ v_0, v_1, v_2, \ldots \\ \cdot \quad \cdot \quad \cdot \\ \cdot \quad \cdot \quad \cdot \\ \cdot \quad \cdot \quad \cdot \end{pmatrix} \tag{8.29}$$

10. Examples

EXAMPLE 8.9. In Example 8.4, we discussed an irreducible aperiodic process with 1-step transition matrix

$$\mathbf{P} = \begin{pmatrix} \frac{3}{4} & \frac{1}{4} \\ \\ \frac{1}{2} & \frac{1}{2} \end{pmatrix}$$

By computation, we found that

$$\mathbf{P}^{(10)} = \begin{pmatrix} .666 & .333 \\ .666 & .333 \end{pmatrix}$$

If we compare this with (8.29), we surmise that the limit vector is $\mathbf{v} = (\frac{2}{3}, \frac{1}{3})$. To check this, we use Theorem 3 and determine the stationary probability vector that satisfies (8.28): $\mathbf{x} = \mathbf{xP}$. Thus

$$\mathbf{x} = \mathbf{x} \begin{pmatrix} \frac{3}{4} & \frac{1}{4} \\ \\ \frac{1}{2} & \frac{1}{2} \end{pmatrix}$$

or

$$x_0 = x_0\left(\frac{3}{4}\right) + x_1\left(\frac{1}{2}\right)$$

$$x_1 = x_0\left(\frac{1}{4}\right) + x_1\left(\frac{1}{2}\right)$$

Rearranged, these equations become

$$\left(\frac{1}{4}\right)x_0 - \left(\frac{1}{2}\right)x_1 = 0$$

$$-\left(\frac{1}{4}\right)x_0 + \left(\frac{1}{2}\right)x_1 = 0$$

Since one equation is simply the negative of the other, the system gives only one condition on the stationary vector (x_0, x_1):

$$\left(\frac{1}{4}\right)x_0 - \left(\frac{1}{2}\right)x_1 = 0$$

or

$$x_1 = \frac{1}{2} x_0 \tag{8.30}$$

Another condition comes from the fact that this is a probability vector, $\sum x_i = 1$. Thus

$$x_0 + x_1 = 1 \tag{8.31}$$

If (8.30) and (8.31) are solved simultaneously, we find that $x_0 = \frac{2}{3}$, $x_1 = \frac{1}{3}$. Thus, by Theorem 3, the limiting vector is given by

$$\mathbf{v} = \mathbf{x} = \left(\frac{2}{3}, \frac{1}{3}\right) \qquad \triangle$$

EXAMPLE 8.10. Suppose that a piece of equipment can be either busy or idle and that, if it is idle, it is either down for repair or waiting for more work. Denote the states "busy," "under repair," and "waiting for more work," by 0, 1, 2. Observe the state of the system on a sequence of successive days, and assume that there is sufficient lack of memory in the system that it can be approximated by a Markov chain with one-step transition matrix

$$\mathbf{P} = \begin{pmatrix} p_{00} & p_{01} & p_{02} \\ p_{10} & p_{11} & p_{12} \\ p_{20} & p_{21} & p_{22} \end{pmatrix} = \begin{pmatrix} .8 & .1 & .1 \\ .2 & .6 & .2 \\ .6 & 0 & .4 \end{pmatrix}$$

Thus, for example, $p_{01} = .1$ means that the probability is .1 that a busy machine will break down; $p_{11} = .6$ means that the probability is .6 that a machine under repair today will still be under repair tomorrow; $p_{21} = 0$ means that an idle machine does not break down.

If we are interested in the long-run proportions of time that the equipment spends in the three states, the limiting distribution must be computed. The system is clearly irreducible. (Each state can be reached from each other state, although not necessarily in one step; it takes at least 2 steps to go from state 2 to state 1.) It is also finite and aperiodic. By Theorem 3, we need only find the unique stationary probability vector by solving $\mathbf{x} = \mathbf{xP}$, $\sum x_i = 1$. Thus

$$x_0 = (.8)x_0 + (.2)x_1 + (.6)x_2$$
$$x_1 = (.1)x_0 + (.6)x_1$$
$$x_2 = (.1)x_0 + (.2)x_1 + (.4)x_2$$

or

$$\begin{aligned} (.2)x_0 - (.2)x_1 - (.6)x_2 &= 0 \\ -(.1)x_0 + (.4)x_1 \qquad\qquad &= 0 \\ -(.1)x_0 - (.2)x_1 + (.6)x_2 &= 0 \end{aligned}$$

Since the third equation can be obtained by adding the first two and multiplying by -1, the third gives no additional information and can be eliminated. The first two equations, together with the fact that $\sum x_i = 1$, will determine the unique solution:

$$\begin{aligned} (.2)x_0 - (.2)x_1 - (.6)x_2 &= 0 \\ -(.1)x_0 + (.4)x_1 \qquad\qquad &= 0 \\ x_0 + \quad x_1 + \quad x_2 &= 1 \end{aligned}$$

From the first two we find that $x_1 = (\frac{1}{4})x_0$, $x_2 = (\frac{1}{4})x_0$ which, when substituted in the last equation, yield

$$x_0 = \frac{2}{3}, \qquad x_1 = \frac{1}{6}, \qquad x_2 = \frac{1}{6}$$

Thus the limiting probability vector is $\mathbf{v} = (\frac{2}{3}, \frac{1}{6}, \frac{1}{6})$, and this gives the long-run proportions of time the system spends in these three states. △

EXERCISES

1. Let $\{X_n\}$ be a Markov chain with state space $\{0, 1, 2\}$, initial probability vector $\mathbf{p}^{(0)} = (\frac{1}{4}, \frac{1}{2}, \frac{1}{4})$, and 1-step transition matrix

$$\mathbf{P} = \begin{pmatrix} 1/4 & 3/4 & 0 \\ 1/3 & 1/3 & 1/3 \\ 0 & 1/4 & 3/4 \end{pmatrix}$$

   (a) Compute $p(0, 1, 1) = \Pr[X_0 = 0, X_1 = 1, X_2 = 1]$.

   (b) Show that $P[X_1 = 1 \text{ and } X_2 = 1 \mid X_0 = 0] = p_{01}p_{11}$.

   (c) Compute $p_{01}{}^{(2)}$.

2. A gambler plays a "fair game" in which the odds are 2 to 1. In other words, at each play he has probability $\frac{1}{3}$ of winning and $\frac{2}{3}$ of losing. If he wins, he wins \$2. If he loses, he loses \$1. Assume that the total dollar resources of the gambler and his opponent are N dollars. If the capital of either player falls below the point where he could pay up if he lost on the next play, the game terminates. Let X_n represent the first player's capital after n plays.
   (a) Set up the 1-step transition matrix of the Markov chain $\{X_n\}$.
   (b) Suppose that the two players agree that if the capital of either one falls to \$1 they will perform the next play at even odds—win or lose \$1 with equal probability. Set up the 1-step transition matrix for this case.
3. In Exercise 1, compute $p_{ij}^{(3)}$ for $i, j = 0, 1, 2$.
4. Let $\{X_n\}$ be a Markov chain with state space $\{0, 1, 2, 3, 4\}$ and 1-step transition matrix $\mathbf{P}$. For each of the following cases, determine the closed communicating classes and absorbing states.

   (a)
$$\mathbf{P} = \begin{pmatrix} 1/2 & 0 & 0 & 1/2 & 0 \\ 1/2 & 1/2 & 0 & 0 & 0 \\ 1/4 & 1/2 & 0 & 0 & 1/4 \\ 0 & 0 & 0 & 1 & 0 \\ 0 & 0 & 1/2 & 1/4 & 1/4 \end{pmatrix}$$

   (b)
$$\mathbf{P} = \begin{pmatrix} 0 & 1/3 & 0 & 1/3 & 1/3 \\ 1/3 & 1/3 & 0 & 1/3 & 0 \\ 0 & 0 & 2/3 & 0 & 1/3 \\ 1/4 & 1/4 & 0 & 1/4 & 1/4 \\ 0 & 0 & 1/3 & 0 & 2/3 \end{pmatrix}$$

5. Show that the random walk example in Section 5, Chapter 7, is periodic.
6. Show that the Markov chain in Exercise 1 is irreducible.
7. Compute the limiting probabilities for the chain in Exercise 1 by finding the stationary probability vector.
8. Discuss the limiting behavior as $n \to \infty$ of the nonirreducible Markov chain in Exercise 4 (a).
9. A certain factory has 2 machines and 1 repair crew. Assume that the probability of any one machine breaking down in a given day is α. Assume that if the repair crew is working on a machine, the probability that they will complete the repair in one more day is β. For simplicity, ignore the possibility of a repair completion or a breakdown taking place except at the end of a day. Let X_n represent the number of machines in operation at the end of the nth day. Assume that the behavior of $\{X_n\}$ can be modeled by a Markov chain.
   (a) Set up the 1-step transition matrix $\mathbf{P}$.

(b) If the system starts out with both machines operating, what is the probability that both will be operating 2 days later?

10. Suppose that we throw one die many times and total the faces as they appear. Let $X_n =$ the total "score" at the nth throw. Is $\{X_n\}$ a Markov process? If it is, what is its one-step transition probability matrix?

11. A simple scheme for predicting the behavior of a stochastic process $\{X_n\}$ is called the "moving average predictor." It is computed by the formula:

$$Y_m = (X_m + X_{m+1} + \cdots + X_{m+n})/n$$

In this computation one always retains the last n observations. When a new observation (X_i) is taken, the observation taken n periods ago is discarded from the average. Hence, Y_m is always the average of n terms representing observations on the stochastic process. Is $\{Y_m\}$ always a Markov process if $\{X_n\}$ is? If $\{X_n\}$ is an independent sequence, is $\{Y_m\}$ a Markov process?

12. The owner of the local one-chair barber shop is thinking of expanding because there seem to be too many people waiting. Observations indicate that, in the time required to cut one person's hair, there may be 0, 1, 2, 3 new arrivals with probability .3, .4, .2, .1, respectively. The shop has a fixed capacity of six people, which includes the one whose hair is being cut. Let X_n be the number of people in the shop at the completion of the nth customer's haircut. Show that $\{X_n\}$ is a Markov chain. Find its one-step transition probability matrix. Determine the "long-run" proportion of time that the shop has 6 people in it; that it has 5 people in it.

13. Suppose that you have conducted a series of tests on a manual dexterity training procedure and find that the following probability matrix describes the course of "correct" and "incorrect" responses.

		$(j+1)$st trial: correct	$(j+1)$st trial: incorrect
jth trial	correct	0.95	0.05
	incorrect	0.01	0.99

(a) What proportion of correct responses would one expect of a "fully trained" trainee?

(b) What proportion of correct responses would one expect of a trainee after five repetitions of the procedure if the initial response is equally likely to be correct or incorrect?

(c) How many repetitions (trials) of this procedure must one perform before the trainee is within 1% of his ultimate performance?

(d) What is the probability that a correct response is obtained for the first time exactly four trials after an incorrect response?

14. Suppose that we have a container with a membrane that separates it into compartments A and B. Initially there are j molecules in compartment A and $a - j$ in compartment B. A *transition* is said to occur whenever a molecule crosses through the membrane (either from A to B or from B to A). Let $X_n =$ the number of molecules in compartment A after n transitions. At each transition, X_n increases or decreases by exactly one molecule. Suppose the probability that a molecule changes compartments is proportional to the number of molecules in the compartment that the molecule leaves. Set up a Markov chain model of this diffusion process. The model discussed here is called the Ehrenfest diffusion model.

15. Suppose that on a final assembly line for automobiles we have this set of rules:

(1) Two convertibles can never follow each other down the line because the work content would unbalance the line.

(2) A station wagon must be followed by a sedan to balance the line.

(3) A sedan can be followed by either a station wagon or a convertible, but not by a sedan. (Only those three models are produced on this line.)

Set up a possible transition matrix for these rules by inserting the letters a, b, c, d, ..., etc., for the values that are not numerically defined by the above rules.

(a) Is the chain irreducible?

(b) What is the probability that after one convertible, the next one occurs on the line one space apart (separated by one other vehicle)? Use your letters a, b, c, d, ..., if they are needed.

(c) If $\mathbf{P}$ is the one-step transition matrix, what interpretation would you give the (ij)th element of $\mathbf{P}^n$, for large n?

(d) If we had a production schedule that was given by a vector of the proportion of each type of auto to be made in a month, how might we determine a probability sequence that met the schedule and the rules of sequencing given in this problem?

(e) Suppose that we had contracted to deliver 10,000 autos at the end of a month. Of these 10,000 ,approximately 50% had to be sedans, 10% had to be station wagons, and 40% had to be convertibles. Set up some probability schedule rule that ensures this production and that still meets constraints 1, 2, and 3 of this problem.

16. A power conveyor is a materials-handling device. Material is placed on carriers that are equally spaced along a constant speed-powered chain. As a loaded carrier passes by a worker, he removes the material from the conveyor, performs

some work on it, and returns it to the conveyor so that it can be moved to the next work station. Because of the constant speed and equal spacing of carriers, it is convenient to measure time in terms of the number of carriers that pass and, hence, to have a discrete variable to measure time.

Suppose that a particular work station has work brought to it by a power conveyor on which every carrier has a single part to be worked on by this station. Suppose further that, in the time between two consecutive carriers, the work station can complete 0, 1, or 2 parts with probability $\frac{3}{8}$, $\frac{1}{2}$, $\frac{1}{8}$, respectively. Suppose further that this work station can accumulate (as temporary storage), at most, one part in addition to a part that is being worked on. If a carrier passes by and the station is empty or has just one part in operation, then the carrier is unloaded and the part is laid aside to be operated on when the one now in progress is completed. If a carrier passes by and the station already has one part in operation and one in storage, then that carrier load bypasses the station and is lost.

Let the states of the work station be 0, 1, 2, representing an empty station, a station with exactly one part in progress, and a station with one part in progress and one in storage, respectively. Let the nth step occur just *before* the arrival of the nth carrier. Then

(a) Show that the one-step transition probability matrix is given by

$$\mathbf{P} = \begin{pmatrix} 5/8 & 3/8 & 0 \\ 1/8 & 1/2 & 3/8 \\ 1/8 & 1/2 & 3/8 \end{pmatrix}$$

(that is, give a logical argument based on the system description and probabilities that will support the use of $\mathbf{P}$ as the transition probabilities for this problem).

(b) Find the proportion of time that the work station spends in each state for the long run.

(c) If the station starts the day empty, what is the probability that the third carrier that arrives will not be able to discharge its load and, hence, bypasses the station.

(d) Find the average length of time between idle times.

(e) What is the probability that the station will go from state 2 to state 0 for the first time in two steps? How can we interpret this probability?*

* In this problem we have "discretized" time and in the process have created some anomalies. For example, if a part is completed at any time within the interval between two carrier passings, the worker is considered to be idle for the entire time. Hence estimates, based on the probabilities obtained, may be in error in any real-life problem in which time is continuous.

17. Suppose that we run an (s, S) inventory system, as follows. At fixed equally spaced instants $t_1 < t_2 < t_3 \ldots$ we make a decision whether to purchase a resupply of inventory. This resupply is immediately delivered. How much is ordered depends on the level of inventory at the instant t_i. If X_i is the on-hand inventory at t_i, then we reorder if $X_i \leq s$ and do not reorder if $s < X_i \leq S$. Whenever an order is placed, it is always for an amount that brings the on-hand inventory up to S.

    Hence X_i takes values $S, S-1, S-2, \ldots 0, -1, -2, \ldots$ where negative values indicate that demand is back-ordered. Backlogged demand is satisfied immediately on receipt of a new supply.

    Suppose that d_i represents the total demand that occurs in (t_{i-1}, t_i) and that d_i is Poisson distributed, with the demand in consecutive periods independent, identically distributed random variables. Thus

$$\begin{aligned} X_{i+1} &= X_i - d_{i+1} \qquad \text{if} \qquad s < X_i \leq S \\ &= S - d_{i+1} \qquad \text{if} \qquad X_i \leq s \end{aligned}$$

    Show that the process $\{X_i\}$ is a Markov process and determine its transition probability matrix $\mathbf{P}$.

18. Suppose that we try to model a sequence of decisions in the following manner.

    In decision-making one collects and interprets data. For many reasons, including collecting the wrong data, we might interpret the data correctly, with probability p, or incorrectly, with probability q, irrespective of the decision. Certainly $p + q = 1$. Having the data, we then make a decision. The decision can be "correct" or "incorrect." Which decision occurs depends, of course, on whether the data were interpreted correctly or not. Suppose also that whether a decision is correct or not depends on whether the previous decision was correct or not. For example, if a correct decision is made, then we would expect that, at the next decision point, the process being controlled was near its control level. Hence, the next decision would be routine and would have a large chance of being correct. On the other hand, an incorrect decision might put the process so far from its control level that drastic action was required which had a large chance of being incorrect.

    Let $X_n = 0$ if a correct decision is made at the nth decision point, and $X_n = 1$ if an incorrect decision is made at the nth decision point.

    Determine the transition probabilities of this decision process:

$$p_{ij} = \Pr[X_n = j \mid X_{n-1} = i], \qquad i, j = 0, 1$$

    Suppose the probability p that the data are interpreted correctly is $\frac{8}{9}$, and the probability q that they are interpreted incorrectly is $\frac{1}{9}$. Suppose that

$$\Pr[X_n = 0 \mid X_{n-1} = 0 \text{ and the data for the } n\text{th decision are interpreted correctly}] = .95$$

$$\Pr[X_n = 1 \mid X_{n-1} = 1 \text{ and the data for the } n\text{th decision are interpreted incorrectly}] = .99$$

$$\Pr[X_n = 0 \mid X_{n-1} = 0 \text{ and the data are interpreted incorrectly}] = 0.5$$

$$\Pr[X_n = 1 \mid X_{n-1} = 1 \text{ and the data are interpreted correctly}] = 0.25$$

From this description determine the Markov chain transition probability matrix **P**.

(a) Suppose that the "mission" is successful if the 4th decision is correct. ("A missile mission is successful if the 4th corrective signal puts it on course.") If the initial decision is correct, what is the probability that the mission is successful for the above calculated probabilities?

(b) If a decision-maker is a "good one" if in the long run he is correct 75% of the time, is the above decision-maker a good one?

(c) If the decision process is to be improved (increase the "long run" probability of a correct decision) should we improve the data interpretation step (that is, change p and q) or improve the decision-maker (that is, change the probabilities $\Pr[X_n = j \mid X_{n-1} = i$ and the data are correct (or incorrect)])? Discuss the relative importance of these two ways of improving matters.

9 • Algebraic Methods Useful in the Study of Markov Chains*

1. Introduction

In the preceding chapters we have observed that properties of interest for a discrete Markov process depend on the powers $\mathbf{P}^n$ of a matrix $\mathbf{P}$, and the limiting values of these powers. For a simple process with few states, we have some hope of raising the matrix to higher powers by straightforward multiplication. For larger matrices this is unwieldy. For large matrices any method is laborious. However, with the use of digital computers there is a possibility of finding the desired matrices with less trouble.

The method of finding powers of matrices that is discussed in this chapter is based on finding a diagonal matrix that is equivalent to the given matrix $\mathbf{P}$. If this matrix can be found, powers of $\mathbf{P}$ will be simply related to powers of the diagonal matrix, which are easy to compute. Later, we shall find that these methods are useful for other purposes as well.

We start with some general remarks. Then we apply the results to several small problems to indicate how the work can be done. The methods used here can be programmed for a digital computer. Existing programs are available for finding the elements of the diagonal matrix, called the *eigenvalues*, and the rows and columns of the transformation matrices, called *eigenvectors*. We do not discuss these programs.

2. Eigenvalues and Eigenvectors

A matrix $\mathbf{A}$ is said to be *similar* to a matrix $\mathbf{\Lambda}$ if there exists a nonsingular matrix $\mathbf{S}$ so that

$$\mathbf{S\,A\,S}^{-1} = \mathbf{\Lambda} \qquad \text{or} \qquad \mathbf{A} = \mathbf{S}^{-1}\,\mathbf{\Lambda}\,\mathbf{S} \qquad \text{or} \qquad \mathbf{SA} = \mathbf{\Lambda\,S} \tag{9.1}$$

* This chapter may be omitted without loss of continuity.

If the matrix $\mathbf{\Lambda}$ is a *diagonal* matrix (entries zero except on the main diagonal), then we say that the matrix $\mathbf{A}$ is *diagonalized* by the relation (9.1). We use the term here to imply that the diagonal elements of $\mathbf{\Lambda}$ are all *distinct*:

$$\mathbf{\Lambda} = \begin{pmatrix} \lambda_{00} & 0 & \dots & 0 \\ 0 & \lambda_{11} & \dots & 0 \\ \cdot & \cdot & \ddots & \cdot \\ \cdot & \cdot & & \cdot \\ \cdot & \cdot & & \cdot \\ 0 & 0 & \dots & \lambda_{N-1,N-1} \end{pmatrix}$$

If the matrix $\mathbf{A}$ can be diagonalized, the diagonal elements, λ_{ii}, of $\mathbf{\Lambda}$ are called *eigenvalues* of $\mathbf{A}$. In terms of the elements of the matrices involved in (9.1), we have

$$\mathbf{s}_i \mathbf{A} = \lambda_{ii} \mathbf{s}_i \tag{9.2}$$

where $\mathbf{s}_i$ is the ith row of the matrix $\mathbf{S}$. Any vector $\mathbf{s}_i$ that satisfies (9.2) is termed a *left eigenvector* of the matrix $\mathbf{A}$ belonging to the eigenvalue λ_{ii}.

We can show that, if $\mathbf{A}$ is an $N \times N$ matrix having N distinct eigenvalues λ_{ii} with corresponding nonzero vectors $\mathbf{s}_i$ that satisfy (9.2), then $\mathbf{A}$ can be diagonalized. Furthermore, any matrix $\mathbf{S}$ having such nonzero eigenvectors as rows will have rank N and satisfy (9.1)

In general, the elements of $\mathbf{S}$ are not determined uniquely by (9.2). Each row of $\mathbf{S}$ is determined up to a multiplicative constant.

The elements of $\mathbf{S}^{-1}$ can be found from $\mathbf{S}\,\mathbf{S}^{-1} = \mathbf{I}$; that is, by inverting the known matrix $\mathbf{S}$, or by rewriting (9.1) in the form:

$$\mathbf{A}\,\mathbf{S}^{-1} = \mathbf{S}^{-1}\,\mathbf{\Lambda}$$

This implies that the *column* vectors of $\mathbf{S}^{-1}$, $\mathbf{s}^j$ must satisfy

$$\mathbf{A}\mathbf{s}^j = \mathbf{s}^j \lambda_{jj} \tag{9.3}$$

Any column vector $\mathbf{s}^j$ that satisfies (9.3) is termed a *right eigenvector* of $\mathbf{A}$.

From (9.1), by using matrix multiplication, it follows that

$$\mathbf{A}\,\mathbf{A} = (\mathbf{S}^{-1}\,\mathbf{\Lambda}\,\mathbf{S})(\mathbf{S}^{-1}\,\mathbf{\Lambda}\,\mathbf{S})$$

or, since $\mathbf{S}^{-1}\,\mathbf{S} = \mathbf{I}$,

$$\mathbf{A}^2 = \mathbf{S}^{-1}\,\mathbf{\Lambda}^2\,\mathbf{S}$$

and by induction

$$\mathbf{A}^n = \mathbf{S}^{-1}\,\mathbf{\Lambda}^n\,\mathbf{S} \tag{9.4}$$

The problem of raising a diagonalizable matrix $\mathbf{A}$ to a power is thus seen to be solvable by finding the matrices $\mathbf{S}$ and $\mathbf{\Lambda}$ that diagonalize $\mathbf{A}$. Since $\mathbf{\Lambda}$ is a diagonal matrix, $\mathbf{\Lambda}^n$ is found in a trivial fashion to be

$$\mathbf{\Lambda}^n = \begin{pmatrix} \lambda_{00}{}^n & 0 & \cdots & 0 \\ 0 & \lambda_{11}{}^n & \cdots & 0 \\ \cdot & \cdot & \cdot & \cdot \\ \cdot & \cdot & \cdot & \cdot \\ \cdot & \cdot & \cdot & \cdot \\ 0 & 0 & \cdots & \lambda_{N-1,N-1}^n \end{pmatrix}$$

Hence

$$\mathbf{A}^n = \mathbf{S}^{-1}\,\mathbf{\Lambda}^n\,\mathbf{S} \tag{9.5}$$

is easy to compute. The problem, of course, is to find $\mathbf{\Lambda}$ and $\mathbf{S}$. In practice, it is often easiest to test if $\mathbf{A}$ is diagonalizable by trying to find its eigenvalues and eigenvectors. We adopt this procedure here.

The eigenvalues can be determined by the following argument. From (9.2) we notice immediately that each eigenvalue λ_{ii}, and corresponding left eigenvector $\mathbf{s}_i$, satisfy

$$\mathbf{s}_i(\mathbf{A} - \lambda_{ii}\mathbf{I}) = \mathbf{0}, \qquad i = 0, 1, \ldots, N-1$$

Since $\mathbf{s}_i$ is a nonzero vector, it follows that whenever $\lambda = \lambda_{ii}$

$$|\mathbf{A} - \lambda\mathbf{I}| = 0 \tag{9.6}$$

($|\mathbf{A}|$ is read as "the determinant of the matrix $\mathbf{A}$") giving an Nth degree polynomial equation whose roots determine the unknown eigenvalues, λ_{ii}. The *eigenvectors* can be found from (9.2) for each eigenvalue above. Equation (9.6) is termed the *characteristic equation* of the matrix $\mathbf{A}$.

Thus the determination of $\mathbf{A}^n$ can be schematicized as follows:

1. Solve (9.6) for the n distinct *eigenvalues*, λ_{ii}. If the N roots are not all distinct *stop*. This method of finding $\mathbf{A}^n$ will not work.
2. Solve (9.2) for a nonzero eigenvector $\mathbf{s}_i$ corresponding to each λ_{ii} for $i = 0, 1, \ldots N-1$.
3. Form $\mathbf{S}$ from the $\mathbf{s}_i$.
4. Find $\mathbf{S}^{-1}$ by
   (a) inverting $\mathbf{S}$, or
   (b) finding the *right eigenvectors* $\mathbf{s}^j$ from (9.3). (Correct proportionality constants must be introduced to assure $\mathbf{S}^{-1}\,\mathbf{S} = \mathbf{I}$.)
5. Find $\mathbf{A}^n$ by (9.4).

3. Examples

EXAMPLE 9.1. To illustrate the preceding remarks, let us consider the tank example (Examples 8.4 and 8.9 of Chapter 8).

We recall that

$$\mathbf{P} = \begin{pmatrix} \frac{3}{4} & \frac{1}{4} \\ \frac{1}{2} & \frac{1}{2} \end{pmatrix}$$

We seek $\mathbf{P}^n$ and lim $\mathbf{P}^n$ as $n \to \infty$. Let us find $\mathbf{P}^n$ by using the methods of this section. (The results of Examples 8.4 and 8.9 will serve as a check on these calculations.)

For this matrix the characteristic equation

$$|\mathbf{P} - \lambda\mathbf{I}| = 0$$

is

$$\begin{vmatrix} \frac{3}{4} - \lambda & \frac{1}{4} \\ \frac{1}{2} & \frac{1}{2} - \lambda \end{vmatrix} = 0$$

When expanded, this gives

$$\left(\frac{1}{2} - \lambda\right)\left(\frac{3}{4} - \lambda\right) - \frac{1}{8} = 0$$

which is the polynomial form of the characteristic equation. Thus the eigenvalues must be the roots of

$$4\lambda^2 - 5\lambda + 1 = 0$$

Hence, $\lambda_{00} = 1$; $\lambda_{11} = \frac{1}{4}$. (We will comment briefly on $\lambda_{00} = 1$ later.) Then from (9.2) we have

$$\mathbf{s}_0 \mathbf{P} = 1\mathbf{s}_0$$

and

$$\mathbf{s}_1 \mathbf{P} = \left(\frac{1}{4}\right)\mathbf{s}_1 \tag{9.7}$$

where $\mathbf{s}_0$ and $\mathbf{s}_1$ are row vectors, say $\mathbf{s}_0 = (s_{00}, s_{01})$; $\mathbf{s}_1 = (s_{10}, s_{11})$. Then we have the two sets of equations

$$(s_{00}, s_{01})\begin{pmatrix} \frac{3}{4} & \frac{1}{4} \\ \frac{1}{2} & \frac{1}{2} \end{pmatrix} = 1(s_{00}, s_{01})$$

with a corresponding equation for $\mathbf{s}_1$. From them, we have

$$\left(\frac{3}{4}\right)s_{00} + \left(\frac{1}{2}\right)s_{01} = s_{00}$$

$$\left(\frac{1}{4}\right)s_{00} + \left(\frac{1}{2}\right)s_{01} = s_{01}$$

or

$$-\left(\frac{1}{4}\right)s_{00} + \left(\frac{1}{2}\right)s_{01} = 0$$

$$\left(\frac{1}{4}\right)s_{00} - \left(\frac{1}{2}\right)s_{01} = 0$$

Note carefully that this system of equations will not determine (s_{00}, s_{01}) uniquely. (The second equation is essentially the same as the first. The equations are not independent.) But we notice that they determine (s_{00}, s_{01}) to within a multiplicative constant. Hence, *at this time* we arbitrarily set $s_{00} = 1$. Then, from the above equations, $s_{01} = \frac{1}{2}$. Thus we have $\lambda_{00} = 1$ and $\mathbf{s}_0 = (1, \frac{1}{2})$. We now find $\mathbf{s}_1$.

From (9.7) we have

$$(s_{10}, s_{11})\begin{pmatrix} \frac{3}{4} & \frac{1}{4} \\ \frac{1}{2} & \frac{1}{2} \end{pmatrix} = \left(\frac{1}{4}\right)(s_{10}, s_{11})$$

or

$$\left(\frac{3}{4}\right)s_{10} + \left(\frac{1}{2}\right)s_{11} = \left(\frac{1}{4}\right)s_{10}$$

$$\left(\frac{1}{4}\right)s_{10} + \left(\frac{1}{2}\right)s_{11} = \left(\frac{1}{4}\right)s_{11}$$

from which we have

$$\left(\frac{1}{2}\right)s_{10} + \left(\frac{1}{2}\right)s_{11} = 0$$

$$\left(\frac{1}{4}\right)s_{10} + \left(\frac{1}{4}\right)s_{11} = 0$$

(Again, notice that we cannot find s_{10}, s_{11} uniquely.) We take $s_{10} = 1$ and, consequently, find that $s_{11} = -1$. Thus we have

$$\mathbf{S} = \begin{pmatrix} 1 & \frac{1}{2} \\ 1 & -1 \end{pmatrix}, \qquad \mathbf{\Lambda} = \begin{pmatrix} 1 & 0 \\ 0 & \frac{1}{4} \end{pmatrix}$$

By inverting $\mathbf{S}$ we find directly that

$$\mathbf{S}^{-1} = \begin{pmatrix} \frac{2}{3} & \frac{1}{3} \\ \frac{2}{3} & -\frac{2}{3} \end{pmatrix}$$

It follows that $\mathbf{P}$ can be represented in the equivalent form

$$\mathbf{P} = \begin{pmatrix} \frac{2}{3} & \frac{1}{3} \\ \frac{2}{3} & -\frac{2}{3} \end{pmatrix} \begin{pmatrix} 1 & 0 \\ 0 & \frac{1}{4} \end{pmatrix} \begin{pmatrix} 1 & \frac{1}{2} \\ 1 & -1 \end{pmatrix}$$

$$= \mathbf{S}^{-1} \quad \cdot \quad \mathbf{\Lambda} \quad \cdot \quad \mathbf{S}$$

which is easily checked by direct multiplication.

It then follows from (9.4) that

$$\mathbf{P}^n = \begin{pmatrix} \frac{2}{3} & \frac{1}{3} \\ \frac{2}{3} & -\frac{2}{3} \end{pmatrix} \begin{pmatrix} 1 & 0 \\ 0 & \left(\frac{1}{4}\right)^n \end{pmatrix} \begin{pmatrix} 1 & \frac{1}{2} \\ 1 & -1 \end{pmatrix}$$

or

$$\mathbf{P}^n = \begin{pmatrix} \frac{2}{3} + \frac{1}{3}\left(\frac{1}{4}\right)^n, & \frac{1}{3} - \frac{1}{3}\left(\frac{1}{4}\right)^n \\ \frac{2}{3} - \frac{2}{3}\left(\frac{1}{4}\right)^n, & \frac{1}{3} + \frac{2}{3}\left(\frac{1}{4}\right)^n \end{pmatrix}$$

If we set $n = 3$, 5 or 10, the results may be checked against the values of $\mathbf{P}^3$, $\mathbf{P}^5$, and $\mathbf{P}^{10}$ that were calculated directly in Example 8.4.

If we let $n \to \infty$ in the above formula, the limiting matrix approached by $\mathbf{P}^n$ can be determined directly from this formula, since $(\frac{1}{4})^n \to 0$:

$$\lim_{n \to \infty} \mathbf{P}^n = \begin{pmatrix} \dfrac{2}{3} & \dfrac{1}{3} \\ \dfrac{2}{3} & \dfrac{1}{3} \end{pmatrix}$$

Thus the limiting matrix has both rows equal to the limiting probability vector $(\frac{2}{3}, \frac{1}{3})$, in agreement with the result that was obtained in a different way in Example 8.9. △

Even in the preceding simple example we can observe some interesting properties of the sequence of matrices $\mathbf{P}^n$ and the eigenvectors and eigenvalues. Notice the following facts in this example.

(i) Each row of $\mathbf{P}^n$ sums to 1 *regardless of n*. Since P^n is a stochastic matrix, of course this must be true.

(ii) Each entry of $\mathbf{P}^n$ involves a constant plus a term that depends on n. The constant term comes directly from the manipulations involving $\lambda_{00} = 1$.

(iii) As $n \to \infty$, each term approaches a limiting value (found in Chapter 8 by using other methods). The terms approach the limit geometrically fast. That is, they approach the limit as fast as $(\frac{1}{4})^n \to 0$.

(iv) The method outlined in this section allows us to determine both the limiting probabilities (v_k) (or the steady state solution), and the transient solution (the solution depending on n).

In component form, we observe that the n-step probabilities are

$$p_{00}{}^{(n)} = \frac{2}{3} + \frac{1}{3}\left(\frac{1}{4}\right)^n$$

$$p_{01}{}^{(n)} = \frac{1}{3} - \frac{1}{3}\left(\frac{1}{4}\right)^n$$

$$p_{10}{}^{(n)} = \frac{2}{3} - \frac{2}{3}\left(\frac{1}{4}\right)^n$$

$$p_{11}{}^{(n)} = \frac{1}{3} + \frac{2}{3}\left(\frac{1}{4}\right)^n$$

We also notice that if the 0th shot is equally likely to hit or to miss, that is, $\mathbf{p}^0 = (\frac{1}{2}, \frac{1}{2})$, then the nth shot probabilities are

$$p_0{}^{(n)} = \frac{2}{3} - \frac{1}{6}\left(\frac{1}{4}\right)^n$$

and

$$p_1{}^{(n)} = \frac{1}{3} + \frac{1}{6}\left(\frac{1}{4}\right)^n$$

These are the elements of the vector $\mathbf{p}^{(n)}$ that are given by

$$\mathbf{p}^{(n)} = \mathbf{p}^{(0)}\,\mathbf{P}^n$$

(v) In the example $\lambda_{00} = 1$ was an eigenvalue. This is always true. It can be shown that every matrix $\mathbf{P}$ of transition probabilities for a Markov chain has $\lambda = 1$ as one of its eigenvalues. In the case of a finite, irreducible, aperiodic Markov chain which we are considering, it can be shown that $|\lambda_{jj}| < 1$ for every other eigenvalue. Thus we can index the eigenvalues in descending order of absolute value:

$$\lambda_{00} = 1 > |\lambda_{11}| \geq |\lambda_{22}| \geq \cdots \geq |\lambda_{N-1,N-1}|$$

We can also show that, roughly, the limiting matrix of $\mathbf{P}^n$ is approached geometrically, at the same rate as $|\lambda_{11}|^n \to 0$ as $n \to \infty$. In the example $\lambda_{11} = \frac{1}{4}$, and we noted in (iii) that the limit was approached geometrically as fast as $(\frac{1}{4})^n \to 0$.

EXAMPLE 9.2. We can make our remarks about the behavior of Example 9.1 more precise for the general problem of chains with 2 states. Thus, suppose that we have a two-state chain with transition matrix

$$\mathbf{P} = \begin{pmatrix} 1-p & p \\ q & 1-q \end{pmatrix}$$

with $0 < p < 1$ and $0 < q < 1$. These conditions insure that the chain is irreducible. The characteristic equation is

$$(1 - p - \lambda)(1 - q - \lambda) - pq = 0$$

from which we obtain

$$\lambda^2 - (2 - p - q)\lambda + (1 - p - q) = 0$$

and

$$\lambda_{00} = 1, \qquad \lambda_{11} = 1 - p - q$$

By using equations (9.2) and (9.3), we have

$$(s_{00}, s_{01})\begin{pmatrix} 1-p & p \\ q & 1-q \end{pmatrix} = (s_{00}, s_{01})$$

and

$$(s_{10}, s_{11})\begin{pmatrix} 1-p & p \\ q & 1-q \end{pmatrix} = (1-p-q)(s_{10}, s_{11})$$

Thus

$$\begin{aligned} (1-p)s_{00} + q\, s_{01} &= s_{00} \\ p\, s_{00} + (1-q)s_{01} &= s_{01} \end{aligned}$$

and

$$\begin{aligned} (1-p)s_{10} + q\, s_{11} &= (1-p-q)s_{10} \\ p\, s_{10} + (1-q)s_{11} &= (1-p-q)s_{11} \end{aligned}$$

We observe that the first set of equations yields the equation

$$-p\, s_{00} + q\, s_{01} = 0$$

and the second set yields

$$q\, s_{10} + q\, s_{11} = 0$$

It follows that for this general problem we have

$$\left(\frac{p}{q}\right)s_{00} = s_{01} \qquad \text{for } \lambda_{00} = 1$$

and

$$s_{10} = -s_{11} \qquad \text{for } \lambda_{11} = 1-p-q$$

Thus, we have

$$\mathbf{S} = \begin{pmatrix} 1 & p/q \\ 1 & -1 \end{pmatrix}$$

In general $\mathbf{S}^{-1}$ can be found by using (9.3) or from $\mathbf{S}\,\mathbf{S}^{-1} = \mathbf{I}$. By either method, we easily find it to be

$$\mathbf{S}^{-1} = \begin{pmatrix} \dfrac{q}{p+q} & \dfrac{p}{p+q} \\[2ex] \dfrac{q}{p+q} & \dfrac{-q}{p^2+q} \end{pmatrix}$$

Then

$$\mathbf{P} = \mathbf{S}^{-1} \boldsymbol{\Lambda} \mathbf{S}$$

$$\mathbf{P} = \frac{1}{p+q} \begin{pmatrix} q & p \\ q & -q \end{pmatrix} \begin{pmatrix} 1 & 0 \\ 0 & 1-p-q \end{pmatrix} \begin{pmatrix} 1 & \dfrac{p}{q} \\ 1 & -1 \end{pmatrix}$$

From it, we obtain the elements of $\mathbf{P}^n$ as

$$p_{00}{}^{(n)} = \frac{1}{p+q} [q + p(1-p-q)^n] = \frac{1}{p+q} [q + p\, \lambda_{11}{}^n]$$

$$p_{01}{}^{(n)} = \frac{1}{p+q} [p - p(1-p-q)^n] = \frac{1}{p+q} [p - p\, \lambda_{11}{}^n]$$

$$p_{10}{}^{(n)} = \frac{1}{p+q} [q - q(1-p-q)^n] = \frac{1}{p+q} [q - q\, \lambda_{11}{}^n]$$

$$p_{11}{}^{(n)} = \frac{1}{p+q} [p + q(1-p-q)^n] = \frac{1}{p+q} [p + q\, \lambda_{11}{}^n]$$

Thus we observe that our remarks at the conclusion of Example 9.1 are true for the general two-state chain.

Furthermore

$$\lim_{n\to\infty} p_{00}{}^{(n)} = \lim_{n\to\infty} p_{10}{}^{(n)} = \frac{q}{p+q}$$

and

$$\lim_{n\to\infty} p_{01}{}^{(n)} = \lim_{n\to\infty} p_{11}{}^{(n)} = \frac{p}{p+q}$$

We notice that the terms converge to these limits as

$$\lambda_{11}{}^n = (1-p-q)^n \to 0 \qquad \triangle$$

4. Difference Equations

Frequently, the analysis of certain problems which concern Markov chains involves the determination of a set of probabilities $p_0, p_1, \ldots, p_N$ that satisfy a system of equations which might take the form

$$a\, p_{k+1} + b\, p_k + c\, p_{k-1} = 0 \qquad \text{for } 0 < k < N \tag{9.8}$$

and that are subject to certain constraints (*initial* and *terminal* conditions) which involve p_0 and p_N. Here a, b, and c represent certain real constants. Thus, for example, equation (7.21) in the Gambler's Ruin example of Chapter 7 is of the form (9.8).

In many cases, these equations can be solved by special tricks, or by iteration and induction, as in the Gambler's Ruin discussion. In other cases, a more general argument is necessary.

Equation (9.8) is termed a *second-order, linear, homogeneous difference equation with constant coefficients*, in the unknowns $\{p_k\}$. It is second-order because the subscripts $k+1$, k, $k-1$ are separated by two, it is linear with constant coefficients for obvious reasons, and it is homogeneous since the right side constant term is zero.

The study of such equations has close analogy with the study of linear differential equations. We can show that (9.8) always has a solution of the form $p_k = r^k$. If this is substituted into (9.8), we find that r must satisfy

$$a\,r^{k+1} + b\,r^k + c\,r^{k-1} = 0$$

or

$$a\,r^2 + b\,r + c = 0 \tag{9.9}$$

Equation (9.9) is termed the *secular equation* of the system. If r is a solution of (9.9), then $p_k = r^k$ is a solution of (9.8). To obtain the general solution, three cases must be considered:

(a) If (9.9) has two distinct real roots, r_1 and r_2, then the general solution of (9.8) is given by

$$p_k = A\,r_1{}^k + B\,r_2{}^k \tag{9.10}$$

where A and B are constants.

(b) If (9.9) has equal real roots, $r_1 = r_2$, then the general solution of (9.8) is given by

$$p_k = A\,r_1{}^k + B\,k\,r_1{}^k \tag{9.11}$$

where as before A and B are constants.

(c) If (9.9) has complex roots, which necessarily must be complex conjugates, $r_1 = r(\cos\theta + i\sin\theta)$, $r_2 = r(\cos\theta - i\sin\theta)$, then the general solution of (9.8) has the form:

$$p_k = r^k(A\cos k\theta + B\sin k\theta) \tag{9.12}$$

where again A and B are constants.

In each case, the constants A and B are determined by the initial and terminal conditions.

This discussion generalizes easily to linear difference equations of higher order than two.

5. Examples

EXAMPLE 9.3 In the Gambler's Ruin example (Section 6 of Chapter 7), equation (7.21), which can be written in the form

$$p\, p_{k+1} - p_k + q\, p_{k-1} = 0, \qquad 0 < k < N \tag{9.13}$$

was solved for the p_k's subject to the initial and the terminal conditions:

$$p_0 = 1, \qquad p_N = 0 \tag{9.14}$$

If we treat this as a linear difference equation of the form (9.8) with the constant coefficients p, -1, q (where $p + q = 1$), the secular equation is

$$pr^2 - r + q = 0$$

or

$$(r - 1)(pr - q) = 0$$

and the roots are $r_1 = 1$, $r_2 = q/p$.

If $p \neq q$, the roots are distinct and case (a) of this section applies. The general solution is given by (9.10)

$$p_k = A + B\left(\frac{q}{p}\right)^k, \qquad k = 0, 1, 2, \ldots, N \tag{9.15}$$

To determine A and B, the conditions (9.14) can be substituted into (9.15) yielding two equations:

$$A + B = 1, \qquad A + B\left(\frac{q}{p}\right)^N = 0$$

They are easily solved to give

$$A = \frac{-(q/p)^N}{1 - (q/p)^N}, \qquad B = \frac{1}{1 - (q/p)^N}$$

If these values are substituted into (9.15), the final solution is found to be

$$p_k = \frac{(q/p)^k - (q/p)^N}{1 - (q/p)^N}, \qquad k = 0, 1, \ldots, N$$

in agreement with (7.26).

If $p = q$, then $r_1 = r_2 = 1$, and case (b) applies. The general solution is given by (9.11) to be

$$p_k = A + Bk, \qquad k = 0, 1, \ldots, N$$

Again, by substituting into (9.14), we find that $A = 1$, $B = -1/N$, and

$$p_k = 1 - \frac{k}{N}, \qquad k = 0, 1, \ldots, N$$

in agreement with (7.27). △

EXAMPLE 9.4. We shall find later that a set of unknown terms $\{q_k\}$ are determined by a difference equation of the form

$$\mu q_{k+1} - (\lambda + \mu) q_k + \lambda q_{k-1} = 0, \qquad k > 0 \tag{9.16}$$

where λ and μ are positive constants. The initial condition that involves q_0 takes the form:

$$\mu q_1 - \lambda q_0 = 0 \tag{9.17}$$

The second condition, necessary to determine the constants involved, comes from the fact that, in the application we have in mind, the terms $\{q_k\}$ form a probability distribution and, hence, satisfy

$$\sum_k q_k = 1 \tag{9.18}$$

The secular equation here is

$$\begin{aligned} \mu r^2 - (\lambda + \mu) r + \lambda &= 0 \\ (r-1)(\mu r - \lambda) &= 0 \\ r_1 = 1, \qquad r_2 &= \frac{\lambda}{\mu} \end{aligned}$$

Thus, if $\lambda \neq \mu$, case (a) applies and the general solution is

$$q_k = A + B\left(\frac{\lambda}{\mu}\right)^k$$

By substituting this into (9.17), we have

$$\begin{aligned} \mu\left[A + B\left(\frac{\lambda}{\mu}\right)\right] - \lambda[A + B] &= 0 \\ (\mu - \lambda) A &= 0 \\ A &= 0 \end{aligned}$$

Thus the solution that satisfies (9.16) and (9.17) is

$$q_k = B(\lambda/\mu). \qquad k \geq 0 \tag{9.19}$$

We now wish to use (9.18) to determine B. However, two possibilities exist here. The range of k may be finite, $k = 0, 1, 2, \ldots, N$, or it may be infinite, $k = 0, 1, 2, \ldots$. Both these cases occur in various applications.

(i) Finite range, $k = 0, 1, 2, \ldots, N$.

In this case (9.18) becomes

$$\sum_{k=0}^{N} q_k = B \sum_{k=0}^{N} \left(\frac{\lambda}{\mu}\right)^k = 1$$

or

$$\frac{B[1 - (\lambda/\mu)^{N+1}]}{1 - (\lambda/\mu)} = 1$$

by using the formula for the sum of $N + 1$ terms of a geometric series. Thus

$$B = \frac{1 - (\lambda/\mu)}{1 - (\lambda/\mu)^{N+1}}$$

and

$$q_k = \frac{[1 - (\lambda/\mu)](\lambda/\mu)^k}{1 - (\lambda/\mu)^{N+1}}, \qquad k = 0, 1, 2, \ldots, N \tag{9.20}$$

gives the required solution. A random variable that has probability density $\{q_k\}$ given by (9.20) is said to have a *truncated geometric distribution.*

(ii) Infinite range, $k = 0, 1, 2, \ldots$

If $\lambda < \mu$, $\lambda/\mu < 1$, then the formula for the sum of a geometric progression can be applied when (9.19) is substituted into (9.18):

$$\sum_{k=0}^{\infty} q_k = B \sum_{k=0} \left(\frac{\lambda}{\mu}\right)^k = 1 \tag{9.21}$$

$$\frac{B}{1 - (\lambda/\mu)} = 1$$

$$B = 1 - \left(\frac{\lambda}{\mu}\right)$$

Thus the solution is

$$q_k = \left[1 - \left(\frac{\lambda}{\mu}\right)\right]\left(\frac{\lambda}{\mu}\right)^k, \qquad k = 0, 1, 2, \ldots$$

A random variable with this probability density is said to have a *modified geometric distribution*. Such random variables were discussed in Section 4 of Chapter 6.

If $\lambda > \mu$, $\lambda/\mu > 1$, the series in (9.21) *diverges*. Thus, (9.21) cannot hold true for any value of B, and it follows that no set of $\{q_k\}$ exists that satisfies (9.16) and (9.17), and that forms a probability distribution [i.e., satisfies (9.18)].

If $\lambda = \mu$, the reader can easily check that no solution exists in the infinite range case (ii). In the finite range case (i), if $\lambda = \mu$, the solution is

$$q_k = \frac{1}{n+1}, \qquad k = 0, 1, 2, \ldots, N \qquad \triangle$$

EXERCISES

1. By using the methods of this chapter, determine the eigenvalues, eigenvector, and $\mathbf{P}^n$ for the transition matrix given by

$$\mathbf{P} = \begin{pmatrix} 1/2 & 1/2 \\ 1 & 0 \end{pmatrix}$$

2. From Exercise 1, determine $\lim_{n\to\infty} \mathbf{P}^n$ and find the number of steps necessary for $\mathbf{P}^n$ to be within 1% of its limiting value. Compute $\mathbf{P}^n$ for $n = 1, 2, \ldots 10$ for each i, j.

3. By using the methods of this chapter, determine the eigenvalues, eigenvectors, and $\mathbf{P}^n$ for each of the following transition matrices:

$$\text{(a)}\quad \mathbf{P} = \begin{pmatrix} 1/2 & 1/4 & 1/4 \\ 1/4 & 1/2 & 1/4 \\ 1/4 & 1/4 & 1/2 \end{pmatrix}; \qquad \text{(b)}\quad \mathbf{P} = \begin{pmatrix} 1/2 & 1/2 & 0 \\ 3/8 & 1/2 & 1/8 \\ 1/4 & 1/2 & 1/4 \end{pmatrix}$$

4. For Exercise 3, find $\lim_{n\to\infty} \mathbf{P}^n$ and find the number of steps necessary for $\mathbf{P}^n$ to be within 1% of its limiting value.

5. For the matrices in Exercises 1 and 3, show that $\mathbf{P}^n$ is a stochastic matrix for every $n > 0$.

10 • Nonirreducible and Nonfinite Markov Chains

1. Introduction

In Chapter 8 we were mainly concerned with finite, irreducible, Markov chains. Nonetheless, chains that do not meet these conditions do come up in applications fairly frequently. For instance, the industrial inspection process in Example 8.2 and the Gambler's Ruin example discussed in Chapter 7 were nonirreducible processes since each had absorbing states. The version of the Gambler's Ruin problem in which play is against an opponent with unlimited cash has an infinite state space, as well as being nonirreducible. The random walk discussed in Chapter 7 has an infinite state space but is irreducible.

In this chapter we will briefly discuss discrete parameter Markov chains that are nonirreducible or nonfinite or both. The student is referred to a more extensive treatise, such as the one by Karlin (see Bibliography) for a more complete discussion.

2. Nonirreducible Markov Chains

Let us recall again the Gamber's Ruin example. The state space contains $N + 1$ points: 0, 1, 2, ..., N, that represent the gambler's possible capital after each play. These states are of two essentially different types: The states 0 and N are *absorbing states*, that is, states which cannot be left once they are entered. The states 1, 2, ..., $N - 1$ are *transient states*, that is, states that are not certain to be returned to, even if the process starts in the state.

This division of the states into two such categories can be extended to any finite, nonirreducible Markov chain. It is necessary, however, to consider the possibility that an absorbing "state" might actually be a *class* of several states with the property that transitions are possible between states within the class, but not from states inside the class to states outside. Such a class is termed a *closed communicating class*. The formal definition is as follows:

DEFINITION. *A* CLOSED COMMUNICATING CLASS *of a discrete parameter Markov chain is a subset C of the state space with the following properties:*

(i) *For each pair of states i and j in C, an eventual transition from state i to state j is possible; that is*

$$p_{ij}^{(n)} > 0$$

for some integer $n \geq 1$.

(ii) *For each state i in C and each state j not in C,*

$$p_{ij}^{(n)} = 0$$

for every integer $n \geq 1$.

Notice that, although a closed communicating class cannot be left, it can possibly be entered by a transition from a state outside the class to one inside the class.

Observe also that any two distinct closed communicating classes C and C' are mutually exclusive, since if $C \cap C'$ contains a state i, eventual transition is possible from any state in C to any state in C' by proceeding first to the state i; hence C and C' must be the same closed communicating class.

DEFINITION. *An* ABSORBING STATE *is a state that by itself forms a closed communicating class.*

DEFINITION. *In a finite, nonirreducible Markov chain, any state that does not belong to a closed communicating class is termed a* TRANSIENT STATE.*

Clearly an irreducible finite Markov chain can have no transient states since the class of *all* states forms a closed communicating class.

Since a process can move from a transient state into a closed communicating class, but not vice versa, the following theorem seems reasonable. Its proof, although comparatively simple, is omitted.

THEOREM. *If a finite Markov chain starts in a transient state* (X_0 *is transient*), *then the chain is certain* (*with probability* 1) *eventually to enter some closed communicating class.* △

The situation can be summed up as follows.

The state space of any finite Markov chain can be subdivided into mutually exclusive classes $C_1, C_2, \ldots, C_1, T$, where $C_1, C_2, \ldots, C_1$ are the closed communicating classes, and T contains all the transient states. Once the process enters

* *Note.* This definition only applies to a *finite* Markov chain. For an infinite chain a more general definition of the term "transient state" is needed. For a discussion, see the treatise by Karlin listed in the Bibliography.

any C_k, it remains there, and within each C_k it acts like an irreducible process, to which the discussion of Chapter 8 applies. Thus, its behavior within each closed communicating class can be analyzed separately (and, of course, if the class consists of just one absorbing state, this analysis is trivial). If the process starts in T, then eventually it must enter one of the C_k's. To analyze the long-run behavior of the process, we must compute the probabilities t_i that a process starting in the transient state i eventually enters the closed communicating class C_k, for each such C_k:

$$t_i = \Pr[X_n \text{ in } C_k \text{ eventually } | X_0 = i] \tag{10.1}$$

for each transient state.

To determine the quantities t_i, let us define new probabilities s_i which represent the probability of a transition from the transient state i into the closed communicating class C_k *in one step*:

$$\begin{aligned} s_i &= \Pr[X_{n+1} \text{ in } C_k | X_n = i] \\ &= \sum_{j \in C_k} p_{ij} \end{aligned} \tag{10.2}$$

where the summation is over all indices j in C_k.

Now, if the chain eventually proceeds from the transient state i into the closed communicating class C_k, it must do so in one of two ways: Either it makes this transition in one step, or it goes to another transient state in its first step and from there eventually proceeds to C_k. Thus we are led to the following system of equations:

$$t_i = s_i + \sum_{j \in T} p_{ij} t_j \tag{10.3}$$

for each transient state i. The summation is here over all transient states j.

The system of linear equations (10.3) can be solved for the probabilities t_i. If the system contains more than one closed communicating class, a similar system can be set up and solved for each class.

3. Examples

EXAMPLE 10.1. The Gambler's Ruin example, (Example 8.1 and Section 6 of Chapter 7) is a nonirreducible process with state space 0, 1, 2, ..., N that contains two closed communicating classes each of which consists of only a single absorbing state, $\{0\}$ and $\{N\}$. Thus the transient states are $T = \{1, 2, \ldots, N-1\}$ and, by the theorem of the preceding section, if the system starts in one of these states, it eventually must pass into state 0 (player ruined) or into state N (opponent ruined).

Let us fix our attention on the "ruin" absorbing state $\{0\}$. The probabilities p_{ij} are given by $p_{i\,i+1} = p$ if $i = 1, 2, \ldots, N-1$; $p_{i\,i-1} = q$ if $i = 1, 2, \ldots, N-1$; and $p_{ij} = 0$ otherwise. The probability s_i is the probability of ruin (entering state 0) in one step from the state i. Thus, $s_1 = q$, $s_i = 0$ for $i = 2, 3, \ldots, N-1$. Hence, the system (10.3) becomes

$$\begin{aligned} t_1 &= q + p\,t_2 \\ t_i &= 0 + q\,t_{i-1} + p\,t_{i+1}, \qquad \text{for} \qquad i = 2, 3, \ldots, N-2 \\ t_{N-1} &= 0 + q\,t_{N-2} \end{aligned}$$

We easily observe that this system is equivalent to the system (7.21). [The system (7.21) appears somewhat simpler, since it uses the simplifying notation $t_0 = 1$, $t_N = 0$, which can be interpreted to mean that ruin is certain starting from state 0 and that ruin is impossible starting from state N. With this notation, the system can be written with the single formula (7.21) instead of by using three formulas.]

The system is solved explicitly in Chapter 7, with the solution given by (7.26) and (7.27). △

EXAMPLE 10.2. A certain piece of electronic equipment can be in any one of 3 states: *Good condition* (state 0), *marginal condition* (state 1), and *burned out* (state 2). At the end of each day of operating and servicing, the condition of the system is recorded. A discrete parameter random process $X_0, X_1, X_2, \ldots$ results, with state space $\{0, 1, 2\}$. Assume that this system can be approximated by a Markov chain with transition matrix:

$$\mathbf{P} = \begin{pmatrix} .7 & .2 & .1 \\ .7 & .3 & .2 \\ 0 & 0 & 1 \end{pmatrix}$$

This is a nonirreducible process with only one absorbing state, state 2, and two transient states, states 0 and 1. By the theorem of the preceding section, the process is certain eventually to enter state 2 (it will eventually burn out). Thus, $t_0 = t_1 = 1$. A more interesting problem is to compute the probability $t_i^{(n)}$ that the equipment will burn out on the nth day, for $n = 1, 2, \ldots$, given that it starts in state i ($i = 0$ or 1):

$$t_i^{(n)} = \Pr[X_n = 2, X_j = 0 \text{ or } 1 \quad \text{for} \quad j < n \mid X_0 = i]$$

We see immediately that $t_0^{(1)} = p_{02} = .1$ and $t_1^{(1)} = p_{12} = .2$.

For $n > 1$, the first transition must be to state 0 or 1, and from there to the *burned out* state 2 in $n - 1$ steps. Thus

$$\begin{aligned} t_0{}^{(n)} &= p_{00}\, t_0{}^{(n-1)} + p_{01}\, t_1{}^{(n-1)} \\ t_1{}^{(n)} &= p_{10}\, t_0{}^{(n-1)} + p_{11}\, t_1{}^{(n-1)} \end{aligned} \tag{10.4}$$

or

$$\begin{aligned} \mathbf{t}^{(n)} = \begin{pmatrix} t_0{}^{(n)} \\ t_1{}^{(n)} \end{pmatrix} &= \begin{pmatrix} p_{00} & p_{01} \\ p_{10} & p_{11} \end{pmatrix} \begin{pmatrix} t_0{}^{(n-1)} \\ t_1{}^{(n-1)} \end{pmatrix} \\ &= \mathbf{A}\, \mathbf{t}^{(n-1)} \end{aligned} \tag{10.5}$$

where $\mathbf{A} = \begin{pmatrix} p_{00} & p_{01} \\ p_{10} & p_{11} \end{pmatrix}$ is the submatrix of $\mathbf{P}$ that involves only the transitions between transient states. Thus, by iteration,

$$\mathbf{t}^{(n)} = \mathbf{A}^{n-1}\, \mathbf{t}^{(1)} \tag{10.6}$$

where the powers of $\mathbf{A}$ can either be computed directly or by using the methods of Chapter 9.

In our example

$$\mathbf{A} = \begin{pmatrix} .7 & .2 \\ .5 & .3 \end{pmatrix} \quad \text{and } \mathbf{t}^{(1)} = \begin{pmatrix} .1 \\ .2 \end{pmatrix}$$

Thus, for instance,

$$\mathbf{t}^{(3)} = \begin{pmatrix} .7 & .2 \\ .5 & .3 \end{pmatrix}^2 \begin{pmatrix} .1 \\ .2 \end{pmatrix} = \begin{pmatrix} .59 & .20 \\ .50 & .19 \end{pmatrix} \begin{pmatrix} .1 \\ .2 \end{pmatrix} = \begin{pmatrix} .099 \\ .088 \end{pmatrix}$$

The probability that a piece of equipment originally in good condition burns out on the third day is $t_0{}^{(3)} = .099$. △

Equations (10.5) and (10.6) generalize immediately to any nonirreducible Markov chain and give the probabilities $t_i{}^{(n)}$ of transition into a given closed communicating class on the nth step, starting in the transient state i.

4. Infinite Irreducible Markov Chains

If $\{X_0, X_1, \ldots\}$ is an irreducible, aperiodic Markov chain with infinite state space $\{0, 1, 2, \ldots\}$, then the discussion in Section 8 of Chapter 8 must be modified. In Section 8, we indicated that, for a *finite* Markov chain, the transition probabilities $p_{ij}{}^{(n)}$ always approach limits v_j as $n \to \infty$, which form a probability vector. In the infinite state case, it might happen that $X_n \to \infty$ as $n \to \infty$, with probability 1, in which case $p_{ij}^{(n)} \to 0$ for every fixed i and j.

Theorems 1 and 2 of Section 8, Chapter 8, apply equally well to the infinite state case. Thus, in an aperiodic, irreducible infinite Markov chain, we know that the limits

$$v_j = \lim_{n \to \infty} p_{ij}^{(n)}$$

all exist, and do not depend on the initial state i. In the infinite state case the possibility remains, however, that these limits might be 0.

DEFINITION. *An irreducible Markov chain is termed* POSITIVE RECURRENT *if*

(a) $v_j = \lim_{n \to \infty} p_{ij}^{(n)} > 0$ *for every* j; *and*

(b) $\sum_j v_j = 1$

(*That is, the vector* $\mathbf{v} = (v_0, v_1, \ldots)$ *forms a probability vector with each entry nonzero.*)

In Section 9, Chapter 8, a probability vector $\mathbf{x} = (x_0, x_1, \ldots)$ was termed a *stationary* vector if it satisfied identity (8.28):

$$x_j = \sum_k x_k p_{kj} \qquad \text{or} \qquad \mathbf{x} = \mathbf{xP} \tag{10.7}$$

Theorem 3 in that section can be modified for infinite state chains as follows.

THEOREM

(a) *An irreducible, aperiodic Markov chain is positive recurrent if, and only if, there exists a unique stationary probability vector* $\mathbf{x}$; *that is, if there exists a unique solution to the identities* (10.7) *that satisfies* $x_j \geq 0$, $\sum x_j = 1$. *In this case the components of* $\mathbf{x}$ *give the limiting probabilities,* $v_j = x_j$ *for each* j.

(b) *If an irreducible, aperiodic Markov chain is not positive recurrent, then*

$$v_j = \lim_{n \to \infty} p_{ij}^{(n)} = 0$$

for every j. △

EXAMPLE 10.3. In Example 8.6, we considered a waiting line process in which X_n represented the number of individuals waiting in line at the time when the nth individual completed his service. The transition matrix had the form:

$$\mathbf{P} = \begin{pmatrix} p_{00} & p_{01} & p_{02} & \cdots \\ p_{10} & p_{11} & p_{12} & \cdots \\ 0 & p_{21} & p_{22} & \cdots \\ 0 & 0 & p_{32} & \cdots \\ \cdot & \cdot & & \\ \cdot & \cdot & \cdot & \\ \cdot & \cdot & \cdot & \end{pmatrix}$$

The zero terms below the dotted line are indicative of the fact that during a single service operation the length of the waiting line cannot decrease by more than one. Here $\{X_n\}$ is an irreducible process with possible states $0, 1, 2, \ldots$. The study of these processes is included in the extensive and useful branch of applied probability termed queueing theory (see Chapter 13).

In one important special case the terms in the matrix $\mathbf{P}$ have the following form:

$$\mathbf{P} = \begin{pmatrix} 1-t & (1-t)t & (1-t)t^2 & \cdots \\ 1-t & (1-t)t & (1-t)t^2 & \cdots \\ 0 & 1-t & (1-t)t & \cdots \\ 0 & 0 & 1-t & \cdots \\ \cdot & \cdot & \cdot & \\ \cdot & \cdot & \cdot & \\ \cdot & \cdot & \cdot & \end{pmatrix} \tag{10.8}$$

or $p_{ij} = (1-t)t^{j-i+1}$ for $i = 1, 2, \ldots$ and $j = i-1, i, i+1, \ldots$; $p_{0j} = (1-t)t^j$ for $j = 0, 1, 2, \ldots$; and $p_{ij} = 0$ otherwise. Here t, $0 < t < 1$, represents a parameter that is related to the relative speed of the server compared to the rate of arrivals to the system.

Direct calculation of the n-step transition probabilities is difficult. However, we can apply our theorem and attempt to solve the system (10.7), which becomes here:

$$\begin{aligned} x_0 &= (1-t)x_0 + (1-t)x_1 \\ x_1 &= (1-t)tx_0 + (1-t)tx_1 + (1-t)x_2 \\ &\cdot \\ &\cdot \\ &\cdot \\ x_j &= (1-t)t^jx_0 + (1-t)\sum_{i=1}^{j+1} t^{j+1-i}x_i, \qquad \text{for} \qquad j = 1, 2, 3, \ldots \end{aligned} \tag{10.9}$$

This system can be solved recursively in terms of x_0 to give

$$x_1 = \left(\frac{t}{1-t}\right) x_0$$

$$x_2 = \left(\frac{t}{1-t}\right)^2 x_0$$

$$\vdots$$

$$x_j = \left(\frac{t}{1-t}\right)^j x_0, \qquad \text{for} \qquad j = 0, 1, 2, \ldots$$

(The student should check this.) In order that $\sum x_j = 1$ we must have

$$x_0 \sum_{j=0}^{\infty} \left(\frac{t}{1-t}\right)^j = 1 \tag{10.10}$$

The condition that the infinite series in (10.10) converge is that $t/(1-t) < 1$, which is equivalent to $t < 1 - t$ or $t < \frac{1}{2}$. If this holds true, we have

$$x_0 \cdot \frac{1}{1 - \left(\dfrac{1}{1-t}\right)} = 1$$

$$x_0 = \frac{1-2t}{1-t}$$

$$x_j = \frac{1-2t}{1-t}\left(\frac{t}{1-t}\right)^j, \qquad \text{for} \qquad j = 0, 1, 2, \ldots \tag{10.11}$$

Thus, by our theorem, if $t < \frac{1}{2}$, the system is positive recurrent and the limiting probabilities v_j are given by (10.11). Formula (10.11) gives an approximation to the probability that j individuals are waiting in line after the process has been going on for a long time.

If $t \geq \frac{1}{2}$, then $t/(1-t) \geq 1$ and the series in (10.10) *diverges*. Thus, no solution to the system (10.9) exists that forms a probability vector. It follows from our theorem that the system is not positive recurrent and that $v_j = \lim_{n\to\infty} p_{ij}^{(n)} = 0$ for all j. The waiting line will tend to grow indefinitely here, and the probability of there being any fixed number j in line will approach zero. △

EXAMPLE 10.4. The random walk discussed in Section 5 of Chapter 7 is a periodic process with period 2. Let us modify this process to admit the possibility of a 0-transition (no change). Assume that the transition probabilities are given by

$$p_{i,i+1} = p, \qquad p_{i,i-1} = q, \qquad p_{i,i} = r, \qquad p_{ij} = 0 \text{ otherwise}$$

for $i, j = 0, \pm 1, \pm 2, \ldots,$ where $p + q + r = 1$. If p, q and r are all nonzero, the process is an irreducible aperiodic Markov chain, with transition matrix:

$$\mathbf{P} = \begin{pmatrix} \ddots & & & \vdots & & & \\ q & r & p & 0 & 0 & 0 & 0 \\ 0 & q & r & p & 0 & 0 & 0 \\ \cdots\ 0 & 0 & q & r & p & 0 & 0\ \cdots \\ 0 & 0 & 0 & q & r & p & 0 \\ 0 & 0 & 0 & 0 & q & r & p \\ & & & \vdots & & & \ddots \end{pmatrix}$$

The results of this section apply, since the state space is countably infinite. although it is indexed by *all* integers $\ldots -2, -1, 0, 1, 2, \ldots$ rather than by only the nonnegative integers $0, 1, 2, \ldots$.

The explicit calculation of the n-step transition probabilities $p_{ij}^{(n)}$ is somewhat complicated. Nonetheless, we can attempt to find the limiting probabilities by solving the system (10.7), which becomes

$$x_j = px_{j-1} + rx_j + qx_{j+1}$$

for $j = 0, \pm 1, \pm 2, \ldots$. Since $r = 1 - p - q$, this system can be written as

$$qx_{j+1} - (p + q)x_j + p_{j-1} = 0 \tag{10.12}$$

for $j = 0, \pm 1, \pm 2, \ldots$. We can solve this difference equation by using the methods of Chapter 9, Section 3. The secular equation is

$$qr^2 - (p + q)r + p = 0$$

which has roots $r = 1$ and p/q. The general solution then is $x_j = A + B(p/q)^j$ if $p \neq q$, ($x_j = A + Bj$ if $p = q$). In order that the solution vector $\mathbf{x}$ be a probability vector, we need $\sum x_j = 1$. However, since the range of the index runs from $-\infty$ to $+\infty$, the series $\sum_{j=-\infty}^{\infty} [A + B(p/q)^j]$ will *diverge* for all p and q, unless $A = B = 0$ in which case $x_j = 0$ for every j.

[If $p = q$, $\sum(A + Bj)$ also diverges.] Thus, in every case, we observe that no solution to the system (10.7) exists which forms a probability vector. By our theorem, we conclude that the system is not positive recurrent, and that $v_j = \lim_{n\to\infty} p_{ij}^{(n)} = 0$ for every j. Notice that the system is not positive recurrent even in the case of a "symmetric" random walk with $p = q$. △

5. Infinite Nonirreducible Chains

The only example of a infinite state, nonirreducible Markov chain that we have discussed thus far is the Gambler's Ruin problem, with play against an infinitely rich opponent (Section 6 of Chapter 7). The process was not irreducible since 0 is an absorbing state. In Chapter 7, the process was analyzed by first placing a bound on the opponent's capital, by studying the resulting finite-state process, and then by using a limiting argument to pass to the infinite state case. The question of the legality of such an argument was ignored. Actually, the argument can be justified rigorously or, alternatively, the infinite-state process can be analyzed directly.

When we studied finite nonirreducible chains, we found it convenient to divide the state space into two categories: the closed communicating classes, and the transient states. In the infinite state situation, a similar division into two categories is also used. However, greater care must be taken in giving definitions, since a process can possibly remain in an infinite class for an infinite number of steps without leaving the class and without necessarily returning to its starting point. The detailed analysis of these processes constitutes one of the most elegant theories in the study of probability. The reader should again refer to the work by Karlin for an exposition of this theory.

EXERCISES

1. Give a precise answer to Exercise 8 of Chapter 8 by using the theorem of Section 2.
2. Consider the Markov chain of Exercise 4(b) in Chapter 8.
   - **(a)** Show that $\{2, 4\}$ is a closed communicating class and that $\{0, 1, 3\}$ are transient states.
   - **(b)** Show that $\lim_{n\to\infty} p_{ij}^{(n)} = v_j$ exists for each j, and does not depend on i, although the chain is not irreducible.
   - **(c)** Compute v_j for $j = 0, 1, 2, 3, 4$.

3. A missile is launched and, as it is tracked, a sequence of course-correction signals are sent to it. Suppose that the system has four states that are labeled as follows:

State 0: *on-course*, no further correction necessary
State 1: *minor deviation*
State 2: *major deviation*
State 3: *abort*, off-course so badly a self-destruct signal is sent

Let X_n represent the state of the system after the nth course correction, and assume that the behavior of $\{X_n\}$ can be modeled by a Markov chain that has the following 1-step transition matrix:

$$\mathbf{P} = \begin{pmatrix} 1 & 0 & 0 & 0 \\ 1/2 & 1/4 & 1/4 & 0 \\ 0 & 1/2 & 1/4 & 1/4 \\ 0 & 0 & 0 & 1 \end{pmatrix}$$

(a) Show that states 0 and 1 are absorbing states, and that states 1 and 2 are transient states.

(b) If the missile starts initially in state 1, compute the probability that it eventually gets on-course.

(c) If the missile has a 50-50 chance of starting in either state 1 or state 2, compute the probability that it eventually gets on-course.

(d) If the missile is equally likely to start in any of the states 0, 1, 2, or 3, compute the probability that it eventually gets on-course.

4. A Markov chain has state space $\{0, 1, 2, 3, 4\}$ and 1-step transition matrix:

$$\mathbf{P} = \begin{pmatrix} 1/2 & 1/2 & 0 & 0 & 0 \\ 1/3 & 2/3 & 0 & 0 & 0 \\ 1/4 & 0 & 1/4 & 1/4 & 1/4 \\ 0 & 1/4 & 1/4 & 1/4 & 1/4 \\ 0 & 0 & 0 & 0 & 1 \end{pmatrix}$$

(a) Determine the closed communicating classes and the transient states.

(b) For each closed communicating class, find the probabilities t_i that the chain will enter that class, given that it starts in the transient state i.

5. In Exercise 2, compute the probabilities $t_i^{(3)}$ that the chain will enter the closed communicating class $\{2, 4\}$ for the first time on the third step, starting in states $i = 0, 1, 3$.
(*Note:* Apply (10.6) by using the appropriate submatrix of $\mathbf{P}$ as the matrix $\mathbf{A}$.)

6. In Exercise 4 compute the probability $t_4^{(3)}$ of entering the closed communicating class $\{0, 1\}$ on the third step, starting in state 3.

7. In Exercise 3 compute the probability that the second course correction puts the missile on course, by assuming initially that it has a 50-50 chance of starting in state 1 or state 2.

8. Exercise 2 of Chapter 8 defines a Gambler's Ruin problem with state space $\{0, 1, 2, \ldots, N\}$ and transition matrix:

$$\mathbf{P} = \begin{pmatrix} 1 & 0 & 0 & 0 & 0 & & \cdots & \cdots \\ 2/3 & 0 & 0 & 1/3 & 0 & & \cdots & \cdots \\ 0 & 2/3 & 0 & 0 & 1/3 & & \cdots & \cdots \\ & & & \ddots & & & & \\ & & & & & \ddots & & \\ 0 & 0 & 0 & 0 & 0 & \cdots & 1 & 0 \\ 0 & 0 & 0 & 0 & 0 & \cdots & 0 & 1 \end{pmatrix}$$

This system has 3 absorbing states: $0, N-1, N$.

(a) Set up the system (10.3) for the probabilities t_i of ruin (entering state 0), by starting from state i.

(b) Solve the system in (a) by using the methods of Section 3, Chapter 9, and show that the solution is

$$t_i = 1 - \frac{3i + (-2)^{1-N}[(-2)^i - 1]}{3N - 2 - (-2)^{1-N}}$$

for $i = 1, 2, \ldots, N-2$.

(c) Discuss the behavior of the solution in (b) as $N \to \infty$.

9. Consider a random walk with a reflecting barrier at $-\frac{1}{2}$. This means that from state 0 the process either proceeds to state 1 (probability p) or bounces from $-\frac{1}{2}$ and ends up by staying at 0 (probability q).

(a) Set up the 1-step transition matrix for this process.

(b) Show that this yields an irreducible aperiodic Markov chain.

(c) Set up the system of equations (10.7) for a stationary vector for the system.

(d) Show that the solution to the system in (c) is a geometric sequence.

(e) Show that the system is positive recurrent if, and only if, $p < q$, and compute the limiting probabilities in this case.

10. Consider the irreducible random walk discussed in Section 5 of Chapter 7

(a) By using formula (7.17) show that

$$p_{ij}^{(n)} = \binom{n}{\frac{n+j-i}{2}} (pq)^{n/2} \left(\frac{p}{q}\right)^{(j-i)/2}$$

whenever $(n+j-i)/2$ is an integer between 0 and n.

(b) Prove that $pq \leq 1/4$ for $0 < p < 1$, $q = 1 - p$.
[*Hint:* Maximize $pq = p(1-p)$ for $0 < p < 1$.]

(c) Show that

$$p_{ij}^{(2n)} \leq \binom{2n}{\frac{2n+j-i}{2}} \frac{1}{2^{2n}} \left(\frac{p}{q}\right)^{(j-i)/2}$$

(d) Show that

$$p_{ij}^{(2n)} \leq \binom{2n}{n} \frac{1}{2^{2n}} \left(\frac{p}{q}\right)^{(j-i)/2}$$

[*Hint:* The middle coefficient in the binomial expansion of $(1+x)^{2n}$ is the largest.]

(e) By using the results of Exercise 15(b) of Chapter 2, show that

$$v_j = \lim_{n \to \infty} p_{ij}^{(2n)} = 0 \qquad \text{for every } i \text{ and } j$$

(f) Conclude that this process has no positive recurrent states.

11 • Continuous Parameter Markov Chains

1. Introduction

We have observed, by examples, that many real-world processes can be modeled by random processes which have the Markov property. In the preceding three chapters our attention was focused on *discrete parameter* situations, in which we dealt with Markovian *sequences* of random variables. In this chapter we discuss processes proceeding in continuous time, for instance, the Poisson process (see Chapter 7). This means that, instead of studying a *sequence* $\{X_n\}$ of random variables, we study a random variable X_t that depends on a parameter t, which can take a continuous range of values (usually all positive real numbers), not just integers. Thus, we are concerned with a *family* of random variables of the form $\{X_t: 0 \leq t < \infty\}$ where t is usually interpreted as *time*.

A random variable is itself a function of sample points and, thus, we say, more precisely, that a continuous parameter random process is a function, $X_t(s)$, of *two* variables, t and s, where t is a continuous time variable and s represents a sample point in some fixed sample space S. This sample space is part of a probability system in which all the statements that we wish to make about the process define *events* and, thus, have probabilities. In this treatment we will ignore the delicate mathematical aspects of this last statement.

As in our discrete-parameter discussion we will confine attention here to *discrete state* processes. This means that, although the parameter t has a continuous range of values, the values (*states*) of X_t itself always lie in some discrete set termed the *state space*. To fix our notation, we shall usually assume that the state space is a set of integers, finite or infinite, of the form $\{0, 1, 2, \ldots, N\}$ or $\{0, 1, 2, \ldots\}$. For example, in the Poisson process example of Section 7, Chapter 7, if X_t represents the number of telephone calls up to time t then, clearly, t can be any positive number, but X_t must be an integer.

A single experiment now results not in a single number, but in an entire *function* X_t. Thus, one experimental result or single sample point defines X_t

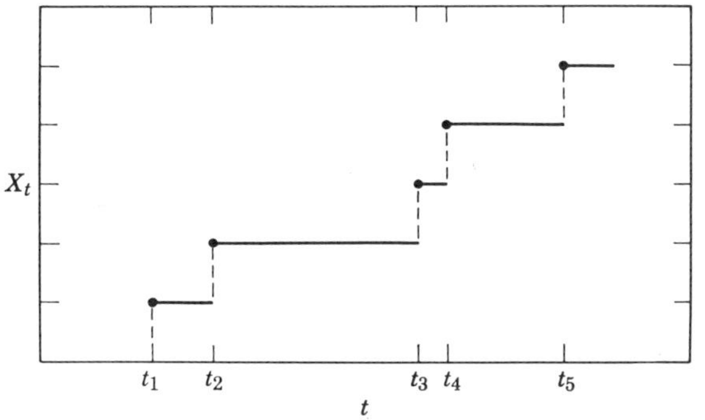

Figure 11.1 One realization of a Poisson process X_t = number of calls up to time t.

for all $t \geq 0$. In the Poisson process example each observation consists of an entire sequence of calls that defines the corresponding counts X_t for all $t \geq 0$. Thus, each sample point will yield a function X_t—a *realization* of the process—which will have a form like the one shown in Figure 11.1. It will take constant integer values between abrupt transitions from state to state. In the Poisson process example these transitions will occur at the times of incoming calls, and X_t will proceed successively through the integer values 0, 1, 2, ...

2. The Markov Property

In previous chapters we discussed the Markov property for certain random sequences (discrete parameter random processes). We found that many real situations can be modeled by these processes, and that the "forgetfulness" property built intc a Markovian model permits fairly simple and complete mathematical analysis. This property can be extended readily to apply to continuous parameter processes.

For our purposes we make the following somewhat imprecise definition.

DEFINITION. X_t *is a continuous-parameter, discrete-state* MARKOV PROCESS (*or* MARKOV CHAIN) *if it has the following Markovian property.*

For any fixed time τ *and state* x, *if* A *is an event that depends only on the values of* X_t *for* $t > \tau$, *and* B *is an event that depends only on the values of* X_t *for* $t < \tau$ *then, conditional on the fact that* $X_\tau = x$, A *and* B *are independent: in symbols*

$$\Pr[A \mid B \text{ and } X_\tau = x] = \Pr[A \mid X_\tau = x] \qquad \triangle \quad (11.1)$$

In words, this states that if the value of X_t is observed at any fixed time τ, then any event A, depending only on the "future" of the process, will be independent of any event B, depending only on the "past" of the process. As far as predicting the chance of A's occurring, the event B can be "forgotten." All information necessary to a determination of the probability of A is given by specifying the "present" state of the process, $X_\tau = x$, and this probability is unaffected by giving any further information that concerns the "past." Loosely speaking, knowledge of the "present" state of the process makes its "future" independent of its "past."

Given three times t', τ, t'' with $t' < \tau < t''$, and three states (integers) x', x, x'', a special case of formula (11.1) is obtained by specializing the events A and B as follows: $A = [X_{t''} = x'']$ and $B = [X_{t'} = x']$. Then the Markov property (11.1) yields

$$\Pr[X_{t''} = x'' \mid X_{t'} = x' \text{ and } X_\tau = x] = \Pr[X_{t''} = x'' \mid X_\tau = x] \tag{11.2}$$

$t' < \tau < t''$.

The probability on the right side of (11.2) has the form

$$\Pr[X_{\tau+t} = j \mid X_\tau = i] \tag{11.3}$$

where i and j are states (integers) and τ and t are times, $\tau \geq 0, t \geq 0$. This probability is termed a *transition probability*, and (11.3) is said to be the *probability of a transition from state i to state j during the time interval from time τ to time $\tau + t$.*

In many important applications, the probability (11.3) does not depend on the *initial time* τ, but only on the *elapsed time* t, and on the *initial* and the *terminal* states i and j. In this case, the process is termed *time homogeneous*. Henceforth, we confine our attention to these time-homogeneous processes. The ideas that are involved are readily generalized to the nonhomogeneous case, if necessary.

We now assume that X_t is a time-homogeneous Markov process and introduce the following more compact notation for the transition probability (11.3):

$$p_{ij}(t) = \Pr[X_{\tau+t} = j \mid X_\tau = i] \tag{11.4}$$

for $i, j = 0, 1, 2, \ldots$ and $t \geq 0$ (and for any initial time τ for which $\Pr[X_\tau = i] \neq 0$).

Let us also introduce the notation

$$p_j(t) = \Pr[X_t = j] \tag{11.5}$$

$j = 0, 1, 2, \ldots$ and $t \geq 0$, for the *state probabilities* at time t.

Clearly, the functions $p_{ij}(t)$ and $p_j(t)$ must satisfy

$$\sum_j p_{ij}(t) = 1 \tag{11.6}$$

for each i and t, and

$$\sum_j p_j(t) = 1 \tag{11.7}$$

for each $t \geq 0$, since at any given time the process must be in *some* state.

By using the Markov properties (11.1) and (11.2) and the basic formulas (2.10) and (2.13) for conditional probabilities, any desired individual or joint probability of the process can be expressed in terms of the transition probabilities $p_{ij}(t)$ and in terms of the *initial probabilities*:

$$p_i = p_i(0) = \Pr[X_0 = i]$$

For example

$$\begin{aligned} p_j(t) = \Pr[X_t = j] &= \sum_i \Pr[X_t = j \text{ and } X_0 = i] \\ &= \sum_i \Pr[X_0 = i] \Pr[X_t = j \mid X_0 = i] \\ &= \sum_i p_i p_{ij}(t) \end{aligned} \tag{11.8}$$

by using formula (2.13). Formula (11.8) can be expressed more compactly by using vector and matrix notation. Introduce the vectors $\mathbf{p}(t)$, $\mathbf{p}$, and the matrix $\mathbf{P}(t)$ as follows:

$$\mathbf{p}(t) = (p_0(t), p_1(t), p_2(t), \ldots) \tag{11.9}$$

$$\mathbf{p} = \mathbf{p}(0) = (p_0, p_1, p_2, \ldots) \tag{11.10}$$

$$\mathbf{P}(t) = (p_{ij}(t)) = \begin{pmatrix} p_{00}(t) & p_{01}(t) & \cdots \\ p_{10}(t) & p_{11}(t) & \cdots \\ \vdots & \vdots & \end{pmatrix} \tag{11.11}$$

If we observe that the right side of (11.8) is simply a term obtained by multiplying a vector and a matrix, this formula can be written in the form:

$$\mathbf{p}(t) = \mathbf{p}\mathbf{P}(t) \tag{11.12}$$

Similarly, joint probabilities can be computed. For example, if $0 < t_1 < t_2$ and if j_1, j_2 are integers,

$$\begin{aligned} \Pr[X_{t_1} = j_1 \text{ and } X_{t_2} = j_2] &= \Pr[X_{t_1} = j_1]\Pr[X_{t_2} = j_2 \mid X_{t_1} = j_1] \\ &= p_{j_1}(t_1)p_{j_1 j_2}(t_2 - t_1) \\ &= [\sum_i p_i p_{ij_1}(t_1)]p_{j_1 j_2}(t_2 - t_1) \end{aligned}$$

If $0 < t_1 < t_2 < t_3$ and if j_1, j_2, j_3 are integers,

$$\begin{aligned}\Pr[X_{t_1} = j_1, X_{t_2} = j_2, X_{t_3} = j_3] \\ &= \Pr[X_{t_1} = j_1, X_{t_2} = j_2]\Pr[X_{t_3} = j_3 \mid X_{t_1} = j_1, X_{t_2} = j_2] \\ &= \Pr[X_{t_1} = j_1, X_{t_2} = j_2]\Pr[X_{t_3} = j_3 \mid X_{t_2} = j_2]\end{aligned}$$

by using the Markov property (11.2). Thus

$$\Pr[X_{t_1} = j_1, X_{t_2} = j_2, X_{t_3} = j_3] = [\sum_i p_i p_{ij_1}(t_1)] p_{j_1 j_2}(t_2 - t_1) p_{j_2 j_3}(t_3 - t_2) \quad (11.13)$$

Formula (11.13) extends to give joint probabilities of any order: For $0 < t_1 < t_2 < \cdots < t_n$,

$$\begin{aligned}\Pr[X_{t_1} = j_1, X_{t_2} = j_2, \ldots, X_{t_n} = j_n] \\ &= [\sum_i p_i p_{ij_1}(t_1)] p_{j_1 j_2}(t_2 - t_1) \cdots p_{j_{n-1} j_n}(t_n - t_{n-1}) \quad (11.14)\end{aligned}$$

The analogy with formula (8.8) in the discrete parameter case is clear.

3. Examples

EXAMPLE 11.1. The Poisson process, discussed in Chapter 7, has been mentioned previously as an example of a continuous parameter Markov chain. In Chapter 7, we showed that, when events occurred randomly in time (for instance telephone calls coming into an exchange), under certain precisely stated hypotheses the number of events (calls) in a time interval of length t had a Poisson distribution, with parameter proportional to t: $\Pr[n \text{ calls in time } t] = e^{-\lambda t}(\lambda t)^n/n!$, $n = 0, 1, 2, \ldots$. If X_t represents the number of events (calls) in the time interval from 0 to t, then we can easily observe that X_t is a Markov chain with infinite state space: 0, 1, 2, ..., and with transition probabilities given by:

$$p_{ij}(t) = \begin{cases} e^{-\lambda t} \dfrac{(\lambda t)^{j-i}}{(j-i)!}, & \text{if } \quad j \geq i \\ 0, & \text{if } \quad j < i \end{cases} \quad (11.15)$$

for $i, j = 0, 1, 2, \ldots$. [Obviously, $p_{ij}(t) = 0$ for $j < i$, since the total number of calls that have come in cannot *decrease* over a positive time interval.]

We also found that for such a process the time *between* events (calls) has a negative exponential distribution with probability density function

$f(t) = \lambda e^{-\lambda t}, \quad t > 0$. This fact is closely connected with the "forgetfulness" property of the negative exponential discussed in Section 7 of Chapter 6. △

EXAMPLE 11.2. A piece of equipment has two possible states: "operating" and "under repair." Let us designate these states by the integers 1 and 0: 1 means "operating," 0 means "under repair." Now consider the random process X_t, where X_t represents the state of the system (0 or 1) at time t. Furthermore, suppose the time that the system remains in the operating state before breaking down is a random variable that satisfies the "forgetfulness" property (Section 7, Chapter 6) independent of the lengths of previous operating-repair cycles. Thus the operating time must have a negative exponential distribution. Suppose also that the length of a repair time has this property. Under these conditions we can show that the random process X_t is a Markov process to which the discussion of this chapter applies. For example, $p_{01}(t)$ would represent the probability that a machine which is under repair at any given time will be operating t units later. △

4. Properties of Transition Probabilities

The transition probabilities $p_{ij}(t)$ of a Markov process clearly have the properties:

$$0 \leq p_{ij}(t) \leq 1 \tag{11.16}$$

for each i, j and $t \geq 0$,

$$\sum_j p_{ij}(t) = 1 \tag{11.17}$$

for each j and $t \geq 0$ [see (10.6)], and

$$p_{ij}(0) = \begin{cases} 1, & \text{if} \quad i = j \\ 0, & \text{if} \quad i \neq j \end{cases} \tag{11.18}$$

The last condition simply states that, if the process is in state i at any given time, it is also in state i zero time units later.

In terms of the matrix $\mathbf{P}(t) = (p_{ij}(t))$, (11.17) states that every row adds up to 1, while (11.18) is equivalent to

$$\mathbf{P}(0) = \mathbf{I} \tag{11.18$'$}$$

$\mathbf{I}$ representing the identity matrix with 1's on the main diagonal and 0's elsewhere.

If $\tau \geq 0$ and $t \geq 0$ are two times, an extremely important identity can be derived by expressing $p_{ij}(t+\tau)$ in terms of the sets of transition probabilities $\{p_{ij}(t)\}$ and $\{p_{ij}(\tau)\}$. We notice that to pass from state i to state j in time $t+\tau$, at time t the process must be in *some* state, say k. Thus

$$p_{ij}(t+\tau) = \sum_k \Pr[X_{t+\tau}=j,\, X_t=k \mid X_0=i] \tag{11.19}$$

The Markov property (11.1) can then be applied as follows:

$$\Pr[X_{t+\tau}=j,\, X_t=k \mid X_0=i] = \frac{\Pr[X_0=i,\, X_t=k,\, X_{t+\tau}=j]}{\Pr[X_0=i]}$$

$$= \frac{\Pr[X_0=i,\, X_t=k]}{\Pr[X_0=i]}\;\frac{\Pr[X_0=i,\, X_t=k,\, X_{t+\tau}=j]}{\Pr[X_0=i,\, X_t=k]}$$

$$= \Pr[X_t=k \mid X_0=i]\,\Pr[X_{t+\tau}=j \mid X_0=i,\, X_t=k]$$

$$= \Pr[X_t=k \mid X_0=i]\,\Pr[X_{t+\tau}=j \mid X_t=k]$$

by using (11.2)

$$= p_{ik}(t)p_{kj}(\tau)$$

Thus (11.19) yields

$$p_{ij}(t+\tau) = \sum_k p_{ik}(t)p_{kj}(\tau) \tag{11.20}$$

for all states i, j, and $t \geq 0$, $\tau \geq 0$. This identity (11.20) is termed the *Chapman-Kolmogorov Equation*, and can be shown to be equivalent to the Markov property (11.1). In words, it can be interpreted as stating that to make a transition from i to j in time $t+\tau$, one must first make a transition to *some* state k in time t and then a transition from k to j in the remaining time τ. This identity allows us to analyse the behavior of the transition probabilities over *long* time intervals in terms of their behavior over *short* time intervals.

Notice that the sum on the right side of (11.20) is of the type obtained when two matrices are multiplied. Thus (11.20) can be written in matrix form as

$$\mathbf{P}(t+\tau) = \mathbf{P}(t)\mathbf{P}(\tau) \tag{11.21}$$

To simplify our discussion we also assume that, considered as functions of t, the functions $p_{ij}(t)$ are sufficiently "smooth." More precisely, we assume that the functions $p_{ij}(t)$ are *continuous* and *differentiable* for $t \geq 0$ with *uniformly bounded derivatives*

$$\left|\frac{dp_{ij}(t)}{dt}\right| < K$$

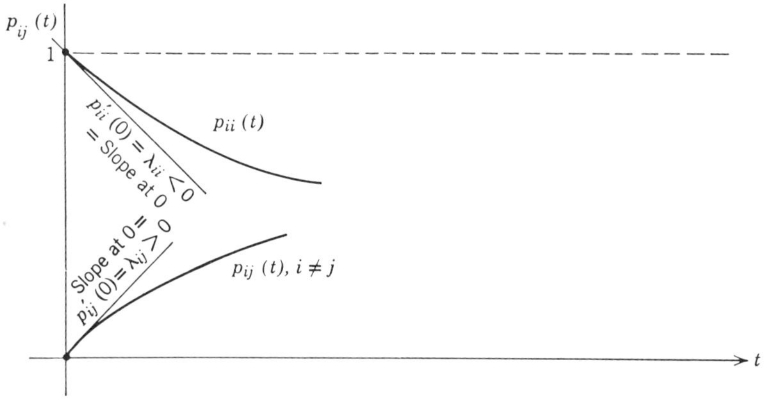

Figure 11.2 Typical graphs for transition probability functions.

for all i, j, and $t \geq 0$. This assumption is much stronger than we actually need, and much of our discussion would hold true under weaker hypotheses.

Graphically speaking, the functions $p_{ij}(t)$ all start at 0 or 1, depending on whether $i \neq j$ or $i = j$, by (11.18), and they always lie between 0 and 1 by (11.16) (Figure 11.2). We also notice that the *slope* (derivative) of $p_{ij}(t)$ at $t = 0$ must be ≥ 0 if $i \neq j$, ≤ 0 if $i = j$.

We state without proof two important properties.

THEOREM 1. *The probability of two or more transitions between states in time Δt is $o(\Delta t)$ as $\Delta t \to 0$.* △

Note. A function $f(t)$ is termed $o(\Delta t)$ if

$$\lim_{\Delta t \to 0} \frac{f(\Delta t)}{\Delta t} = 0$$

In other words, $f(\Delta t)$ approaches zero so rapidly as $\Delta t \to 0$, that, even when divided by the small quantity Δt, the ratio *still* approaches zero.

THEOREM 2. *The time that the process stays in a given state (time between transitions) is a random variable that has negative exponential distribution.* △

A proof of this theorem can be constructed by showing that the Markov property implies that the length of time the process stays in a given state must have the "forgetfulness" property of the negative exponential discussed in Section 7 of Chapter 6.

5. The Rate Matrix and the Kolmogorov Differential Equations

Let us examine more closely the behavior of the transition probability functions $p_{ij}(t)$ for small values of t. If $i \neq j$, we have noted that $p_{ij}(t)$ starts at 0 when $t = 0$, and initially moves with an upward slope while, if $i = j$, it starts at 1 and initially moves with a downward slope. Let us denote this initial slope by

$$\lambda_{ij} = \left.\frac{dp_{ij}(t)}{dt}\right|_{t=0} \tag{11.22}$$

where $\lambda_{ij} \geq 0$ if $i \neq j$, and $\lambda_{ii} \leq 0$. The quantities λ_{ij} can be collected into a matrix $\mathbf{\Lambda}$ that is termed the *rate matrix* of the process

$$\mathbf{\Lambda} = (\lambda_{ij}) = \begin{pmatrix} \lambda_{00} & \lambda_{01} & \lambda_{02} & \cdots \\ \lambda_{10} & \lambda_{11} & \lambda_{12} & \cdots \\ \lambda_{20} & \lambda_{21} & \lambda_{22} & \cdots \\ \vdots & \vdots & \vdots & \end{pmatrix} \tag{11.23}$$

This matrix turns out to be most important in describing the properties of the process.

In case $i \neq j$, the definition of a derivative yields

$$\lambda_{ij} = \left.\frac{dp_{ij}(t)}{dt}\right|_{t=0} = \lim_{\Delta t \to 0} \frac{p_{ij}(\Delta t) - 0}{\Delta t}$$

Hence, for small values of Δt, we have approximately

$$p_{ij}(\Delta t) \approx \lambda_{ij}\,\Delta t \tag{11.24}$$

or, more precisely,

$$p_{ij}(\Delta t) = \lambda_{ij}\,\Delta t + o(\Delta t) \tag{11.25}$$

where again $o(\Delta t)$ refers to a term so small that, even when divided by Δt, it still approaches zero as $\Delta t \to 0$,

$$\lim_{\Delta t \to 0} \frac{o(\Delta t)}{\Delta t} = 0$$

Note. This $o(\Delta t)$ notation is very useful to describe a term of *smaller order* than Δt. Since addition or multiplication by a scalar does not change the property of approaching zero, even when divided by Δt, this small-o symbol satisfies useful identities such as $o(\Delta t) + o(\Delta t) = o(\Delta t)$ and $ao(\Delta t) = o(\Delta t)$ for all constants a.

If $i = j$, we must modify our procedure slightly:

$$\lambda_{ii} = \left.\frac{dp_{ii}(t)}{dt}\right|_{t=0} = \lim_{\Delta t \to 0} \frac{p_{ii}(\Delta t) - 1}{\Delta t}$$

Hence

$$p_{ii}(\Delta t) \approx 1 + \lambda_{ii}\,\Delta t \tag{11.26}$$

for small Δt or, precisely,

$$p_{ii}(\Delta t) = 1 + \lambda_{ii}\,\Delta t + o(\Delta t) \tag{11.27}$$

Formulas (11.25) and (11.27) are useful in determining the values of the λ_{ij}'s in applications. They are illustrated in Figure 11.3.

Recall formula (11.17):

$$\sum_j p_{ij}(t) = 1$$

If we differentiate each side of this identity with respect to t and set $t = 0$, we obtain

$$\sum_j \lambda_{ij} = 0 \tag{11.28}$$

for each i. This derivation is valid if the state space is finite and, under the assumptions that we have made, it can also be justified in the infinite case with the sum in (11.28) being an infinite series.

Since $\lambda_{ij} \geq 0$ for $i \neq j$, it follows from (11.28) that the one negative term

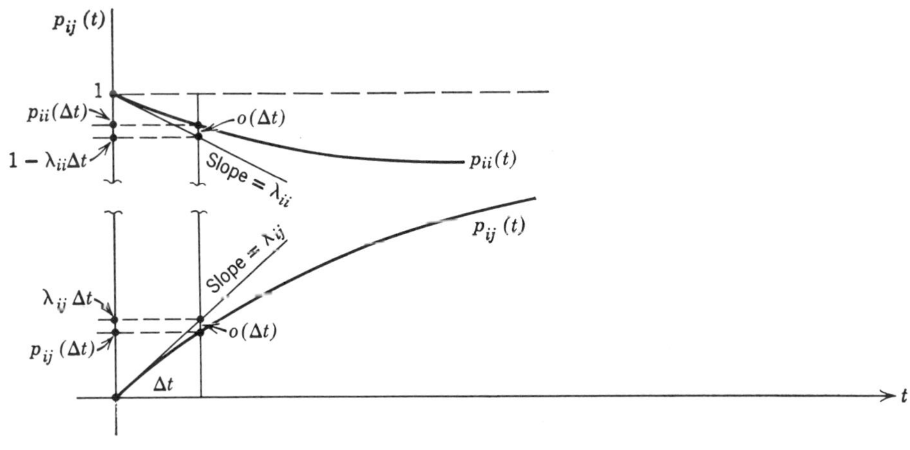

Figure 11.3 An approximation of $p_{ij}(t)$ near $t = 0$ by a linear function.

λ_{ii} must cancel all the positive terms. Thus, the matrix Λ has the following properties.

(i) The off-diagonal terms are ≥ 0; the diagonal terms are ≤ 0.
(ii) In each row the sum of the off-diagonal terms is equal in magnitude to the diagonal term.

Hence, in applications, if all the λ_{ij} terms with $i \neq j$ are known, properties (i) and (ii) allow one to compute the diagonal terms λ_{ii}.

Now let us return to the Chapman-Kolmogorov equation (11.20) and let us differentiate each side with respect to τ:

$$\frac{\partial p_{ij}(t+\tau)}{\partial \tau} = p_{ij}'(t+\tau) = \sum_k p_{ik}(t) \frac{dp_{kj}(\tau)}{d\tau}$$

This term-by-term differentiation is valid even in the infinite state case under our assumptions. If we then set $\tau = 0$, we obtain

$$p_{ij}'(t) = \frac{dp_{ij}(t)}{dt} = \sum_k p_{ik}(t)\lambda_{kj} \tag{11.29}$$

for $i, j = 0, 1, 2, \ldots$. Formula (11.29) represents a system of linear differential equations which, theoretically at least, can be solved for the unknown functions $p_{ij}(t)$. The initial conditions necessary to the solution are given by (11.18):

$$p_{ij}(0) = \begin{cases} 1, & \text{if} \quad i = j \\ 0, & \text{if} \quad i \neq j \end{cases}$$

The equations (11.29) are termed the *Kolmogorov Forward System of Differential Equations* of the Markov process. By studying their solutions, one can determine the complete behavior of the transition probabilities $p_{ik}(t)$ in terms of the *rate constants* λ_{ij}.

If the Chapman-Kolmogorov equation is differentiated with respect to t instead of τ, and then t is set equal to 0, another system of differential equations with the same initial conditions, is obtained.

$$\frac{dp_{ij}(\tau)}{d\tau} = \sum_k \lambda_{ik} p_{kj}(\tau) \tag{11.30}$$

for $i, j = 0, 1, 2, \ldots$. This is termed the *Kolmogorov Backward System of Differential Equations*. In all the cases in which we are interested, both of these systems have the same solution (although this is not always true under more general hypotheses). We use whichever system is simplest in each application that we consider.

The two Kolmogorov systems (11.29) and (11.30) can be written in matrix form as

$$\frac{d\mathbf{P}(t)}{dt} = \mathbf{P}(t)\mathbf{\Lambda} \tag{11.29'}$$

$$\frac{d\mathbf{P}(\tau)}{d\tau} = \mathbf{\Lambda}\mathbf{P}(\tau) \tag{11.30'}$$

while the initial condition is

$$\mathbf{P}(0) = \mathbf{I}$$

The analogy is striking between (11.29′) or (11.30′) and the simple ordinary differential equation $dy/dt = \lambda y$, with $y(0) = 1$. This equation of exponential growth has solution $y = e^{\lambda t}$, which suggests that we seek a solution of (11.29′) or (11.30′) in the form:

$$\mathbf{P}(t) = e^{\mathbf{\Lambda} t} \tag{11.31}$$

This indeed *is* the solution to these systems, under our assumptions, provided that we define properly what is meant by an exponential with a matrix exponent. In analogy with the series expansion $e^{\lambda t} = \sum_{n=0}^{\infty} \lambda^n t^n/n!$, the proper *definition* of $e^{\mathbf{\Lambda} t}$ turns out to be

$$e^{\mathbf{\Lambda} t} = \mathbf{I} + \sum_{n=1}^{\infty} \frac{\mathbf{\Lambda}^n t^n}{n!} \tag{11.32}$$

where the sum of a series of matrices is the matrix having entries equal to the sum of the corresponding entries in the terms.

Formulas (11.31) and (11.32) provide a formal solution to the problem of finding the transition matrix $\mathbf{P}(t)$ in terms of the rate matrix $\mathbf{\Lambda}$. In practice, this method often is cumbersome in direct calculations, and alternate methods must be sought.

6. Examples

EXAMPLE 11.3. The *Poisson process* discussed in Chapter 7 and in Section 3 of this chapter is a continuous parameter Markov process with state space: 0, 1, 2, ... By differentiating formula (11.15) and setting $t = 0$, we find that the rate constants λ_{ij} are given by

$$\begin{aligned}
\lambda_{i\,i+1} &= \lambda, && \text{for} \quad i = 0, 1, 2, \ldots \\
\lambda_{ii} &= -\lambda, && \text{for} \quad i = 0, 1, 2, \ldots \\
\lambda_{ij} &= 0, && \text{for all other } i, j
\end{aligned}$$

$$\mathbf{\Lambda} = \begin{pmatrix} -\lambda & \lambda & 0 & 0 & \cdots \\ 0 & -\lambda & \lambda & 0 & \cdots \\ 0 & 0 & -\lambda & \lambda & \cdots \\ \vdots & \vdots & \vdots & \vdots & \end{pmatrix} \tag{11.33}$$

Alternately the λ_{ij}'s can be determined by using (11.25). We observed in Chapter 7 that the time between events (calls) in such a Poisson process has a negative exponential distribution with probability density function $\lambda e^{-\lambda t}$. Since $p_{i\,i+1}(\Delta t)$ is the probability of an event occurring within time Δt, we have

$$\begin{aligned}
p_{i\,i+1}(\Delta t) &= \int_0^{\Delta t} \lambda e^{-\lambda t} dt = 1 - e^{-\lambda \Delta t} \,* \\
&= 1 - \left(1 - \lambda \Delta t + \frac{\lambda^2 (\Delta t)^2}{2} \mp \cdots \right) \\
&= \lambda \Delta t + (\Delta t)^2 \left(-\frac{\lambda^2}{2} \mp \cdots \right) \\
&= \lambda \Delta t + o(\Delta t)
\end{aligned} \tag{11.34}$$

where the last term is $o(\Delta t)$, since it has a factor of $(\Delta t)^2$ and, hence, approaches zero as $\Delta t \to 0$ even when divided by Δt.

By comparison with (11.25), we find that $\lambda_{i\,i+1} = \lambda$.

Since $p_{ij}(\Delta t)$ is the probability of $j - i$ events in time Δt, we notice that $p_{ij}(\Delta t) = 0$ if $i > j$, and that $p_{ij}(\Delta t) = o(\Delta t)$ if $j - i \geq 2$ (by Theorem 1 of Section 4). Comparison with (11.25) gives $\lambda_{ij} = 0$ in both these cases. This determines all the elements of the matrix $\mathbf{\Lambda}$ with the exception of the diagonal elements λ_{ii}, which are obtained by using the fact that each row sums to zero.

Formula (11.34) is illustrated in Figure 11.4.

In Section 7 of Chapter 6 the mean of a random variable that has the negative exponential probability density function of the form $\lambda e^{-\lambda t}$ was

* We have ignored the possibility of 2 or more events in time Δt. By Theorem 1 of Section 4 this would have probability $o(\Delta t)$ and, hence, would not affect the final result (11.34).

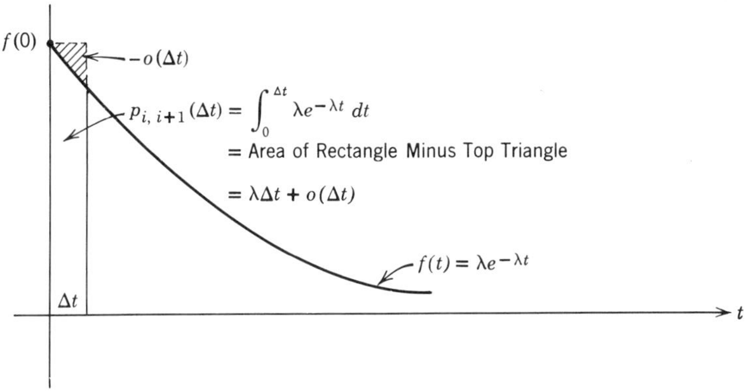

Figure 11.4 Pr[1 event in time Δt] $\approx \lambda\, \Delta t$ for negative exponential inter-event times.

shown to be $1/\lambda$ (formula 6.36). Thus (11.33) can be written in the form

$$p_{i\ i+1}(\Delta t) = \frac{1}{\beta} \Delta t + o(\Delta t) \tag{11.35}$$

where β is the mean time *between* events.

By using (11.33), the Kolmogorov Forward System of differential equations becomes

$$p_{ii}'(t) = -\lambda\, p_{ii}(t)$$
$$p_{ij}'(t) = -\lambda\, p_{ij}(t) + \lambda\, p_{ij-1}(t), \qquad j = i+1, i+2, \ldots$$

This system is essentially the same as the system studied in Section 7 of Chapter 7 (formulas (7.36) and (7.37)), but uses a slightly different notation. It can be solved in the same way, with initial conditions given by (11.18). The solution is (11.15). $\triangle$

EXAMPLE 11.4 THE TWO-STATE PROCESS. Consider the simplest possible nontrivial example of a random process as one that has only two states, designated 0 and 1. For instance, a machine could be considered as either in state 1 ("operating" state) or in state 0 ("under repair" state). As we have mentioned previously, if the length of stay in each state has a negative exponential distribution, independent of the lengths of previous stays, then the "forgetfulness" property of the negative exponential can be used to show that the process is Markovian.

Suppose that the length of an operating period has probability density function $\mu e^{-\mu t}$, $t > 0$, and that the length of a repair period has probability density function $\lambda e^{-\lambda t}$, $t > 0$. The parameters λ and μ can be interpreted by noting that $1/\lambda$ is the *mean* length of a repair period and that $1/\mu$ is the *mean* length of an operating period. Thus these quantities can be estimated by observing the *average* operating times and repair times over a lengthy period of observation.

Now, $p_{01}(t)$ represents the probability of a machine, which initially is under repair, being in operation t units later. Thus, $p_{01}(\Delta t)$ is the probability of a repair being completed in time Δt, ignoring the possibility of 2 or more transitions in time Δt as a term of smaller order, $o(\Delta t)$, (Theorem 1 of Section 4). Thus the same argument that we used in deriving formula (11.34) applies:

$$\begin{aligned} p_{01}(\Delta t) &= \int_0^{\Delta t} \lambda e^{-\lambda t} dt = 1 - e^{-\lambda \Delta t} \\ &= \lambda \Delta t - \frac{\lambda^2(\Delta t)^2}{2} \pm \cdots \\ &= \lambda \Delta t + o(\Delta t) \end{aligned}$$

By comparison with (11.25), we have $\lambda_{01} = \lambda$. Similarly, $p_{10}(\Delta t) = \mu\, \Delta t + o(\Delta t)$, $\lambda_{10} = \mu$. Since there are only 2 states, $\mathbf{\Lambda}$ is a 2×2 matrix, and the diagonal terms λ_{00}, λ_{11} are determined, since each row sums to 0:

$$\mathbf{\Lambda} = \begin{pmatrix} -\lambda & \lambda \\ \mu & -\mu \end{pmatrix}$$

The Kolmogorov Forward Equations are

$$\begin{aligned} p_{i0}'(t) &= -\lambda p_{i0}(t) + \mu p_{i1}(t) \\ p_{i1}'(t) &= \quad \lambda p_{i0}(t) - \mu p_{i1}(t) \end{aligned} \tag{11.36}$$

$i = 0, 1$. Since $p_{i0}(t) + p_{i1}(t) = 1$, $p_{i1}(t) = 1 - p_{i0}(t)$, and the first of these equations can be written as a simple first-order linear differential equation

$$p_{i0}'(t) + (\lambda + \mu)p_{i0}(t) = \mu$$

which has general solution:

$$p_{i0}(t) = \frac{\mu}{\lambda + \mu} + Ce^{-(\lambda+\mu)t}$$

The constant C can be found by using the fact that $p_{i0}(0) = 1$ if $i = 0$, 0 if $i = 1$. The final solutions for $p_{i0}(t)$ and $p_{i1}(t) = 1 - p_{i0}(t)$ are

$$p_{00}(t) = \frac{\mu + \lambda e^{-(\lambda+\mu)t}}{\lambda + \mu}, \qquad p_{01}(t) = \frac{\lambda - \lambda e^{-(\lambda+\mu)t}}{\lambda + \mu} \tag{11.37}$$

$$p_{10}(t) = \frac{\mu - \mu e^{-(\lambda+\mu)t}}{\lambda + \mu}, \qquad p_{11}(t) = \frac{\lambda + \mu e^{-(\lambda+\mu)t}}{\lambda + \mu}$$

[Notice that the substitution $p_{i1}(t) = 1 - p_{i0}(t)$ changes the second of the equations (11.36) into a form identical to the first and, hence, the solution that uses the second equation would be the same.] The graphs of the transition probabilities $p_{ij}(t)$ are sketched in Figure 11.5. △

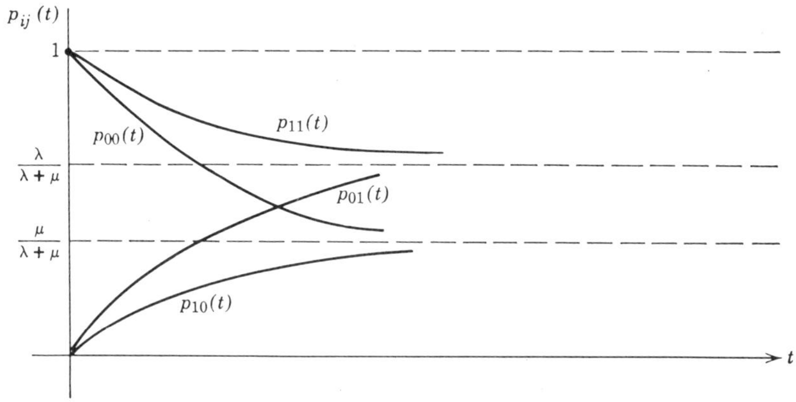

Figure 11.5 Transition probabilities for a 2-state Markov process.

EXAMPLE 11.5. MACHINERY BREAKDOWN. The situation of the previous example can be generalized. Suppose that we have N identical machines, operating independently, serviced by a single repair crew. If a machine breaks down while another is being repaired, it must wait its turn before repairs can start. As before, assume that *each* machine has an operating time that is a negative exponentially distributed random variable with mean $1/\mu$. Also assume that the time required to repair a machine is negative exponentially distributed with mean $1/\lambda$.

Let X_t be the number of machines in operating condition at time t. Thus X_t is a Markov chain with state space: 0, 1, 2, ..., N.

Next we must determine the transition rates λ_{ij}.

First notice the probability that two breakdowns (or a breakdown and a repair completion) will occur exactly simultaneously is clearly 0. It then follows from Theorem 1 of Section 4 that $p_{ij}(\Delta t) = o(\Delta t)$ whenever the subscripts i and j differ by more than 1. Thus, $\lambda_{ij} = 0$ when $j \neq i - 1$, i, $i + 1$.

Let us consider the breakdown and repair sides of the problem separately. Now, $p_{i\,i+1}(\Delta t)$ represents the probability of one machine being repaired during a time interval of length Δt (and, hence, the number of operating machines increases from i to $i+1$). The same argument we used in the previous example applies. Again, by ignoring the possibility of more than one transition (for example, two repairs and one further breakdown in time Δt) as a term of order $o\,(\Delta t)$, we have

$$
\begin{aligned}
p_{i\,i+1}(\Delta t) &= \int_0^{\Delta t} \lambda\, e^{-\lambda t}\, dt = 1 - e^{-\lambda \Delta t} \\
&= \lambda\, \Delta t - \lambda^2 \frac{(\Delta t)^2}{2} \mp \cdots \\
&= \lambda\, \Delta t + o(\Delta t)
\end{aligned}
$$

for $i = 0, 1, 2, \ldots, N-1$. Thus

$$\lambda_{i\,i+1} = \lambda, \qquad \text{for} \quad i = 0, 1, 2, \ldots, N-1 \tag{11.38}$$

To determine $\lambda_{i\,i-1}$ we must estimate $p_{i\,i-1}(\Delta t)$. This is the probability of one machine breaking down when i are operating. The probability of any given machine breaking down is $\int_0^{\Delta t} \mu\, e^{-\mu t}\, dt = 1 - e^{-\mu \Delta t} = \mu \Delta t + o(\Delta t)$. The probability of exactly one out of i machines breaking down is i times this, $i[\mu \Delta t + o(\Delta t)]$, *minus* the probability of two or more breaking down. By Theorem 1 of Section 4 this latter probability is $o(\Delta t)$. Thus

$$
\begin{aligned}
p_{i\,i-1}(\Delta t) &= i[\mu \Delta t + o(\Delta t)] + o(\Delta t) \\
&= i\mu \Delta t + o(\Delta t)
\end{aligned}
$$

and

$$\lambda_{i\,i-1} = i\mu \qquad \text{for } i = 1, 2 \ldots, N \tag{11.39}$$

Since $\lambda_{ij} = 0$ unless $j = i-1$, i or $i+1$, and the rows of $\mathbf{\Lambda}$ sum to zero, the terms λ_{ii} can be calculated from (11.38) and (11.39):

$$
\begin{aligned}
\lambda_{ii} &= -(\lambda + i\mu) \qquad \text{for} \qquad i = 0, 1, 2, \ldots, N-1 \\
\lambda_{NN} &= -N\mu
\end{aligned}
$$

$$
\mathbf{\Lambda} = \begin{pmatrix}
-\lambda & \lambda & 0 & 0 & \cdots & 0 \\
\mu & -(\lambda+\mu) & \lambda & 0 & \cdots & 0 \\
0 & \mu & -(\lambda+2\mu) & \lambda & \cdots & 0 \\
\vdots & \vdots & \vdots & \vdots & & \\
0 & 0 & 0 & 0 & \cdots & -N\mu
\end{pmatrix} \tag{11.40}
$$

The explicit solution of the Kolmogorov differential equations for this $\mathbf{\Lambda}$ is somewhat more complicated. A possible method is indicated in Section 8. △

EXAMPLE 11.6. A SIMPLE QUEUEING MODEL. Consider a situation in which individuals or demands arrive at a certain service point in random sequence, line up, and wait their turn at this point. Then, successively, they have some service operation performed on them, and they depart. Such a situation is termed a single-server queue. Examples readily come to mind that involve supermarket checkout lines, telephone traffic, computer demands, factory production lines, and inventories, in addition to many others. Assume that interarrival times and service times are all independent random variables, and that the times between arrivals have negative exponential distributions with mean $1/\lambda$ (or, equivalently, that the arrivals form a Poisson process with parameter λ). Assume likewise that the service times also have negative exponential distributions with mean $1/\mu$.

Now let X_t represent the length of the queue or waiting line at time t. Under our assumptions X_t is a Markov process with state space : 0, 1, 2, ... The probability of an arrival in time Δt is $\int_0^{\Delta t} \lambda\, e^{-\lambda t}\, dt = \lambda\, \Delta t + o(\Delta t)$,and the probability of a service completion in time Δt is $\mu\, \Delta t + o(\Delta t)$ (assuming that some individual is actually being served at the beginning of this time interval). Thus we have $p_{i,i+1}(\Delta t) = \lambda\, \Delta t + o(\Delta t)$, $i = 0, 1, 2, \ldots$, and $p_{i,i-1}(\Delta t) = \mu\, \Delta t + o(\Delta t)$, $i = 1, 2, 3, \ldots,$ and hence

$$\begin{aligned} \lambda_{i,i+1} &= \lambda, \qquad i = 0, 1, 2, 3 \ldots \\ \lambda_{i,i-1} &= \mu, \qquad i = 1, 2, 3, \ldots \end{aligned} \tag{11.41}$$

By Theorem 1 of Section 4, $\lambda_{ij} = 0$ otherwise for $i \neq j$, and consequently we have necessarily

$$\begin{aligned} \lambda_{00} &= -\lambda \\ \lambda_{ii} &= -(\lambda + \mu), \qquad i = 1, 2, 3, \ldots \end{aligned}$$

in order that the rows of $\mathbf{\Lambda}$ sum to 0.

$$\mathbf{\Lambda} = \begin{pmatrix} -\lambda & \lambda & 0 & 0 & \cdots \\ \mu & -(\lambda + \mu) & \lambda & 0 & \cdots \\ 0 & \mu & -(\lambda + \mu) & \lambda & \\ \vdots & \vdots & \vdots & \vdots & \end{pmatrix} \tag{11.42}$$

The Kolmogorov Forward Equations are

$$\begin{aligned} p_{i0}'(t) &= -\lambda\, p_{i0}(t) + \mu\, p_{i1}(t) \\ p_{ij}'(t) &= \lambda\, p_{i,j-1}(t) - (\lambda + \mu) p_{ij}(t) + \mu\, p_{i,j+1}(t), \qquad j = 1, 2, \ldots, \end{aligned} \tag{11.43}$$

for $i = 0, 1, 2, \ldots$. The solution of this system requires certain special techniques (see Cox and Smith). $\triangle$

EXAMPLE 11.7. THE COMPOUND POISSON PROCESS. A Markov process X_t of wide applicability is one which stays constant between points of change that occur at times given by a Poisson process with parameter λ, but where the amount of each change is given by a positive random variable, independent of the change times, and not always 1 as in the simple Poisson process. These situations occur frequently—for example, orders to a factory (orders are not always for the same dollar amount), claims on an insurance company, certain radioactive disintegrations, and many others.

Given that an increase in X_t occurs at time t, assume that the amount of the increase is one of the integers 1, 2, 3, ... with probabilities q_1, q_2, $q_3, \ldots, \sum q_i = 1$, with the increases being independent of each other and of the times of increase.

Under these circumstances X_t is termed a *compound Poisson process.*

The probability $p_{ij}(\Delta t)$ of a change from state i to state j, $i < j$, in time Δt is obtained by multiplying the probability of any change, $\lambda\, \Delta t + o(\Delta t)$, by the probability q_{j-i} that the change is of the amount $j - i$. Thus

$$p_{ij}(\Delta t) = q_{j-i}\, \lambda\, \Delta t + o(\Delta t), \qquad \text{for} \qquad i < j$$

and

$$\begin{aligned} \lambda_{ij} &= \lambda\, q_{j-i}, && \text{for} \quad i < j \\ \lambda_{ij} &= 0, && \text{for} \quad i > j \\ \lambda_{ii} &= -\lambda \sum_j q_{j-i} = -\lambda \end{aligned} \tag{11.44}$$

$$\boldsymbol{\Lambda} = \begin{pmatrix} -\lambda & \lambda q_1 & \lambda q_2 & \cdots \\ 0 & -\lambda & \lambda q_1 & \cdots \\ 0 & 0 & -\lambda & \cdots \\ \vdots & \vdots & \vdots & \end{pmatrix}$$

The Kolmogorov Forward Equations are

$$p_{ij}'(t) = -\lambda[p_{ij}(t) - \sum_{k=i}^{j-1} q_{j-k}\, p_{ik}(t)] \tag{11.45}$$

for $j = i, i + 1, i + 2, \ldots$, and for each fixed i.

A further analysis of the functions $p_{ij}(t)$ is facilitated by using the concept of a probability generating function that is discussed in Chapter 5, Section 13. For each t introduce the following generating function

$$\psi = \psi_i(t, z) = \sum_j p_{ij}(t) z^j \tag{11.46}$$

for $i = 0, 1, 2, \ldots, t \geq 0, |z| \leq 1$. Define also the generating function for the q_k distribution as

$$q(z) = \sum_k q_k z^k$$

Now

$$\begin{aligned}\frac{\partial \psi}{\partial t} &= \sum_j p_{ij}'(t)\, z^j \\ &= -\lambda\left[\sum_{j=i}^{\infty} p_{ij}(t) z^j - \sum_{j=i+1}^{\infty} \sum_{k=i}^{j-1} q_{j-k}\, p_{ik}(t) z^j\right]\end{aligned}$$

by using (11.45). The first summation yields ψ again. The double summation adds over all values of the indices k and j that satisfy $i \leq k < j < \infty$, with i fixed. If the order of summation is reversed, we obtain

$$\begin{aligned}\frac{\partial \psi}{\partial t} &= -\lambda\left[\psi - \sum_{k=i}^{\infty} \sum_{j=k+1}^{\infty} q_{j-k}\, p_{ik}(t) z^j\right] \\ &= -\lambda\left[\psi - \sum_{k=1}^{\infty} p_{ik}(t) z^k \sum_{j=k+1}^{\infty} q_{j-k}\, z^{j-k}\right] \\ &= -\lambda\left[\psi - \psi \sum_{n=1}^{\infty} q_n z^n\right] \\ &= -\lambda[\psi - \psi q(z)] \\ \frac{\partial \psi}{\partial t} &= -\lambda \psi\, [1 - q(z)]\end{aligned} \tag{11.47}$$

Now (11.47) is a simple linear differential equation of the form $\partial\psi/\partial t =$ constant$\cdot\psi$. (It is only formally a partial differential equation.) If $t = 0$, ψ has initial condition

$$\psi_i(0, z) = \sum_j p_{ij}(0) z^j = z^i$$

since $p_{ii}(0) = 1$, $p_{ij}(0) = 0$ for $i \neq j$. The solution of (11.47), subject to this initial condition, is easily found to be

$$\psi = \psi_i(t, z) = z^i\, e^{-\lambda t[1 - q(z)]} \tag{11.48}$$

For any given $q(z)$, (11.48) can be expanded in powers of z and the functions $p_{ij}(t)$ can be determined as the coefficients of these powers by using (11.46).

In case each increase of X_t is one unit, then $q_1 = 1$, $q_k = 0$ for $k > 1$, $q(z) = z$, and

$$\psi = e^{-\lambda t(1-z)}$$

provided that the initial state is assumed to be $i = 0$. By comparison with (6.24) we notice that, in this case, X_t has a Poisson distribution with mean λt. Under these circumstances, the compound Poisson process specializes to an ordinary Poisson process. △

7. The Birth-and-Death Process

All of the examples in the previous section, except the last, shared the property that changes in X_t occurred only in unit steps, up or down; each change in the value of X_t was either $+1$ or -1. Only in the compound Poisson process, among the examples discussed in Section 6, could X_t change by 2 or more. In that process, changes could be 1, 2, 3, ... with probabilities q_1, q_2, q_3, ... If X_t can only change by $+1$ or -1 at any instant, the process is said to be a *birth-and-death* process. These processes have many applications and an extensively worked-out theory. They are characterized as continuous parameter Markov processes with state space: 0, 1, 2, ..., and with rates that satisfy $\lambda_{ij} = 0$ if i and j differ by 2 or more. Thus the rate matrix of such a process has the form

$$\mathbf{\Lambda} = \begin{pmatrix} -\lambda_0 & \lambda_0 & 0 & 0 & \cdots \\ \mu_1 & -(\lambda_1 + \mu_1) & \lambda_1 & 0 & \cdots \\ 0 & \mu_2 & -(\lambda_2 + \mu_2) & \lambda_2 & \cdots \\ 0 & 0 & \mu_3 & -(\lambda_3 + \mu_3) & \cdots \\ \vdots & \vdots & \vdots & \vdots & \end{pmatrix} \tag{11.49}$$

The Kolmogorov Forward Equations are

$$p_{i0}'(t) = -\lambda_0 p_{i0}(t) + \mu_1 p_{i1}(t)$$
$$p_{ij}'(t) = \lambda_{j-1} p_{i\,j-1}(t) - (\lambda_j + \mu_j) p_{ij}(t) + \mu_{j+1} p_{i\,j+1}(t), \qquad i = 1, 2, \ldots$$

Historically this structure of transition probabilities probably arose in connection with bacterial studies. One interpretation with regard to these studies is the following.

Suppose that we are observing a colony of bacteria. At each instant the colony can increase by one member. The probability of an increase depends on how many members there are in the colony, and, in an interval of length Δt,

$$p_{i,i+1}(\Delta t) = \lambda_i \Delta t + o(\Delta t)$$

is the probability of an increase, given that the population is of size i at the start of the interval. This increase is called a *birth*.

Furthermore, suppose that members of the colony die. At each instant the colony can decrease by one because of a death. The probability of a death in an interval of length Δt depends on how many are present at the start of the interval, and it is given by

$$p_{i,i-1}(\Delta t) = \mu_i \Delta t + o(\Delta t)$$

The biological questions involve finding the transition and state probabilities as functions of t, or finding the probability that the colony eventually dies out, or finding the probability that its size exceeds some limit or, if it does eventually die out, estimating how long it takes to do so.

Questions like the ones asked of the bacterial population are often asked of other systems that have nothing to do with biological populations, such as the birth-and-death examples 11.3 to 11.6 in Section 6.

8. Algebraic Methods for Solving Kolmogorov Equations

The algebraic methods discussed in Chapter 9 can often be used to assist in solving the Kolmogorov Forward or Backward Differential Equations for a *finite-state* Markov process.

We recall that a formal solution to these systems is given by (11.31), $\mathbf{P}(t) = e^{\mathbf{\Lambda}t}$, but the difficulty comes in computing $e^{\mathbf{\Lambda}t} = \mathbf{I} + \sum_{n=1}^{\infty} \mathbf{\Lambda}^n t^n/n!$ for a general matrix $\mathbf{\Lambda}$. In the case of a *diagonal* matrix, such an exponential is easily computed. If

$$\mathbf{D} = \begin{pmatrix} d_0 & 0 & 0 & \dots & 0 \\ 0 & d_1 & 0 & \dots & 0 \\ 0 & 0 & d_2 & \dots & 0 \\ \vdots & \vdots & \vdots & \ddots & \vdots \\ 0 & 0 & 0 & \dots & d_N \end{pmatrix}$$

is a diagonal matrix, then clearly

$$\mathbf{D}^n = \begin{pmatrix} {d_0}^n & 0 & \dots & 0 \\ 0 & {d_1}^n & \dots & 0 \\ \vdots & \vdots & \ddots & \vdots \\ 0 & 0 & \dots & {d_N}^n \end{pmatrix}$$

is also diagonal. Hence each entry in the matrix

$$e^{\mathbf{D}t} = \mathbf{I} + \sum_{n=1}^{\infty} \mathbf{D}^n \frac{t^n}{n!}$$

can be obtained by adding the corresponding entries:

$$e^{\mathbf{D}t} = \begin{pmatrix} 1 + \sum_{n=1}^{\infty} d_0{}^n \frac{t^n}{n!}, & 0 & , \dots \\ 0 & , \quad 1 + \sum_{n=1}^{\infty} d_1{}^n \frac{t^n}{n!} & , \dots \\ \vdots & \vdots & \dots \end{pmatrix}$$

$$= \begin{pmatrix} e^{d_0 t} & 0 & 0 & \dots & 0 \\ 0 & e^{d_1 t} & 0 & \dots & 0 \\ \vdots & \vdots & \vdots & & \\ 0 & 0 & 0 & \dots & e^{d_N t} \end{pmatrix} \tag{11.50}$$

Thus the computation of $e^{\mathbf{D}t}$ for a diagonal matrix $\mathbf{D}$ is particularly simple.

In Section 2 of Chapter 9, we stated that any matrix that has distinct eigenvalues [satisfying (9.2) or (9.3)] can be diagonalized, and that the corresponding diagonal matrix will have the eigenvalues as diagonal elements.

The identity $\sum_j \lambda_{ij} = 0$ can be written in the form:

$$\mathbf{\Lambda} \begin{pmatrix} 1 \\ 1 \\ 1 \\ \vdots \\ 1 \end{pmatrix} = \begin{pmatrix} 0 \\ 0 \\ 0 \\ \vdots \\ 0 \end{pmatrix} = \begin{pmatrix} 1 \\ 1 \\ 1 \\ \vdots \\ 1 \end{pmatrix} . 0$$

Thus, by comparison with (9.3), we observe that 0 is one eigenvalue of the rate matrix $\mathbf{\Lambda}$. Let us assume that all eigenvalues of $\mathbf{\Lambda}$ are distinct. Thus, by the results of Section 2 of Chapter 9, there exists a nonsingular matrix S so that

$$\mathbf{\Lambda} = \mathbf{S}^{-1}\,\mathbf{D}\,\mathbf{S} \tag{11.51}$$

where $\mathbf{D}$ has the eigenvalues of $\mathbf{\Lambda}$ as diagonal elements

$$\mathbf{D} = \begin{pmatrix} 0 & 0 & 0 & \dots & 0 \\ 0 & d_1 & 0 & \dots & 0 \\ 0 & 0 & d_2 & \dots & 0 \\ \vdots & \vdots & \vdots & & \vdots \\ 0 & 0 & 0 & \dots & d_N \end{pmatrix}$$

We can also show that for such a rate matrix $\mathbf{\Lambda}$, the nonzero eigenvalues $d_1, d_2, \dots, d_N$ are either *negative* real numbers, or complex numbers with *negative* real parts.

Now

$$\begin{aligned} \mathbf{\Lambda}^n &= (\mathbf{S}^{-1}\,\mathbf{D}\,\mathbf{S})^n \\ &= (\mathbf{S}^{-1}\,\mathbf{D}\,\mathbf{S})(\mathbf{S}^{-1}\,\mathbf{D}\,\mathbf{S})(\mathbf{S}^{-1} \cdot \dots \cdot (\mathbf{S}^{-1}\,\mathbf{D}\,\mathbf{S}) \\ &= \mathbf{S}^{-1}\,\mathbf{D}^n\,\mathbf{S} \end{aligned}$$

since in all the intervening terms $\mathbf{S}\,\mathbf{S}^{-1}=\mathbf{I}$. Thus

$$
\begin{aligned}
\mathbf{P}(t) = e^{\mathbf{\Lambda}t} &= \mathbf{I} + \sum_{n=1}^{\infty} \mathbf{\Lambda}^n \frac{t^n}{n!} \\
&= \mathbf{I} + \sum_{n=1}^{\infty} \mathbf{S}^{-1}\mathbf{D}^n\mathbf{S}\frac{t^n}{n!} \\
&= \mathbf{S}^{-1}\left[\mathbf{I} + \sum_{n=1}^{\infty} \mathbf{D}^n \frac{t^n}{n!}\right]\mathbf{S} \\
&= \mathbf{S}^{-1} e^{\mathbf{D}t}\,\mathbf{S} \\
&= \mathbf{S}^{-1}\begin{pmatrix} 1 & 0 & 0 & \dots & 0 \\ 0 & e^{d_1 t} & 0 & \dots & 0 \\ 0 & 0 & e^{d_2 t} & \dots & 0 \\ \vdots & \vdots & \vdots & \vdots & \\ 0 & 0 & 0 & \dots & e^{d_N t} \end{pmatrix}\mathbf{S} \qquad (11.52)
\end{aligned}
$$

It follows that once the eigenvalues $d_1, d_2, \ldots, d_N$ and the corresponding eigenvectors that give the matrices $\mathbf{S}$ and $\mathbf{S}^{-1}$ are computed, by using the methods of Chapter 9, then (11.52) provides an explicit solution for the transition matrix $\mathbf{P}(t)$.

EXERCISES

1. Let X_t be the Poisson process in Example 11.1. By using (11.15) show that
   (a) $p_{ij}(\Delta t) = o(\Delta t)$, whenever $|j-i|>1$
   (b) $\sum_{k=i}^{j} p_{ik}(t)p_{kj}(s) = p_{ij}(t+s)$, whenever $i<j,\ t\geq 0, s\geq 0$
2. Let X_t be a Markov chain with state space $\{0, 1\}$ and transition probabilities

$$p_{ij}(t) = \frac{1+(-1)^{i+j}\,e^{-2t}}{2}, \qquad \text{for} \qquad i, j = 0, 1$$

   Show that properties (11.16), (11.17), (11.18), and (11.20) for a Markov process are satisfied.
3. Determine the rate matrix $\mathbf{\Lambda}$ for the process of Exercise 2, and show that (11.29) and (11.30) hold true.
4. Two communications satellites are placed in orbit. The lifetime of a satellite is negative exponential with mean $1/\mu$. If one fails, a replacement is sent up. The time necessary to prepare and to send up a replacement is negative exponential

with mean $1/\lambda$. Let X_t = the number of satellites in orbit. Assume that X_t is a Markov process with state space $\{0, 1, 2\}$.

(a) Show that the rate matrix $\mathbf{\Lambda}$ is given by

$$\mathbf{\Lambda} = \begin{pmatrix} -\lambda & \lambda & 0 \\ \mu & -(\lambda+\mu) & \lambda \\ 0 & 2\mu & -2\mu \end{pmatrix}$$

(b) Set up the Kolmogorov Forward Equations for this process.

(c) Solve the system in (b) to find the transition probabilities $p_{ij}(t)$. (Optional.)

5. Write and solve the Kolmogorov Backward Equations for the Poisson process, Example 11.3 of Section 6. [*Hint.* Remember that $p_{ij}(t) = 0$ if $i > j$.]

6. Show for Example 11.4 of Section 6 that $p_{10}(\Delta t) = \mu \Delta t + o(\Delta t)$.

7. For Example 11.4 of Section 6, find the Backward Kolmogorov Equations, and solve for $p_{ij}(t)$.

8. Write the Kolmogorov Forward and Backward Equations for Example 11.5 of Section 6.

9. In Example 11.5 of Section 6, set up the Forward Equations assuming that there are R identical repair crews ($R < N$), each with the same service rate λ. Each crew services at most one machine at a time, and the crews do not cooperate with each other. That is, if i machines are down for repairs, then for $i < R$, $R - i$ crews stand idle awaiting a new breakdown and i crews perform service on the down machines.

10. In the birth-and-death process suppose that the "birth rate" λ_i is inversely proportional to one more than the number in the population, and the "death rate" μ_i is constant, $\mu_i = \mu$. Write the Kolmogorov Forward Equations.

11. In the simple queueing model of Example 11.6 of Section 6, suppose that the waiting line can be at most K individuals (including the one in service). Write the Kolmogorov Forward Equations.

12. Suppose that the time between arrivals of jobs to a shop is negative exponentially distributed. Each job requires Y units of service time, where Y is a binomially distributed random variable on the integers $0, 1, 2, \ldots, N$. Determine the probability generating function for the total work load that enters the shop up to time t. (*Hint.* See Example 11.7 of Section 6.)

13. In certain queueing models one can approximate the behavior of the system by assuming that a birth-and-death model is appropriate, with

$$\begin{aligned} \lambda_{i,i+1} &= \lambda \\ \lambda_{i,i-1} &= i\mu, \qquad \lambda_{ij} = 0, \text{ otherwise, } i \neq j \\ \lambda_{0,0} &= -\lambda \end{aligned}$$

Write the Kolmogorov Forward Equations. One interpretation of this model is that the arrival process is a Poisson process. The service unit contains an unlimited number of identical servers, each having service times negative exponentially distributed with mean $1/\mu$.

14. Suppose that arrivals to a service center are composed of exactly 2 units. Each arrival unit is serviced one at a time. With this single change, write the Kolmogorov Forward Equations for Example 11.6 of Section 6.

15. The birth-and-death process of Section 7 is called a *pure birth process* if $\mu_i = 0$ for every i. The Poisson process is an example of this process. Suppose that $\lambda_i = i\lambda$ for $i = 0, 1, 2, \ldots$. This process would occur in a system, such as a biological population, in which the probability of a birth is proportional to the number of individuals present to act as parents. Assume that initially $X = n$. Show that

$$\Pr[X_t = i] = p_{ni}(t) = \binom{i-1}{n-1} e^{-n\lambda t}(1 - e^{-\lambda t})^{i-n}, \qquad \text{for } i \geq n$$

16. A birth-and-death process is called a *pure death process* if $\lambda_i = 0$ for every i. Suppose that $\mu_i = \mu$ for $i = 1, 2, 3, \ldots$, and that initially $X_0 = n$. Show that $\Pr[X_t = i] = p_{ni}(t)$ has a binomial distribution with parameter $p = e^{-\mu t}$, for $i = 0, 1, 2, \ldots, n$.

17. An insurance company might be faced with the following problem. The time between accidents is a negative exponentially distributed random variable with parameter λ_1. The dollar cost to the insurance company of each accident is a random variable with a Poisson density function with parameter λ_2. One might be interested in determining the probability density function for total claim value in some fixed interval. Explain how this could be done.

18. Find the transition probabilities $p_{ij}(t)$ for the two-state process, discussed in Example 11.4 of Section 6, by using the methods of Section 8.

    [Determine the eigenvalues of $\mathbf{\Lambda}$, and a 2×2 matrix $\mathbf{S}$ so that $\mathbf{\Lambda} = \mathbf{S}^{-1}\mathbf{D}\,\mathbf{S}$, and then use (11.52).] Compare with (11.37).

19. Use the methods of Section 8 to determine the transition probabilities $p_{ij}(t)$ for the machinery repair Example 11.5 of Section 6 in the case $\lambda = 2$, $\mu = 1$, and $N = 2$.

20. Answer Exercise 19 for general λ and μ, in the case $N = 2$.

12 • Limiting Distribution of Continuous Parameter Markov Processes

1. Introduction

In Chapter 11, we derived a system of differential equations, the Kolmogorov equations which, as we observed, completely defined the transition probabilities of the continuous parameter Markov process (with the initial conditions, $\mathbf{P}(0) = \mathbf{I}$). Our job in that chapter was to show the basic probability structure of the Markov process, and to present several examples of how these processes arise naturally in real-life processes. Except for the simplest examples, such as the Poisson process and the two-state process in Examples 11.3 and 11.4 of Section 6 of that chapter, we did not attempt to solve the differential equations explicitly. This was not an oversight on our part for two reasons. First, the solution of the Kolmogorov equation for all but the simplest cases often requires methods that are beyond the scope of this book. (Although one general method which is applicable to finite-state processes was indicated in Section 8.) Second, in many engineering problems one's primary interest is to study the behavior of the solutions for large values of t. As it develops, simple methods are available for studying this behavior that do not require solving the differential equations themselves explicitly. In this chapter we shall explore and exploit these methods to get limiting values of the probabilities as $t \to \infty$ and, hence, approximate values for large t.

In effect, our approach follows rather closely the approach of Chapter 8, Section 8, where we studied Markov chains with a discrete parameter. We recall from our discussion that there were cases where the limiting probabilities exist but that there were also cases in which these limits did not themselves form a probability distribution. Thus, for example, in Chapter 8 we observed that, if the Markov chain was finite, irreducible, and aperiodic, then there always existed a limiting distribution with the property that

$$\lim_{n\to\infty} p_{ij}^{(n)} = v_j, \qquad \text{not depending on } i$$

$$\mathbf{v} = \mathbf{vP}$$

$$0 \le v_j \le 1$$

$$\sum_j v_j = 1$$

Our job in this chapter is to study the behavior of the $p_{ij}(t)$ for $t \to \infty$, to try to understand under what conditions these limits exist as probabilities (that is, under what conditions a limiting distribution exists for the continuous parameter Markov chain), and to find ways of determining these distributions. Of course, one way to find the limiting distribution, is to solve the Kolmogorov differential equations in Chapter 11 and then to compute $\lim_{t\to\infty} p_{ij}(t)$. It is this method that we want to avoid, because of the difficulty of determining $p_{ij}(t)$. Proofs of some of the assertions in this chapter require analysis well beyond the scope of this book. In these cases we will state the results without proof. The interested student can consult one of the listed references on Markov chains with a continuous parameter for more details.

Thus, in this chapter we wish to explore the following problem.

If $p_j(t) = \Pr[X_t = j]$, where X_t is a continuous, parameter Markov process:

(a) Do the limits

$$q_j = \lim_{t\to\infty} p_j(t) \tag{12.1}$$

exist? If so, how do they depend on the initial probability distribution

$$\{p_i = p_i(0)\}?$$

(b) How can these limits be computed?

(c) Do the quantities $\{q_j\}$ form a probability distribution? That is, does $\sum q_j$ equal 1?

2. The existence of the $\{q_i\}$

It can be shown that, under the assumptions of Chapter 11, all of the limits q_j defined by (12.1) do, in fact, exist, and that $0 \le q_j \le 1$ for each j.

In general these limits will depend on the initial value, X_0, (or on the probability distribution of X_0). For example, if X_t represents the number of bacteria in a sealed culture at time t, then the long-run behavior of X_t will depend crucially on whether the initial value of X_0 is 0 or not. In this case, 0 is termed an *absorbing state*, since if the process ever enters this state it never leaves it (see Chapter 8).

There is one important situation in which the initial distribution does not affect the limiting probabilities. This occurs when the process is *irreducible.*

In analogy with the discrete parameter case, a continuous parameter Markov process is termed *irreducible* if, for each i and j, the transition probability $p_{ij}(t)$ is strictly positive for some t. That is, a chain is irreducible if there is a positive probability of eventually reaching any state j from any state i.

THEOREM. *Under our assumptions, if the Markov process X_t is irreducible, then the limits q_j exist and always have the same values, regardless of the initial distribution at time* 0. △

Since $p_{ij}(t)$ can be interpreted as simply $p_j(t) = \Pr[X_t = j]$ provided that the initial value X_0 is assumed to be equal to i,($p_i = 1$, $p_k = 0$ for $k \neq i$) it follows that the statement that the $\{q_j\}$ exist and are independent of the initial distribution is equivalent to the formula:

$$\lim_{t\to\infty} p_{ij}(t) = q_j\,, \qquad \text{for all } i \tag{12.2}$$

3. Determination of $\{q_i\}$—the Normal Equations

It seems reasonable to hope that if the functions $p_{ij}(t)$ approach the constants q_j as $t \to \infty$, then the derivatives of these functions should approach 0 (since the derivative of a constant is 0). This is indeed true, although a rigorous proof is quite deep.

THEOREM. *Under our assumptions*

$$\lim_{t\to\infty} \frac{dp_{ij}(t)}{dt} = 0 \tag{12.3}$$

for each i and j. △

If one lets $t \to \infty$ in the Kolmogorov Forward Equations (11.29) and formally uses (12.2) and (12.3), the following system of linear homogeneous equations for the probabilities q_i is obtained (the *normal equations*):

$$\sum_k q_k \lambda_{kj} = 0 \qquad \text{for} \qquad j = 0, 1, 2, \ldots \tag{12.4}$$

In matrix form this system can be written as

$$\mathbf{q\Lambda} = \mathbf{0} \tag{12.5}$$

where $\mathbf{q} = (q_0, q_1, q_2, \ldots)$. Even for infinite state spaces (12.4) can be justified rigorously. (The proof that the limit of the infinite series is the series of limits requires great care here.)

Thus, we observe that the limits q_j do, in fact, exist and are independent of the initial conditions, if the Markov chain is irreducible. We also can hope that in these cases $\{q_j\}$ can be determined by solving the system (12.4) or (12.5). This gives a partial answer to (a) and (b) of Section 1. Question (c) must still be explored.

4. The Existence of a Limiting Probability Distribution

A homogeneous system of equations like (12.4) or (12.5) always has the trivial solution $\mathbf{q} = (0, 0, 0, \ldots)$ as one possible solution. If another solution exists, then an infinite number of solutions can be found by multiplying by scalars. Thus some additional information is needed to uniquely determine the q_j's. This additional information is provided in the following theorem.

If X_t is an irreducible Markov process, then it can be shown that *either* the limiting probabilities are identically 0, $q_j \equiv 0$ for all j, *or* that they are all positive and form a probability distribution, $q_j > 0$ for all j and $\sum q_j = 1$. In the second case, we say that the process is *positive recurrent*.

Combining the results stated in this section with the results of the previous section, we can show the following.

THEOREM. *Let X_t be an irreducible Markov process that satisfies our assumptions.*

(i) *If the system* (12.4) *or* (12.5) *has the trivial solution* (0, 0, 0, ...) *as the only solution for which $\sum q_j < \infty$, then $q_j \equiv 0$ for all j.*

(ii) *If the system* (12.4) *or* (12.5) *has exactly one solution*

$$\mathbf{q} = (q_0, q_1, q_2, \ldots)$$

in which each $q_j > 0$ and $\sum q_j = 1$, then the system is positive recurrent, and this solution gives the limiting probabilities.

(iii) *If the state space is finite, case* (ii) *must occur.* △

Suppose that the process X_t does have a limiting distribution $\{q_j\}$ which is itself a probability distribution. If that limiting distribution is then taken to be the initial distribution at $t = 0$, $p_i = p_i(0) = q_i$, it follows that

$$p_{ij}(t) \equiv q_j$$

for every $t \geq 0$. That is, the transition probability functions are all constants.

We notice, of course, that (iii) is just a restatement of Theorem 3 in Chapter 8 for the continuous parameter case, and that (i) and (ii) are restatements of the theorem of Section 4, Chapter 10.

5. Irreducible Examples

In Section 6 of Chapter 11, Examples 11.4, 11.5, and 11.6, as well as the general birth-and-death model in Section 7, are irreducible, provided that all the parameters involved are nonzero. The student should convince himself of the truth of this assertion. We shall illustrate how the results of the preceding section can be applied to obtain the limiting stationary distribution in these cases. The following examples refer to the example numbers in Section 6 of Chapter 11. Notice that examples 11.4 and 11.5 have finite state spaces, and hence are known to be positive recurrent by the Theorem of the preceding section.

EXAMPLE 12.1. THE MACHINE OPERATION CYCLE (See Example 11.4). In this example the limiting probabilities can be computed explicitly by using formulas (11.37) for the $p_{ij}(t)$.

$$\begin{aligned} q_0 &= \lim_{t\to\infty} p_{i0}(t) = \frac{\mu}{\lambda+\mu} \\ q_1 &= \lim_{t\to\infty} p_{i1}(t) = \frac{\lambda}{\lambda+\mu} \end{aligned} \tag{12.6}$$

These results could also be found by solving the system (12.4), which in this case becomes

$$\begin{aligned} -\lambda q_0 + \mu q_1 &= 0 \\ \lambda q_0 - \mu q_1 &= 0 \end{aligned}$$

Each equation is equivalent to $q_1 = (\lambda/\mu)q_0$, and hence $q_0 + q_1 = [(\lambda+\mu)/\mu]q_0$. Thus (12.6) is easily seen to be the only solution of this system that satisfies $q_i > 0$, $q_0 + q_1 = 1$. By (ii) of the preceding theorem the system is positive recurrent with these limiting probabilities. The approach of the $p_{ij}(t)$ to their limiting values is illustrated in Figure 11.5. △

EXAMPLE 12.2. THE MACHINERY BREAKDOWN EXAMPLE WITH SEVERAL MACHINES. (See Example 11.5). In this case the system of normal equations (12.4) becomes

$$-\lambda q_0 + \mu q_1 = 0$$

$$\lambda q_{i-1} - (\lambda + \mu i)q_i + \mu(i+1)q_{i+1} = 0, \quad \text{for} \quad i = 1, 2, \ldots, N-1$$

$$\lambda q_{N-1} - \mu N\, q_N = 0$$

From the first equation we have

$$q_1 = \left(\frac{\lambda}{\mu}\right) q_0$$

The second equation (with $i = 1$) can be written as

$$\begin{aligned} q_2 &= \frac{1}{2\mu}\left[(\lambda + \mu)q_1 - \lambda q_0\right] \\ &= \frac{1}{2\mu}\left[(\lambda + \mu)\left(\frac{\lambda}{\mu}\right) q_0 - \lambda q_0\right] \\ &= \frac{1}{2\mu}\left[\left(\frac{\lambda^2}{\mu}\right) q_0\right] \\ &= \frac{1}{2}\left(\frac{\lambda}{\mu}\right)^2 q_0 \end{aligned}$$

This can now be substituted in the equation with $i = 2$, etc., and the terms q_3, q_4, ... solved for recursively in terms of q_0. The general result is

$$q_i = \frac{1}{i!}\left(\frac{\lambda}{\mu}\right)^i q_0, \qquad \text{for} \qquad i = 0, 1, 2, \ldots, N \tag{12.7}$$

(The student should check this.) To find q_0 we simply set

$$\sum q_i = \left[\sum \frac{1}{i!}\left(\frac{\lambda}{\mu}\right)^i\right] q_0 = 1,$$

and find

$$q_0 = \frac{1}{\sum_{i=0}^{N} \frac{1}{i!}\left(\frac{\lambda}{\mu}\right)^i} \tag{12.8}$$

By the theorem in the preceding section, (12.7) and (12.8) do, in fact, give the limiting probabilities for this positive recurrent system. △

Formulas (12.7) and (12.8) together constitute the *Erlang loss formula*, a formula of wide application in congestion and waiting-line theory, where situations frequently occur that can be modeled by a Markov process with rate matrix of the form considered in this example.

EXAMPLE 12.3. THE SIMPLE QUEUEING MODEL (see Example 11.6). In this case the normal equations (12.4) take the form

$$-\lambda q_0 + \mu q_1 = 0$$
$$\lambda q_{i-1} - (\lambda + \mu) q_i + \mu q_{i+1} = 0, \qquad \text{for} \qquad i = 1, 2, 3, \ldots \tag{12.9}$$

As in the preceding example, we can solve recursively for $q_1, q_2, q_3, \ldots$ in terms of q_0. The result is

$$q_i = (\lambda/\mu)^i q_0, \qquad \text{for} \qquad i = 0, 1, 2, \ldots \tag{12.10}$$

(The student should check this.) In order that $\sum q_i = 1$, we must have

$$q_0 \sum_{i=0}^{\infty} \left(\frac{\lambda}{\mu}\right)^i = 1 \tag{12.11}$$

Two cases must now be considered:

(i) $\lambda < \mu$. In this case the series in (12.11) converges and can be evaluated by using the formula for the sum of a geometric series:

$$q_0 \cdot \frac{1}{1 - \lambda/\mu} = 1$$

$$q_0 = 1 - \frac{\lambda}{\mu}$$

$$q_i = \left(1 - \frac{\lambda}{\mu}\right)\left(\frac{\lambda}{\mu}\right)^i, \qquad \text{for} \qquad i = 0, 1, 2, \ldots \tag{12.12}$$

Formula (12.12) gives the only solution of the system (12.9) that satisfies $q_i > 0$, $\sum q_i = 1$, and hence by (ii) of the theorem of the preceding section, the system is positive recurrent with limiting probabilities given by (12.12).

(ii) $\lambda \geq \mu$. In this case the series in (12.11) diverges, and thus equation (12.11) cannot be satisfied for any choice of q_0. The series $\sum q_j$ will diverge for any choice of $q_0 > 0$. By (i) of the theorem of the preceding section, it follows that

$$q_j \equiv 0 \qquad \text{for all } j$$

and thus that the system is not positive recurrent.

By formula (12.12) we observe that in the positive recurrent case, $\lambda < \mu$, the limiting distribution of X_t is *modified geometric* with common ratio $p = \lambda/\mu$ (see Section 4 of Chapter 6). It will have mean and variance.

$$\frac{p}{1-p} = \frac{\lambda}{\mu - \lambda} \qquad \text{and} \qquad \frac{p}{(1-p)^2} = \frac{\lambda\mu}{(\mu - \lambda)^2}$$

Note that the results of this example were also obtained in Example 9.2, Section 3, Chapter 9 by using different methods. △

EXAMPLE 12.4. THE GENERAL BIRTH-AND-DEATH MODEL. This process defined in Section 7 of Chapter 11 is irreducible only if all parameters are nonzero. For example, if $\lambda_0 = 0$, then the state 0 becomes an absorbing state which once entered can never be left, $p_{0j}(t) = 0$ for every $j > 0$; (no "birth" from state 0 is possible). Let us assume that all the parameters λ_i and μ_i are nonzero.

For this case the system (12.4) becomes

$$-\lambda_0 q_0 + \mu_1 q_1 = 0 \tag{12.13}$$

$$\lambda_{j-1} q_{j-1} - (\lambda_j + \mu_j) q_j + \mu_{j+1} q_{j+1} = 0, \quad \text{for} \quad j = 1, 2, 3, \ldots \tag{12.14}$$

Equation (12.14) can be written as

$$\mu_{j+1} q_{j+1} - \lambda_j q_j = \mu_j q_j - \lambda_{j-1} q_{j-1} \tag{12.15}$$

By (12.13), $\mu_1 q_1 - \lambda_0 q_0 = 0$. If this is substituted into (12.15) with $j = 1$, we find that $\mu_2 q_2 - \lambda_1 q_1 = 0$. Substituting again, we obtain by successive substitution

$$\mu_{j+1} q_{j+1} - \lambda_j q_j = 0 \qquad \text{for each } j = 0, 1, 2, \ldots$$

or

$$q_{j+1} = \frac{\lambda_j}{\mu_{j+1}} q_j \tag{12.16}$$

Successive application of (12.16) yields

$$q_n = \frac{\lambda_0 \lambda_1 \ldots \lambda_{n-1}}{\mu_1 \mu_2 \ldots \mu_n} q_0, \qquad n = 1, 2, \ldots \tag{12.17}$$

If the series

$$S = 1 + \frac{\lambda_0}{\mu_1} + \frac{\lambda_0 \lambda_1}{\mu_1 \mu_2} + \cdots + \frac{\lambda_0 \lambda_1 \cdots \lambda_{n-1}}{\mu_1 \mu_2 \cdots \mu_n} + \cdots \tag{12.18}$$

is convergent, then by (ii) of the theorem of the preceding section the system is positive recurrent with limiting probabilities given by (12.17) with

$$q_0 = \frac{1}{S}$$

If the series is divergent, then case (i) applies and the limiting probabilities are

$$q_i = 0 \qquad \text{for} \qquad i = 0, 1, 2, \ldots \quad \triangle$$

From this example we obtain the very important and useful theorem:

THEOREM. *In an irreducible birth-and-death process with* $\lambda_i > 0$, $i = 0$, 1, 2, ... *and* $\mu_i > 0$, $i = 1, 2, \ldots$, *a limiting stationary distribution exists if the sum* (12.18) *converges.* $\triangle$

6. An Interpretation of the $\{q_i\}$

In many practical problems for which a limiting distribution exists, we interpret q_j as the probability that the system will be in state j a "long time" after $t = 0$. As a consequence, we talk of q_j as the *long-run probability* of being in state j, or the *steady state probability* for state j. In many cases, we can show that $p_{ij}(t)$ approaches q_j at a fairly rapid rate as $t \to \infty$. Thus, in many practical applications, we assume that these limiting probabilities are, in fact, good measures of a system's behavior even if the system is observed for modest t-values instead of for large t. Of course, from our study, we do not know the rate at which a particular process approaches its steady state probabilities, and hence care must be exercised in this interpretation of the q_j.

There is another interpretation of the $\{q_j\}$ that make them useful in application. Our approach to the $\{q_j\}$ shows that they are limiting *state* probabilities. That is, for large t, q_j is the probability of *being in state* j (or alternatively the $\{q_j\}$ are the limiting transition probabilities). With a deeper study of the theory of stochastic processes q_j can also be interpreted as *the proportion of time that the system spends in state* j, over the long run. This is an interesting relation. What we are asserting is that state probabilities can be interpreted as proportions of time. The limiting probability of being in state j is equal to the proportion of time spent in the state. Clearly these two concepts are quite different, and we might not expect, in general, that probabilities with respect to states were in any way related to the length of time that a state is occupied. It is of practical value (and theoretically verifiable) that the relation exists. As one consequence, the relation implies that to estimate empirically the probabilities q_j we can proceed in one of two ways. First we could observe many different realizations of the Markov process, long after their commencements. From these many observations we could form a relative frequency ratio (as we did in Chapter 2) of the number of observations in state j relative to the total number of observations. The resulting proportion of observations of j would then serve as an estimate of q_j. Alternatively, we could observe only *one* realization of the Markov process over a long period of time. From this observation we could form the relative frequency ratio of the total time spent in state j relative to the total time of observation. The resulting *proportion* of time in state j would also serve as an estimate of q_j. The importance of this result is that in many applied problems we have available a reasonably long series of observations on a single realization of the process. But many different series of observations are not available, because of the expense of time consumed in collecting this amount of data. Nonetheless, from our remark, we can still estimate the q_j from this one series of observations by using the proportion of time the process spends in state j.

7. Some Nonirreducible Processes

One way in which a process can fail to be irreducible is to have one or more *absorbing states* [that is, states i for which $p_{ij}(t) = 0$ for every $j \neq i$], or *absorbing classes* of states. An absorbing class C is a class of states such that $p_{ij}(t) > 0$ for each $i, j \in C$, and some $t > 0$, while $p_{ij}(t) = 0$ for $i \in C, j \notin C$. If such states or classes exist, then the existence of a unique limiting distribution cannot be guaranteed, since the behavior of the process as $t \to \infty$ might depend crucially on whether or not it started in an absorbing state.

For a nonirreducible process, by a deeper analysis one can show that as $t \to \infty$ one of two things must happen to X_t: either it eventually enters one of the absorbing states or classes (is *absorbed*), or it eventually grows larger and larger, $X_t \to \infty$. Of course, if the state space is finite only the first possibility exists.

8. Examples

EXAMPLE 12.5. Suppose that we consider the birth-and-death model, section 7 of Chapter 11, with

$$\lambda_0 = 0$$

$$\lambda_i = i\lambda, \qquad \text{for} \qquad i = 1, 2, 3, \ldots$$

$$\mu_i = i\mu, \qquad \text{for} \qquad i = 1, 2, 3, \ldots$$

In this case, X_t could be interpreted as the number of bacteria or individuals present in a certain sealed environment; λ and μ are the per unit birth and death rates. The terms $i\lambda$ and $i\mu$ indicate that the probability of a birth or death is proportional to the number present. We observe that state 0 is an absorbing state. Thus, as $t \to \infty$, either X_t eventually enters state 0 (the population becomes extinct), or $X_t \to \infty$. The important question here is: Which of these possibilities happen, given that the population starts with i individuals? Thus we wish to compute

$$\alpha_i = \Pr[X_t \text{ is absorbed in } 0 \mid X_0 = i] = \Pr[\text{population becomes extinct} \mid \text{initially } i \text{ individuals}]$$

Clearly, $\alpha_0 = 1$.

For absorption into 0 to occur, starting with state i, in a small time interval Δt the process will either proceed from state i to state $i+1$ and then eventually be absorbed, or it will proceed from state i to state $i-1$ and then eventually be absorbed, or it will stay in state i and eventually be absorbed. Thus

$$\alpha_i = i\lambda\Delta t\, \alpha_{i+1} + i\mu\Delta t\, \alpha_{i-1} + [1 - i(\lambda+\mu)\Delta t]\alpha_i + o(\Delta t)$$

for $i = 1, 2, 3, \ldots$. It follows that

$$i(\lambda+\mu)\Delta t\, \alpha_i = i\lambda\Delta t\, \alpha_{i+1} + i\, \mu\Delta t\, \alpha_{i-1} + o(\Delta t)$$

or, dividing by $i\Delta t$ and letting $\Delta t \to 0$,

$$(\lambda+\mu)\alpha_i = \lambda\alpha_{i+1} + \mu\alpha_{i-1}, \qquad \text{for } i = 1, 2, 3, \ldots$$

$$\alpha_i = \frac{\lambda}{\lambda+\mu}\alpha_{i+1} + \frac{\mu}{\lambda+\mu}\alpha_{i-1}, \qquad \text{for } i = 1, 2, 3, \ldots$$

This system of equations is identical with the system discussed in the Gambler's Ruin example in Section 6 of Chapter 7, with $p = \lambda/(\lambda+\mu)$, $q = \mu/(\lambda+\mu)$, and $N \to \infty$. Thus the solution is given by (7.28)

$$\alpha_i = \begin{cases} (\mu/\lambda)^i & \text{if} \quad \mu < \lambda \\ 1 & \text{if} \quad \mu \geq \lambda \end{cases} \tag{12.19}$$

Thus, extinction of the population is *certain* if the death rate exceeds *or is equal to* the birth rate. If the birth rate exceeds the death rate, extinction is not *certain*, but is *possible* with probability given by (12.19). △

EXAMPLE 12.6. In the Poisson process no absorbing states exist, yet the process is not irreducible since only transitions to higher states are possible. Thus, necessarily, $X_t \to \infty$ as $t \to \infty$, and $p_{ij}(t) \to 0$ for every i, j. (This can be seen directly by using formula (11.15):

$$q_j = \lim_{t\to\infty} p_{ij}(t) = \lim_{t\to\infty} e^{-\lambda t} \frac{(\lambda t)^{j-i}}{(j-i)!} = 0$$

for all $j \geq i$. Notice that $p_{ij}(t) \equiv 0$ for $j < i$, so that $q_j = \lim_{t\to\infty} p_{ij}(t) = 0$ in this case as well.) This agrees with our intuition. As $t \to \infty$, the number of calls coming in would increase without bound, so that the probability of any fixed number of calls coming in would approach zero. △

EXERCISES

1. For the process of Exercises 2 and 3, Chapter 11, find the limiting probabilities $q_j = \lim_{t\to\infty} p_{ij}(t)$ directly, and also by using the normal equations. Show that the same results are obtained.
2. Find the limiting probabilities for the process of Exercise 4, Chapter 11. For what proportion of time in a year will the communications system be out of action, with no satellites in operation?
3. Find the limiting probabilities for the machinery repair problem discussed in Exercise 9 of Chapter 11.
4. Consider the process discussed in Exercise 10 of Chapter 11. This can be considered as a queueing process in which arrivals are discouraged by a long queue (arrival rate inversely proportional to $i + 1$, where i is number present). Show that this process is always positive recurrent, unlike the simple queueing example, and determine its long-run probabilities.
5. Consider the queueing process with an unlimited number of servers discussed in Exercise 13 of Chapter 11.
   - **(a)** Show that this process is positive recurrent for all λ and μ.
   - **(b)** Show that the limiting distribution is Poisson.
6. Consider the queueing process with batch arrivals (2 at a time) discussed in Exercise 14 of Chapter 11.
   - **(a)** Under what conditions is this positive recurrent?
   - **(b)** Compute the limiting probabilities.
7. In the machinery breakdown Example 12.2, show that the mean and variance of the limiting distribution are

$$\left(\frac{\lambda}{\mu}\right)(1 - q_N) \qquad \text{and} \qquad \left(\frac{\lambda}{\mu}\right)(1 - q_N)\left(\frac{\lambda}{\mu} q_N + 1\right) - \left(\frac{\lambda}{\mu}\right)^2 q_{N-1}$$

13 • Introduction to Queueing Theory

1. Introduction

One of the useful areas of application of the theory of Markov processes is in the study of waiting lines. Queues (waiting lines) abound in practical situations. The history of the development of queueing theory is found in fields as seemingly diverse as hospital management and time-shared computer system design. The earliest use of queuing theory was in the design of telephone systems. In a queueing system one supposes that there are "customers" arriving to a "server" for service. The time between arrivals is a random variable and the time to perform service is a random variable. Because of this random character, it is impossible to guarantee that service completions coincide exactly with customer arrivals. Consequently, there are times when the server completes his service on a customer and finds no one else available to work on. Thus he is idle until the next customer arrives. But there are also times when a customer arrives and finds the server already occupied with some previous arrival. This customer must then wait his turn or leave. In early telephone studies, physical storage devices for letting calls wait were not available, and hence in those studies one assumed that customers who arrived during the time when a server (trunk line) was busy were lost. They left without waiting. In some present day applications (e.g. inventory control theory), there are physical storage devices that are large and, hence, one can reasonably suppose that customers who arrive when the server is busy wait their turn for service. They queue up and form a waiting line or backlog of work to be done.

There are broadly four problems of interest in the study of waiting lines. We can ask how long an arriving customer will wait. This *waiting time*, of course, is a random variable, and we would be interested in its probability density or distribution function. The sequence of waiting times of the successive customers forms a stochastic process with a discrete parameter space and continuous state space. A second stochastic process is called the *busy period.* Here one is interested in determining the probability density function for the time from the commencement of service of a customer who terminates an idle time until the server becomes free for the first time after this. The third problem is the *queue-length problem*. In this we are concerned with the number of customers waiting for service (including the ones that are being served) at any given time. In this chapter we are primarily concerned with this third problem. A fourth problem would be the study of the *departure process*, which describes the behavior of the stream of customers that are leaving the system after completing service. This would be of particular interest if these individuals then were to go to another queueing system for still another service operation.

Many of the examples discussed in Chapters 8, 10, 11, and 12 can be considered as queueing processes. The machine repair Examples 11.4 and 11.5 can be interpreted as queueing problems in which the queue consists of machines awaiting repair, and the servers are the repair crews. Other situations involve the demands on a computer system, the demands for long-distance telephone lines, the customers at banks and toll booths, the scheduling of patients at hospitals and clinics, and the arrival of airplanes at terminal facilities, etc. (See Figure 13.1.)

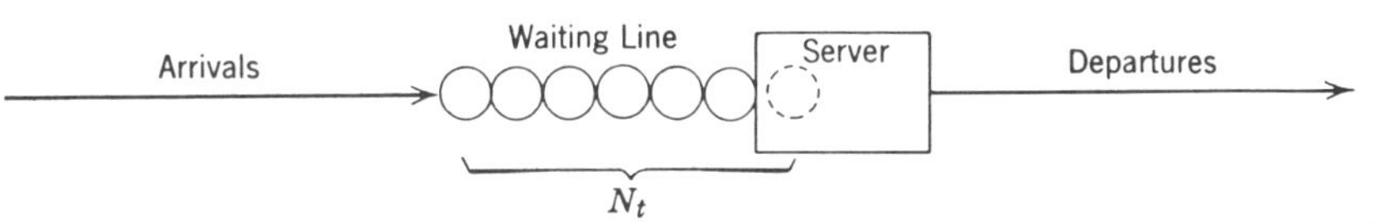

Figure 13.1 A schematic queueing system.

2. Definitions

The description of any queueing system requires a discussion of three parts: the *arrival process*, which describes the way in which individuals join the queue; the *queue discipline*, which describes the way in which choices are made for service (first come first served, last come-first served, random choice of an individual to be served, etc.); and the *server mechanism* which describes the length of time each service operation requires. Since in this chapter we are concerned primarily with queue lengths rather than with waiting times of individuals, we can ignore the effects of different types of queue disciplines.

Consider the sequence $X_1, X_2, X_3, \ldots$ of *interarrival times*. Thus X_1 is the time until the first individual arrives at the system, and X_2 is the time between the first and second arrivals, etc. This random sequence constitutes a discrete parameter random process, each realization of which defines an entire sequence of arrivals to the system. The time T_n of the nth arrival is given by

$$T_n = X_1 + X_2 + \cdots + X_n \tag{13.1}$$

Instead of describing the arrivals in terms of their interarrival times $\{X_n\}$, they can also be described by the process $Y_t =$ the total number of arrivals up to time t, $t \geq 0$. Thus, Y_t is a continuous parameter random process termed the *arrival process from time* 0. The event "n arrivals before time t," is equivalent to "the nth arrival is before time t but the $(n+1)$st arrival is after time t." In symbols:

$$[Y_t = n] = [T_n \leq t < T_{n+1}]$$

Thus the Y_t process is determined once the sequence of interarrival times $X_1, X_2, \ldots$ is known, and conversely. The Y_t process proceeds successively through the states 0, 1, 2, ..., and hence is termed a *counting process*.

More generally, let $Y_{\tau, t}$ represent the number of arrivals *between* times τ and t, $0 \leq \tau \leq t$. Thus, $Y_t = Y_{0, t}$; $Y_{\tau, t}$ is termed the *arrival process from time* τ.

The service mechanism is specified by giving the random sequence $Z_1, Z_2, \ldots$ of service times of the successive individuals served. In a way similar to the one used in defining the arrival processs Y_t in terms of the interarrival times, we also can define a continuous parameter process by describing the service mechanism. Some care is needed, however, since the server will be idle for certain time intervals when the queue is empty. Thus, for each $\tau \geq 0$, we define the *service process from time* τ, $Z_{\tau, t}$, as giving the number of service completions between time τ and time t, $\tau \leq t$, *with the clock being stopped whenever the server is idle*. Thus, $Z_{\tau, t}$ counts the number of service completions between times τ and t, assuming that an individual is always available to enter the service system immediately after each service completion.

The random sequences $\{X_n\}$ and $\{Z_n\}$ or, equivalently, the processes $Y_{\tau, t}$ and $Z_{\tau, t}$, determine the behavior of the waiting line.

Let N_t represent the queue length at time t, that is, the total number of individuals either waiting for service or being served at time t. Thus N_t is a continuous parameter random process with state space $\{0, 1, 2, \ldots\}$. This is the random process that we wish to discuss.

In a complete discussion of queueing systems one should classify systems according to the *number of servers* making up the service system and the maximum number of individuals who can be accommodated in the queue (*storage size*).

3. The Simple Markovian Queue

In general, the process N_t is not a Markov process. In certain important special cases, however, it does have the Markov property and the discussion in Chapters 11 and 12 can be applied.

Let us consider a single-server, unlimited-storage queueing system with the following properties:

(i) *The probability of an arrival in a time interval of length* Δt *is* $\lambda\Delta t + o(\Delta t)$.

(ii) *The probability of a service completion in a time interval of length* Δt *is* $\mu\Delta t + o(\Delta t)$, *subject only to the condition that the queue is not empty at the start of the interval,* $N_t > 0$.

(iii) *The probability of more than one arrival or service completion in a time interval of length* Δt *is* $o(\Delta t)$.

(iv) *The events in* (i) *and* (ii) *are independent of each other and of the past history of the system.*

From properties (i) to (iv) we can show rather readily that the interarrival times $\{X_n\}$ and service times $\{Z_n\}$ satisfy the "forgetfulness" property that characterizes the negative exponential distribution. The sequence $\{X_n\}$ is a sequence of independent random variables that each have density $\lambda e^{-\lambda x}$, $x > 0$; the sequence $\{Z_n\}$ is a sequence of independent random variables that each have density $\mu e^{-\mu z}$, $z > 0$; and these two sequences are independent of each other. The processes $Y_{\tau, t}$ and $Z_{\tau, t}$ are Poisson processes of the type discussed in Section 7 of Chapter 7.

In view of the "forgetfulness" property of the exponential, the probability of an arrival or a departure in some future time interval is unaffected by the past history of the process, given the queue length at some fixed time t, $N_t = n$. Thus, N_t represents a continuous parameter Markov process, termed a *simple Markovian queue*. Such a system is particularly easy to analyze and serves as a good approximation to many real-world queueing situations.

From properties (i) to (iii) it follows that the rate matrix for this Markov chain has the form:

$$\mathbf{\Lambda} = \begin{pmatrix} -\lambda & \lambda & 0 & 0 & \cdots \\ \mu & -(\lambda+\mu) & \lambda & 0 & \cdots \\ 0 & \mu & -(\lambda+\mu) & \lambda & \cdots \\ \cdot & \cdot & \cdot & \cdot & \\ \cdot & \cdot & \cdot & \cdot & \\ \cdot & \cdot & \cdot & \cdot & \end{pmatrix}$$

Thus this process is the one given in Example 11.6, with limiting properties discussed in Example 12.3.

If $\lambda < \mu$ and $t \to \infty$, then by (12.12) the state probabilities approach those of a random variable that has the modified geometric distribution discussed in Section 4 of Chapter 6, with parameter $p = \lambda/\mu$,

$$p_n(t) = \Pr[N_t = n] \to q_n = \left(1 - \frac{\lambda}{\mu}\right)\left(\frac{\lambda}{\mu}\right)^n \tag{13.2}$$

for $n = 0, 1, 2, \ldots$. The mean and variance of N_t approach those of the modified geometric given by (6.17) and (6.18):

$$E[N_t] \to \frac{\lambda/\mu}{1 - \lambda/\mu} = \frac{\lambda}{\mu - \lambda} \tag{13.3}$$

$$\operatorname{Var}[N_t] \to \frac{\lambda/\mu}{(1 - \lambda/\mu)^2} = \frac{\lambda\mu}{(\mu - \lambda)^2} \tag{13.4}$$

If $\lambda \geq \mu$, then the system is not positive recurrent. The queue length N_t grows beyond any bound as $t \to \infty$, with probability 1. Thus, no limiting probability distribution exists. It is important to notice that this happens even when $\lambda = \mu$. This implies that any attempt to exactly balance the queue by making the service rate exactly equal to the arrival rate will inevitably lead to an unstable queue. Basically this is because random downward fluctuations in queue length are bounded, since the queue length cannot be less than zero, while no such bound exists on random upward fluctuations.

4. The Multiple-Server Markovian Queue

Let us assume that the same situation as in the previous section holds true except that the system has r identical serving units operating in parallel, rather than just 1. Properties (i), (iii), and (iv) still hold true; however, property (ii) must be modified somewhat. If the number of individuals in the system does not exceed r, $N_t = i \leq r$, then the probability of a departure in time Δt is approximately proportional to i, as well as to Δt. This follows since, by neglecting $o(\Delta t)$ terms, the probability of a service completion in time Δt if 2 individuals are being served is twice what it would be if only 1 individual were to be served, etc. Thus (ii) must be replaced by the following.

(ii)′ *Given that $N_t = i$, the probability of a service completion in the time interval $(t, t + \Delta t)$ is $i\mu\Delta t + o(\Delta t)$ if $i \leq r$, and is $r\mu\Delta t + o(\Delta t)$ if $i > r$.*

Now the departures after time t clearly depend on the value of N_t, but not on the past history. Thus the Markovian property still holds true. From (i) and (ii)′ the rate matrix $\mathbf{\Lambda}$ is found to be

$$\mathbf{\Lambda} = \begin{pmatrix} -\lambda & \lambda & 0 & 0 & \cdots & & \\ \mu & -(\lambda+\mu) & \lambda & 0 & \cdots & & \\ 0 & 2\mu & -(\lambda+2\mu) & \lambda & \cdots & & \\ & & \ddots & \ddots & \ddots & & \\ & & & r\mu & -(\lambda+r\mu) & \lambda & \cdots \\ & & & & r\mu & -(\lambda+r\mu) & \lambda \\ & & & & & \ddots & \ddots \end{pmatrix}$$

We see that N_t is a birth-and-death process of the type that is discussed in Example 12.4, with $\lambda_i=\lambda \quad \mu_i=i\mu$ for $i\leq r$, $\mu_i=r\mu$ for $i>r$. The series (12.18) becomes

$$S=1+\frac{\lambda}{\mu}+\frac{\lambda^2}{2\mu^2}+\cdots+\frac{\lambda^r}{r!\mu^r}+\frac{\lambda^r}{r!\mu^r}\sum_{i=1}^{\infty}\frac{\lambda^i}{(r\mu)^i} \tag{13.5}$$

This infinite series will converge if, and only if, $\lambda<r\mu$. In this case, the sum can be evaluated by using the formula for the sum of a geometric series, although the resulting expression is somewhat complicated. By (12.17) and the theorem of section 4, Chapter 12, we conclude the following.

If $\lambda<r\mu$,

$$p_n(t)=\Pr[N_t=n]\to\begin{cases}\dfrac{\lambda^n}{n!\mu^nS}, & \text{if } n\leq r\\[2ex] \dfrac{\lambda^n}{r!r^{n-r}\mu^nS}, & \text{if } n>r\end{cases} \tag{13.6}$$

as $t\to\infty$ where S is given by (13.5). Limiting formulas for the mean and variance can also be derived, but they are rather complex.

If $\lambda\geq r\mu$ the queue length grows beyond any bound with probability 1 as $t\to\infty$.

5. A Simple Markov Queue with Bounded Storage

Consider a simple one-server Markovian queue of the type discussed in Section 3, with the additional restriction that there is space for only K individuals in the system. If the queue reaches this size, then additional arrivals are turned away until the number waiting drops below K. The assumptions made in Section 3 still apply with the single exception that property (*i*) must be modified as follows.

(i)′ *Given that $N_t < K$, the probability of an addition to the queue in the time interval $(t, t + \Delta t)$ is $\lambda\Delta t + o(\Delta t)$. Given that $N_t = K$, the probability of an addition in this interval is $o(\Delta t)$.*

Properties (i)′ and (ii) to (iv) imply that N_t is a Markov chain having the finite state space $\{0, 1, 2, \ldots, K\}$ and transition rate matrix

$$\mathbf{\Lambda} = \begin{pmatrix} -\lambda & \lambda & 0 & 0 & \ldots & 0 & 0 & 0 \\ \mu & -(\lambda+\mu) & \lambda & 0 & \ldots & 0 & 0 & 0 \\ 0 & \mu & -(\lambda+\mu) & \lambda & \ldots & 0 & 0 & 0 \\ \cdot & \cdot & \cdot & \cdot & & \cdot & \cdot & \cdot \\ \cdot & \cdot & \cdot & \cdot & & \cdot & \cdot & \cdot \\ \cdot & \cdot & \cdot & \cdot & & \cdot & \cdot & \cdot \\ 0 & 0 & 0 & 0 & \ldots & \mu & -(\lambda+\mu) & \lambda \\ 0 & 0 & 0 & 0 & \ldots & 0 & \mu & -\mu \end{pmatrix}$$

Since the chain is irreducible and finite, it must be positive recurrent for all values of λ and μ. The limiting probabilities are easily found, by using (12.17), to be

$$p_n(t) = \Pr[N_t = n] \to \left(\frac{1 - \lambda/\mu}{1 - (\lambda/\mu)^{K+1}}\right)(\lambda/\mu)^n \tag{13.7}$$

for $n = 0, 1, 2, \ldots, K$.

6. Independent Interarrival and Service Times

Assume that the sequence $\{X_n\}$ of interarrival times forms a sequence of independent, identically distributed random variables. We assume also that there is only 1 server, and that the sequence $\{Z_n\}$ of service times also forms a sequence of independent, identically distributed random variables. Finally, we assume that the terms of the $\{X_n\}$ sequence are independent of the terms of the $\{Z_n\}$ sequence. Under these conditions, we say that the process N_t is a *general independent input and service queueing process*. The notation GI/G/1 is often used to describe such a process. (GI stands for *general independent input*, G for *general service*, and 1 for the *number of servers*). For r servers the notation GI/G/r is used. However, in this case, the independent, identically distributed sequence of service times refers to each individual server, not to the overall system. This notation allows us to specify a whole heirarchy of different queueing systems by specializing the probability distribution of the interarrival times $\{X_n\}$ or the service times $\{Z_n\}$. Many real-world queueing systems fall into these categories.

The simple Markovian queue discussed in Section 3 has this form. The interarrival times $\{X_n\}$ are independent, and each has a negative exponential density: $\lambda e^{-\lambda x}$ for $x > 0$. Similarly, the service times $\{Z_n\}$ are independent, and each has also a negative exponential density: $\mu e^{-\mu z}$ for $z > 0$. This situation is symbolized by the notation M/M/1. Here M stands for *Markovian* and refers to negative exponential interarrival times and service times (or, equivalently, to a Poisson arrival process and a Poisson service process). Similarly the r-server case discussed in Section 4 can be symbolized by M/M/r.

It can be shown that the M/M/1 and M/M/r queues are the *only* general independent input and service queueing processes for which N_t is Markovian.

7. The M/G/1 Queue. Embedded Chains

Let us now illustrate how we might handle a non-Markov queueing process of the type discussed in the previous section. Assume that we have a single-server queue in which the arrivals form a Poisson process. That is, the inter-arrival times are negative exponential with density: $\lambda e^{-\lambda x}$ for $x > 0$. Suppose that the service times $\{Z_n\}$ each have some fixed probability density function $f(z)$, $z > 0$. Again, we also assume that all interarrival and service times are mutually independent. If $f(z)$ is not specified, this queue is said to be of type M/G/1.

Such a queueing system results if arrivals come in essentially at random. For example, telephone calls arrive at an exchange, but the length of each call has a density, illustrated in Figure 13.2, which might appear related more to a gamma density than to a negative exponential.

Suppose that N_t is observed for some fixed time t, $N_t = n$. If $n > 0$, then some individual is being serviced at this time. Clearly, information about when this service operation began assists us in predicting when service will be completed on this individual (unless the service time is negative exponential and has the "forgetfulness" property). This information can be obtained by looking at the past history of the process—particularly, at the time that the previous individual completed his service. Thus the future behavior is not independent of the past; N_t does not satisfy the Markov property.

However, *if we happened to observe N_t at the instant after a departure has*

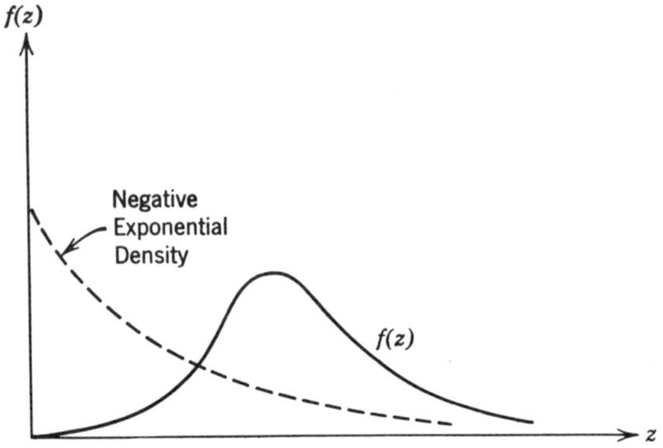

Figure 13.2 A possible service time density $f(z)$.

taken place from the queue, then we know that the next service operation is just starting, and no other information about the past will tell us any more about that operation. Thus the length of this service operation *is* independent of past history. Since the interarrival times are negative exponential, prior information does not help with them either. These observations suggest the following construction:

Let $\tau_1, \tau_2, \tau_3, \ldots$ represent the random sequence of *departure times*. If we consider the queue length N_t only at these times, we obtain a *discrete-parameter* random sequence

$$\{N_{\tau_k}\} = \{N_{\tau_1}, N_{\tau_2}, N_{\tau_3}, \ldots\}^*$$

This sequence is termed an *embedded chain* since it is extracted from the continuous parameter process N_t by observing it only at a discrete sequence of times. The observations in the preceding paragraph imply that this embedded chain $\{N_{\tau_k}\}$ is a Markovian sequence. Thus, instead of working with the entire continuous-parameter process N_t, which is non-Markovian and complicated, we extract from it a discrete parameter process $\{N_{\tau_k}\}$, which is Markovian, and study this simpler process.

$\{N_{\tau_k}\}$ has state space $\{0, 1, 2, \ldots\}$. We need to determine the 1-step transition probabilities:

$$p_{ij} = \Pr[N_{\tau_{k+1}} = j \mid N_{\tau_k} = i]$$

If $j < i - 1$ then clearly $p_{ij} = 0$, since the queue cannot decrease by more than 1 in an interval having only 1 departure.

* N_{τ_k} represents the number left behind after the departure at time τ_k. The departing individual is not counted.

If $j \geq i-1$ and $i > 0$, then p_{ij} is the probability of the queue length increasing from i to j during one service time. Since one individual departs, $j-i+1$ must arrive during this interval. The number of arrivals during a *fixed* time interval has a Poisson distribution. Thus the conditional density of the total number M of arrivals during the $(k+1)$st service time Z_{k+1}, given that $Z_{k+1}=z$, is

$$\begin{aligned} f(m \mid z) &= \Pr[m \text{ arrivals in time } z] \\ &= \Pr[M=m \mid Z_{k+1}=z] = e^{-\lambda z}\frac{(\lambda z)^m}{m!} \end{aligned}$$

for $m = 0, 1, 2, \ldots$. The joint density of M and Z_{k+1} is

$$f(m, z) = f(m \mid z) f(z) = e^{-\lambda z}\frac{(\lambda z)^m}{m!}\cdot f(z)$$

for $m = 0, 1, 2, \ldots$ and $z > 0$. This is a mixed situation that combines a discrete and a continuous random variable, but the analysis of Chapter 4, Section 8, still applies. The density g_m of M alone is obtained by integrating $f(m, z)$ with respect to z [see (4.22)]:

$$g_m = \Pr[M=m] = \int_0^\infty e^{-\lambda z}\frac{(\lambda z)^m}{m!} f(z)\, dz \tag{13.8}$$

It follows that

$$p_{ij} = g_{j-i+1} = \int_0^\infty e^{-\lambda z}\frac{(\lambda z)^{j-i+1}}{(j-i+1)!} f(z)\, dz$$

for $j \geq i-1$ and $i > 0$.

If $i = 0$, then we must modify the argument slightly. This means that after time τ_k the queue is empty. One must wait for an individual to arrive before the service mechanism can start up again, and this first individual to arrive will be the $(k+1)$st departure. He will leave behind a queue that contains all the arrivals during his service time. Thus, if $i = 0$,

$$p_{0j} = g_j = \int_0^\infty e^{-\lambda z}\frac{(\lambda z)^j}{j!} f(z)\, dz \tag{13.9}$$

for $j = 0, 1, 2, \ldots$, and hence the 0th row of the transition matrix coincides with the 1st row.

We conclude that the 1-step transition matrix for the embedded chain $\{N_{\tau_k}\}$ has the form

$$\mathbf{P} = \begin{pmatrix} g_0 & g_1 & g_2 & g_3 & \cdots \\ g_0 & g_1 & g_2 & g_3 & \cdots \\ 0 & g_0 & g_1 & g_2 & \cdots \\ 0 & 0 & g_0 & g_1 & \cdots \\ \cdot & \cdot & \cdot & \cdot & \\ \cdot & \cdot & \cdot & \cdot & \\ \cdot & \cdot & \cdot & \cdot & \end{pmatrix} \tag{13.10}$$

where g_m is given by (13.8).

The chain $\{N_{\tau_k}\}$ is irreducible, and its distribution can be found by solving the system $\mathbf{v} = \mathbf{vP}$ to find the stationary vector $\mathbf{v}$ (see Section 4 of Chapter 10). We will show that a stationary probability vector exists provided that

$$\lambda m < 1 \tag{13.11}$$

where $m = E[Z_k] = \int_0^\infty z f(z)\, dz$. By the theorem of Section 4, Chapter 10, the chain is positive recurrent if, and only if, (13.11) holds true. If $\lambda m \geq 1$, then $N_{\tau_k} \to \infty$ (and hence $N_t \to \infty$) with probability 1.

Most required information about the behavior of the original process N_t as $t \to \infty$ can be obtained by studying the behavior of the embedded chain $\{N_{\tau_k}\}$ as $k \to \infty$.

The equation $\mathbf{v} = \mathbf{vP}$ for the stationary vector $\mathbf{v}$ can be written in component form by using (13.10)

$$v_i = v_0 g_i + \sum_{k=1}^{i+1} v_k g_{i+1-k} \tag{13.12}$$

for $i = 0, 1, 2, \ldots$.

To solve (13.12), we use the method of generating functions. Define the function:

$$V(\xi) = \sum_{i=0}^{\infty} v_i \xi^i \tag{13.13}$$

If we can determine this function and show it to be convergent for $\xi = 1$, with $V(1) = \sum v_i = 1$ and coefficients satisfying (13.12), this will assure us that the system is positive recurrent. The limiting probabilities v_i can be found as the coefficients of the ξ^i-terms in the series expansion of $V(\xi)$.

To find $V(\xi)$, multiply each side of (13.12) by ξ^i and sum, for $i = 0, 1, 2, \ldots$.

$$V(\xi) = \sum_{i=0}^{\infty} v_i \xi^i = v_0 \sum_{i=0}^{\infty} g_i \xi^i + \sum_{i=0}^{\infty} \sum_{k=1}^{i+1} v_k g_{i+1-k} \xi^i$$

$$= v_0 \sum_{i=0}^{\infty} g_i \xi^i + \xi^{-1} \sum_{k=1}^{\infty} v_k \xi^k \sum_{i=k-1}^{\infty} g_{i+1-k} \xi^{i+1-k} \; *$$

$$= v_0 \sum_{i=0}^{\infty} g_i \xi^i + \xi^{-1}[V(\xi) - v_0] \sum_{j=0}^{\infty} g_j \xi^j$$

$$= [v_0(1 - \xi^{-1}) + \xi^{-1} V(\xi)] G(\xi) \tag{13.14}$$

where

$$G(\xi) = \sum_{i=0}^{\infty} g_i \xi^i \tag{13.15}$$

If we solve (13.14) for $V(\xi)$ we obtain

$$V(\xi) = \frac{v_0(\xi - 1)G(\xi)}{\xi - G(\xi)} \tag{13.16}$$

By the definition (13.9) of the terms g_i,

$$G(\xi) = \sum_{i=0}^{\infty} g_i \xi^i = \sum_{i=0}^{\infty} \xi^i \int_0^{\infty} e^{-\lambda z} \frac{(\lambda z)^i}{i!} f(z)\, dz$$

$$= \int_0^{\infty} e^{-\lambda z} \sum_{i=0}^{\infty} \frac{(\lambda z \xi)^i}{i!} f(z)\, dz$$

$$= \int_0^{\infty} e^{-\lambda z} \cdot e^{\lambda z \xi} f(z)\, dz$$

$$= \int_0^{\infty} e^{\lambda(\xi - 1)z} f(z)\, dz$$

$$= M[\lambda(\xi - 1)]$$

where

$$M(\theta) = E[e^{\theta z_k}] = \int_0^{\infty} e^{\theta z} f(z)\, dz$$

is the moment generating function of the service distribution. Thus (13.16) can be written in the form:

$$V(\xi) = \frac{v_0(\xi - 1) M[\lambda(\xi - 1)]}{\xi - M[\lambda(\xi - 1)]} \tag{13.17}$$

* The double summation here is over all indices i and k that satisfy $1 \leq k \leq i + 1 < \infty$ and we have rearranged the order of summation.

To find v_0 we let $\xi \to 1$, and hence $V(\xi) \to V(1) = 1$ in (13.17). Now $M[\lambda(\xi - 1)] \to M(0) = 1$. To compute the limit of the right side of (13.17) we must use l'Hôpital's rule.

$$\lim_{\xi \to 1} V(\xi) = 1 = v_0 \lim_{\xi \to 1} \frac{\xi - 1}{\xi - M[\lambda(\xi - 1)]}$$

$$= v_0 \cdot \frac{1}{1 - \lambda M'(0)}$$

Thus $v_0 = 1 - \lambda M'(0)$. But

$$M'(0) = \int_0^\infty z f(z)\, dz$$

$$= E[Z_k] = m$$

We conclude that

$$v_0 = 1 - \lambda m \tag{13.18}$$

Clearly v_0 will be a probability, and $V(\xi)$ will not be identically zero, only if $1 - \lambda m > 0$. But this is precisely condition (13.11). If (13.11) holds true, it can be shown by simple analytic methods that (13.17) exists and that it defines a convergent series for all $|\xi| \leq 1$. The final solution for $V(\xi)$ is

$$V(\xi) = \frac{(1 - \lambda m)(\xi - 1)M[\lambda(\xi - 1)]}{\xi - M(\lambda(\xi - 1))} \tag{13.19}$$

for $|\xi| \leq 1$, $\lambda m < 1$. In any given special case $M(\theta)$ can be computed and hence $V(\xi)$ can be determined by (13.19). The limiting probabilities v_i can then be found by expanding in powers of ξ.

8. Examples

EXAMPLE 13.1. Customers arrive at a certain airline counter in a terminal at a mean rate of 1 per minute. The manager wants to adjust his staff so that the probability of more than 5 customers waiting for service is no more than 5%. At what rate should he be prepared to service customers?

For convenience, assume that arrivals are completely random, so that the interarrival time is negative exponential, and that the service times are also negative exponential. Thus we have a simple Markovian queue, as discussed in Section 3. Arrivals form a Poisson process with mean rate λ per unit time. The hypothesis is that $\lambda = 1$. We wish to determine the mean service rate μ so that the long-run probability of a queue > 5 does not exceed .05.

$$\sum_{n=6}^{\infty} q_n \leq .05$$

By (13.2) this becomes

$$\left(1 - \frac{1}{\mu}\right) \sum_{n=6}^{\infty} \left(\frac{1}{\mu}\right)^n \leq .05$$

or

$$\frac{1}{\mu^6} \leq .05$$

$$\mu^6 \geq 20$$

$$\mu \geq 20^{1/6} = 1.65$$

Thus the facility must be prepared to handle customers at the mean rate of at least 1.65 per minute (or with mean service time $1/1.65 = .61$ minutes per customer) to meet the requirement. This rate is substantially greater than the mean arrival rate. When $\lambda = 1$ and $\mu = 1.65$, $q_0 = 1 - \lambda/\mu = 1 - 1/1.65 = .39$, so the manager must be prepared to accept the fact that his staff will be standing idle 39% of the time for the stated requirement to be met.

By varying the length of line that he can permit and by comparing it with the cost of providing the desired service, analyses like the one above permit the manager to achieve the best balance between cost and service. $\wedge$

EXAMPLE 13.2. In the preceding example the manager decides to arrange his staff so that there are two service positions at the counter, each with mean rate μ. How would this effect the value of the combined service rate (which will now be 2μ) that is necessary to meet the stated condition?

The discussion of Section 4 now applies with $r = 2$, $\lambda = 1$. By (13.5) and (13.6)

$$S = 1 + \frac{1}{\mu} + \frac{1}{2\mu^2} + \frac{1}{2\mu^2}\sum_{i=1}^{\infty}\left(\frac{1}{2\mu}\right)^i$$

$$= 1 + \frac{1}{\mu} + \frac{1/(2\mu^2)}{1 - 1/(2\mu)}$$

$$= 1 + \frac{1}{\mu} + \frac{1}{2\mu^2 - \mu}$$

$$= \frac{2\mu + 1}{2\mu - 1}$$

$$q_0 = \frac{1}{S} = \frac{2\mu - 1}{2\mu + 1}$$

$$q_1 = \frac{2\mu - 1}{\mu(2\mu + 1)}$$

$$q_n = \frac{2\mu - 1}{2^{n-1}\mu^n(2\mu + 1)} \qquad \text{for } n \geq 2$$

$$\sum_{n=6}^{\infty} q_n = \frac{2\mu - 1}{2^5\mu^6(2\mu + 1)} \cdot \frac{1}{1 - 1/(2\mu)} = \frac{1}{2^4\mu^5(2\mu + 1)}$$

This probability will be $\leq .05$ provided that

$$2^4\mu^5(2\mu + 1) \geq 20$$

This inequality can be solved to give $\mu \geq .856$. The necessary combined service rate is $2\mu = 1.712$, which exceeds the single rate of 1.65 found previously. Thus, splitting the service system does not improve matters. A still greater service capacity would be required then for a single channel. △

EXAMPLE 13.3. Answer the question raised in Example 13.1, if the system is to be designed so that the *mean* queue length should not exceed 2.

By (13.3) the limiting mean queue length is $\lambda/(\mu - \lambda) = 1/(\mu - 1)$. We can solve the inequality

$$\frac{1}{\mu - 1} \leq 2$$

and obtain $\mu \geq 1.5$. Thus a mean service rate of only 1.5 (less than the 1.65 found in example 13.1) will guarantee that the mean length queue is only 2, although there will then be a probability greater than .05 that the length will actually exceed 5. △

EXERCISES

1. Use (13.2) to show that if $\{X_1, X_2, \ldots\}$ is a sequence of identically-distributed, negative exponential random variables, then Y_t has a Poisson distribution for each t. (*Hint.* See Chapter 6, Section 8.)

2. Under the conditions of Exercise 1 show that $Y_{\tau,t}$ has a Poisson distribution for arbitrary $\tau \leq t$.

3. If $\{X_1, X_2, \ldots\}$ is a renewal process (see Chapter 7, Section 3) and the X_i have a gamma density function, determine the probability density function for

$$T_n = \sum_{i=1}^{n} X_i$$

(*Hint.* See Chapter 5, Section 12, and Chapter 6, Section 8.)

4. Consider the multiple server queue with $r = \infty$. Determine $\lim_{t\to\infty} p_n(t)$ and give the conditions under which the process is positive recurrent.

5. In Exercise 4 determine the mean and variance of the queue size for the limiting distribution.

6. Determine the limiting probabilities v_j, by using (13.19), for the special case of negative exponentially distributed service times. Compare your answer with (6.13) and infer that the queue-length probabilities are the ones given by (13.2). Compare it with the result obtained directly in Example 10.4. (*Note.* The limiting probabilities here are identical, but the two stochastic processes N_t and N_{τ_k} are clearly different.)

7. By using (13.19), show that the mean limiting queue length for the M/G/1 queue is given by

$$\lim_{t\to\infty} E[N_t] = \lambda m + \frac{(\lambda m)^2 + \lambda^2 \operatorname{Var}[Z_k]}{2(1 - \lambda m)}$$

[*Hint.* The mean $\sum iv_i = V'(1) = \lim_{\xi\to 1} V'(\xi)$, and l'Hôpital's rule must be used in evaluating this limit.] The mean queue-length formula is called the *Pollaczek-Khintchine* formula, and is extremely useful in queueing systems design.

8. By using the formula in Exercise 7, show that the mean queue-length is minimized for fixed λm whenever the variance of the service time is 0. Describe the service process for which this holds true, and give the mean queue length.

9. In the bounded storage example (Section 5), if a customer arrives when the queue is at full capacity K, that customer is turned away (lost). One can show that with probability $\lambda/(\lambda + \mu)$ the next event in this queueing system will be

an arrival, while with probability $\mu/(\lambda+\mu)$ the next event will be a departure. Determine the probability that the next arriving customer will be turned away.

10. In Example 13.1 it was shown that the probability of the server being idle was .39 if one wished to have a probability of less than .05 that the queue-length exceeded 5. Suppose that in that example the server was a computer that had to be kept running and that one could tolerate no more than 5% idle time resulting from the lack of work. By using the values given in Example 13.1, determine the queue-length probabilities and the mean and variance of the queue length for this system.

11. In Example 13.2 it was argued that we could change the performance of the system by adding another server. Clearly, for fixed λ, say $\lambda = 1$ as in that example, we could increase the service rate of one server or add another server. Suppose that we could speed up the servicing process so that the mean service time was $\frac{1}{2}$ of its previous value or that we could add another identical server. Compare q_0 and $\sum_{n=6}^{\infty} q_n$ for these two possible systems. Also compare the means and variances of the queue length for the two systems.

12. *The* G/M/1 *Queue*. Consider a single server queue with negatively exponentially distributed service times, for which the interarrival times $\{X_n\}$ have some fixed probability density function $g(x)$, $x > 0$. The process N_t is not Markovian for this queue. Discuss why.

13. (Exercise 12 continued) Define $T_1, T_2, \ldots, T_n$ to be the times at which arrivals occur. Verify that $\{N_{T_k}\}$ is a Markov chain with a discrete parameter space and a countable state space.

14. (Exercise 12 continued) Let $p_{ij} = \Pr[N_{T_{k+1}} = j \mid N_{T_k} = i]$ be the 1-step transition probabilities for the *embedded Markov chain* $\{N_{T_k}\}$. Notice that $p_{ij} = 0$ if $j > i + 1$ Determine p_{ij} for $j \leq i + 1$ in terms of $g(x)$ and the mean service rate μ.

14 • Further Properties of Stochastic Processes

1. Introduction

Most of our previous discussion of random processes has been aimed at processes that are (i) discrete state space processes and (ii) Markovian. Even with these restrictions we have found that these processes have wide applicability. In this book we have only begun the study of such processes. The serious student will find that the range of the theory and applications of discrete-state Markov processes goes far beyond what is covered here.

There also exists a substantial theory for analyzing and applying random processes that do not satisfy properties (i) and (ii) above. In this chapter we suggest some directions in which such generalizations can proceed, with primary emphasis on continuous state and continuous parameter random processes.

2. Notation and Definitions

If X_t is a random process, then X_t is a random variable for each fixed t. Define $F(x; t)$ to be the distribution function of this random variable:

$$\Pr[X_t \leq x] = F(x; t)$$

This is the *first-order probability distribution function* of the process. This function depends on both x and the parameter t.

If $F(x; t)$ can be differentiated (piecewise-continuously) with respect to x, then we have the *first-order probability density function*:

$$f(x; t) = \frac{\partial F(x; t)}{\partial x}$$

We can also define a second order or, in general, an *nth order distribution function*, as

$$F(x_1, x_2, \ldots, x_n; t_1, t_2, \ldots, t_n) = \Pr[X_{t_1} \leq x_1, X_{t_2} \leq x_2, \ldots, X_{t_n} \leq x_n]$$

for any set $t_1, t_2, \ldots, t_n$ of values of t. This probability is a function of t_1, t_2, t_n and $x_1, x_2, \ldots, x_n$. The joint density function, if it exists, is then the mixed nth derivative of this distribution function with respect to the n variables $x_1, \ldots, x_n$,

$$f(x_1, x_2, \ldots, x_n; t_1, t_2, \ldots, t_n) = \frac{\partial^n F}{\partial x_1 \, \partial x_2 \ldots \partial x_n}$$

Analogously, for discrete state processes the discrete probability density function $f(x_1, \ldots, x_n; t_1, \ldots, t_n)$ is simply defined as the probability of the event $[X_{t_1} = x_1, X_{t_2} = x_2, \ldots, X_{t_n} = x_n]$.

A stochastic process is completely specified as far as simple probability information is concerned by giving the joint distribution function F for all finite sets of t_i. This specification is usually extremely unwieldy and, hence, in practice one attempts to relate real systems to stochastic processes that are completely described by low-order distribution functions. Surprisingly, many important processes are completely specified by knowledge or first- or second-order distribution functions only.

For example, Markov processes are completely specified by knowledge of their second-order joint probabilities, since they suffice to determine the transition probabilities, or rates, and the initial probabilities and, hence, the entire process.

Since X_t is a random variable for each fixed value of t, one can consider the first moment

$$\mu(t) = E[X_t] = \int_{-\infty}^{\infty} x f(x; t) \, dx$$

[or $\sum_x x f(x; t)$, in the discrete case]. Notice that this expectation will be a function of t (but not of x).

Similarly we can consider higher order moments and covariances. One particularly important expectation is the second-order mixed moment

$$R(t_1, t_2) = E[X_{t_1} \cdot X_{t_2}] = \int_{-\infty}^{\infty} \int_{-\infty}^{\infty} x_1 x_2 f(x_1, x_2; t_1, t_2) \, dx_1 \, dx_2$$

[or $\sum_{x_1} \sum_{x_2} x_1 x_2 f(x_1, x_2; t_1, t_2)$ in the discrete case]. $R(t_1, t_2)$ is termed the *autocorrelation function* of the process. Many important properties of the process are related to this function. In particular, if we set $t_1 = t_2 = t$, then

$$R(t, t) = E[X_t^2] = \sigma^2(t) + [\mu(t)]^2$$

where $\sigma^2(t) = E\{[X_t - \mu(t)]^2\}$ is the *variance* of X_t.

3. Some General Properties of a Random Process

A random process can be classified in many ways.

(**a**) X_t is termed a *strictly stationary* process if its joint distribution function is unaffected by translations of the time axis

$$F(x_1, \ldots, x_n; t_1 + h, \ldots, t_n + h) = F(x_1, \ldots, x_n; t_1, \ldots, t_n)$$

for every h. Roughly speaking, this means that the joint probabilities depend only on the intervals *between* the times at which the process is to be observed, and not on when the sequence of observations starts.

(**b**) Random changes of the form $X_{t+h} - X_t$, for fixed $h > 0$, are termed *increments* of the process. If each set of increments, corresponding to a nonoverlapping collection of time intervals $(t_i, t_i + h)$, is mutually independent, then X_t is said to be a process with *independent increments*. If, in addition, the increment $X_{t+h} - X_t$ has a distribution that depends only on h, and not on t, the process X_t is said to have *stationary, independent increments*. When discussing these processes, for convenience, we usually make the additional assumption that $X_0 \equiv 0$, that is, the process initially starts in state 0.

In Example 4.22 and in the random walk of Chapter 7, Section 5, a process of partial sums was formed by adding the terms of a set of independent, identically distributed random variables that each take only the values ± 1. In the same way, a discrete parameter process $\{S_n\}$ can be formed from any sequence $\{X_n\}$ of independent, identically distributed random variables,

$$S_n = X_1 + X_2 + \ldots + X_n$$

for $n = 1, 2, 3, \ldots$. If this is done, then clearly the increments of this process are

$$S_{n+h} - S_n = X_{n+1} + X_{n+2} + \ldots + X_{n+h}$$

for integers h. Furthermore the $\{X_i\}$ are mutually independent and, hence, the increments

$$\begin{aligned}&\{S_{n+h_1} - S_n; S_{n+h_1+h_2} - S_{n+h_1}; \ldots\}\\ &= \{X_{n+1} + X_{n+2} + \ldots + X_{n+h_1}; X_{n+h_1+1} + X_{n+h_1+2} + \ldots + X_{n+h_1+h_2}; \ldots\}\end{aligned}$$

are independent since the sums involve nonoverlapping sets of X_i's. Furthermore, since the $\{X_i\}$ are identically distributed, the distribution function of each increment depends only on the number h of terms in the sum, and not on n. That is,

$$S_{n+h} - S_n = X_{n+1} + X_{n+2} + \ldots + X_{n+h}$$

has the same distribution function as

$$S_{r+h} - S_r = X_{r+1} + X_{r+2} + \ldots + X_{r+h}$$

for each r, since the distribution of the sum depends only on the fact that there are h terms in each sum, and not on the fact that they are the particular terms given. Hence, the process $\{S_n\}$ is a discrete parameter process with stationary, independent increments.

An example of a continuous parameter process with stationary, independent increments is the Poisson process, discussed in Chapters 7, 11, and 12. Continuous state-space examples are important in the theory of noise, prediction, Brownian movement, and many other areas of application.

Some hint as to why processes with stationary and independent increments are particularly important is obtained by considering the sequence of times $0, t/n, 2t/n, \ldots, nt/n = t$, for each fixed t and n. We denote these times by $t_i = it/n$ for $i = 0, 1, 2, \ldots, n$, and write

$$X_t = X_{t_n} = [X_{t_n} - X_{t_{n-1}}] + [X_{t_{n-1}} \mp \ldots + [X_{t_1} - X_{t_0}]$$

$$= \sum_{i=1}^{n} [X_{t_i} - X_{t_{i-1}}]$$

(We have assumed $X_0 = 0$.) Then we observe that the sum has the form $\sum_{i=1}^{n} X_i$ where the X_i are independent, identically distributed random variables. Hence, all the results applicable to those sums, including extensions of the theory of random walks, can be used to discuss the behavior of X_t.

For $s \geq 0$, $t \geq 0$, write

$$X_{t+s} = X_{t+s} - X_s + X_s - X_0$$

Since $X_{t+s} - X_s$ and $X_s - X_0 = X_s$ are independent, we can use formula (5.27) to obtain

$$\text{Var}[X_{t+s}] = \text{Var}[X_{t+s} - X_s] + \text{Var}[X_s]$$

But $X_{t+s} - X_s$ has the same distribution, and hence, the same variance, as X_t. Thus the above formula can be written in the form:

$$\text{Var}[X_{t+s}] = \text{Var}[X_t] + \text{Var}[X_s] \tag{14.1}$$

for $t \geq 0$, $s \geq 0$. If we let $f(u) = \text{Var}[X_u]$, then (14.1) takes the form:

$$f(t+s) = f(t) + f(s), \qquad \text{for } s \geq 0, t \geq 0 \tag{14.2}$$

It is a simple exercise in mathematical analysis to show that the *only* nonnegative functions $f(t)$ that satisfy the functional equation (14.2) are the linear functions

$$f(t) = at$$

where a is a nonnegative constant.

[The proof of this is particularly simple if $f(t)$ is assumed to be differentiable. By setting $s = 0$ in (14.2), we find that $f(0) = 0$. Then write

$$\frac{df(t)}{dt} = \lim_{\Delta t \to 0} \frac{f(t + \Delta t) - f(t)}{\Delta t}$$

$$= \lim_{\Delta t \to 0} \frac{f(t) + f(\Delta t) - f(t)}{\Delta t}$$

by (14.2)

$$= \lim_{\Delta t \to 0} \frac{f(\Delta t) - f(0)}{\Delta t}$$

$$= f'(0)$$

Denote the constant $f'(0)$ by a. Then

$$f(t) = \int_0^t \frac{df(t)}{dt}\, dt = \int_0^t a\, dt = at]$$

In this way we obtain the following important formula for processes with stationary independent increments.

$$\mathrm{Var}[X_t] = at \tag{14.3}$$

where $a = \mathrm{Var}[X_1]$ is a nonnegative constant (actually a positive constant unless $X_t \equiv 0$). Thus, variance is a linear function of time.

Also we have

$$\begin{aligned}\mathrm{Var}[X_t] &= \mathrm{Var}[X_t - X_0] \\ &= \mathrm{Var}[X_{t+s} - X_s] \\ &= \mathrm{Var}[X_{t+s}] + \mathrm{Var}[X_s] - 2\,\mathrm{Cov}[X_{t+s}, X_s]\end{aligned}$$

using the fact that $X_{t+s} - X_t$ has the same distribution as $X_t - X_0$, by stationarity, and also using formula (5.26).

Thus, the covariance between X_{t+s} and X_t is

$$\mathrm{Cov}[X_{t+s}, X_s] = \frac{1}{2}\{\mathrm{Var}[X_s] + \mathrm{Var}[X_{t+s}] - \mathrm{Var}[X_t]\}$$

$$= \frac{1}{2}\{as + a(t+s) - at\}$$

$$= as$$

for $s \geq 0$, $t \geq 0$. If t_1 and t_2 are arbitrary positive numbers, then by setting the smaller of t_1 and t_2 equal to s and the larger equal to $t + s$, we obtain the formula

$$\text{Cov}[X_{t_1}, X_{t_2}] = a \min\{t_1, t_2\} \tag{14.4}$$

another important formula for processes with stationary, independent increments. ($\min\{t_1, t_2$) means the smaller of the two values t_1 and t_2.)

As mentioned previously, the Poisson processs is an example of a process with stationary independent increments. To determine this, let X_t be a Poisson process. That is, events occur at random times according to the assumptions of Section 7, Chapter 7, and X_t counts the number of events that occur in the time interval from 0 to t. Thus, $X_0 = 0$. Furthermore, since $X_{t_i} - X_{t_{i-1}}$ counts the number of events in the time interval (t_{i-1}, t_i), the assumption of independence of nonoverlapping intervals (the Chapman-Kolmogorov equation) made in Chapter 7 implies that the set

$$X_{t_1} - X_{t_0}, X_{t_2} - X_{t_1}, \ldots, X_{t_n} - X_{t_{n-1}}$$

is an independent set for $t_0 < t_1 < \ldots < t_n$. Since we also assumed in Chapter 7 that the number of events that occur in a time interval depends only on the length of that interval, it follows that the distribution of $X_{t_2} - X_{t_1}$, for $t_1 < t_2$, depends only on $t_2 - t_1$. In fact, this distribution is Poisson with mean $\lambda(t_2 - t_1)$ and probability density function:

$$e^{-\lambda(t_2-t_1)} \frac{[\lambda(t_2 - t_1)]^x}{x!}, \qquad x = 0, 1, 2, \ldots$$

Thus, all conditions are satisfied to assure that X_t is a process with stationary independent increments. Notice that it is *not* a stationary process; the distribution of X_{t+h} is not the same as that of X_t.

(c) The concept of a process with independent increments can be generalized as follows.

The process X_t is said to be a process with *orthogonal increments* if, for each sequence $t_0 < t_1 < \cdots < t_n$, the set of increments

$$X_{t_1} - X_{t_0}, X_{t_2} - X_{t_1}, \ldots, X_{t_n} - X_{t_{n-1}}$$

are *uncorrelated* (rather than independent), that is

$$\text{Cov}[X_{t_{i+1}} - X_{t_i}, X_{t_{j+1}} - X_{t_j}] = 0, \qquad \text{whenever } i \neq j$$

If in addition we assume the increments $X_{t_{i+1}} - X_{t_i}$ have mean and variance depending only on $t_{i+1} - t_i$, then we can show formulas (14.3) and (14.4) hold true for processes of this type as well.

(**d**) If, for any n and any set $(t_1, \ldots, t_n)$ with $t_1 < t_2 < \cdots < t_n$, the process has the property that

$$\Pr\{X_{t_n} \leq x_n \mid X_{t_{n-1}}, X_{t_{n-2}} \ldots X_{t_1}\} = \Pr\{X_{t_n} \leq x_n\}$$

then, clearly, the state of the process at any time in no way depends on the past. Hence the process is termed an *independent process.*

(**e**) The concept of a stationary process can be generalized as follows. A random process X_t, $-\infty < t < \infty$, is said to be *stationary in the wide sense* if its autocorrelation function

$$\mathrm{Cov}[X_{t_1} X_{t_2}]$$

is a function of the difference $|t_2 - t_1|$ only, and not of t_1 and t_2 individually;

$$\mathrm{Cov}[X_{t_1}, X_{t_2}] = R(t) \tag{14.5}$$

where $t = |t_2 - t_1|$. We also add the condition (which sometimes can be eliminated) that

$$\mu(t) = E[X_t] = \mu = \text{constant}$$

for each t, $-\infty < t < \infty$.

Let us also assume that the covariance $\mathrm{Cov}[X_{t_1}, X_{t_2}]$ approaches zero, when t_1 and t_2 are widely separated, at such a rate that the following integral converges:

$$\int_{-\infty}^{\infty} |R(t)|\, dt < \infty$$

Under these conditions we can show that $R(t)$ can be represented in the form

$$R(t) = 2\int_0^{\infty} \cos t\lambda\, f(\lambda)\, d\lambda \tag{14.6}$$

where the function $f(\lambda)$ is known as the *spectral density function* of the process. It is given by

$$f(\lambda) = \frac{1}{\pi}\int_0^{\infty} \cos t\lambda\; R(t)\, dt$$

Under these conditions, we can show that there exists a process Y_s (possible complex-valued) *with orthogonal increments* so that

$$E[\,|Y_s|^2] = \int_0^s f(\lambda)\, d\lambda$$

and the X_t process can be represented in the form

$$X_t = \int_{-\infty}^{\infty} e^{its}\, dY_s + \mu \tag{14.7}$$

with an appropriate definition of this integral.

Formula (14.7) is known as the *spectral decomposition* of the process X_t in terms of a process with orthogonal increments. This decomposition was first suggested by certain ideas of linear filters in electrical engineering. These ideas have led to a vast theory extending, specializing, and generalizing the ideas of this section. This theory has had a tremendous impact on many other branches of applied and pure mathematics.

4. Normal Processes; The Wiener Process

A random process X_t is termed a *normal* process if, for each sequence of times, $t_1 < t_2 < \cdots < t_n$, the random variables $X_{t_1}, X_{t_2}, \ldots, X_{t_n}$ have a joint normal distribution (see Example 4.12).

If we assume that the process X_t

(i) is normal
(ii) has stationary independent increments
(iii) has expectation $\mu(t) = E[X_t] = 0$ for all $t \geq 0$

then X_t is termed a *Wiener process.*

This process has many applications, particularly in electrical engineering and in the theory of Brownian motion.

Since a Wiener process has stationary independent increments, its autocorrelation function is given by (14.4):

$$\operatorname{Cov}[X_{t_1}, X_{t_2}] = a \min\{t_1, t_2\}$$

A multivariate normal distribution is completely determined by its first and second moments: means, variances, and covariances. Since for a Wiener process the means are all zero by property (iii), it follows that all joint distributions are determined by the autocorrelation function.

5. Ergodicity

In the practical application of the theory of stochastic processes we may have available few realizations of the process—often just one. The question then arises as to how well this one realization represents the entire process. Other questions that concern the measurements taken with respect to this realization and what these measurements mean with respect to the process itself are important practical matters. We would hope that the one realization is "typical" of the process, in some manner, so that statistical estimates, made by using the data, reliably estimate parameters of the process, such as the mean, the correlation function, or certain joint functions. Clearly we cannot expect every stochastic process to be so well behaved that one realization is "typical." The word *ergodic* is frequently used in various senses to describe a process in which a single realization normally yields sufficient information about the process. Of course, we must give a more precise discussion of the word than the preceding vague sentence.

A process can be ergodic in many respects, such as "ergodic in the mean," or "ergodic in distribution."

(a) **Ergodic in the Mean.** We assume that the process X_t is wide sense stationary and define the *time average*

$$\frac{1}{2T}\int_{-T}^{T} X_t\,dt = \eta_T$$

for some fixed interval $(-T, T)$. Notice that this quantity η_T is a random variable that depends on the particular realization of the process X_t observed. The practical question then arises: What can be said about $\mu = E[X_t]$ from one measurement or observed value of η_T? More particularly, we would like to have η_T as an *unbiased* estimate of μ, $E[\eta_T] = E[X_t] = \mu$ (μ constant since X_t is stationary). $E[\eta_T]$ can be shown to satisfy this relation. The question is: Is just one observation of η_T close enough to μ to give a good estimate of μ? The general answer to this question is deeper than we can explore here. The result is that, if the variance of the estimate $\sigma_{\eta_T}^2 \to 0$ for $T \to \infty$, then $\lim_{T\to\infty} \eta_T = \mu$ with probability 1. In this case, we say that the stochastic process X_t is *ergodic in the mean.* For practical purposes, then, if we have a single long-time realization of the stochastic process, its time average η_T is very nearly equal to its population average or mean μ.

From the Chebyshev inequality we can estimate "how much" the time average deviates from the "population average." Thus

$$\Pr[|\eta_T - \mu| < \varepsilon] \geq 1 - \frac{\sigma_{\eta_T}^2}{\varepsilon^2}$$

In words, this says that the probability of large deviation between the "time average η_T" and the "population average μ" is small. The probability associated with deviations of size ε depends inversely on ε^2 and the variance of η_T. If the process is ergodic in mean, and $\sigma^2_{\eta_T} \to 0$ as $T \to \infty$, then

$$\lim_{T \to \infty} \Pr[|\eta_T - \mu| < \varepsilon] = 1$$

However, this is a much weaker statement than one which says that $\eta_T \to \mu$ with probability 1.

(b) Ergodicity in Distribution. There is yet another important type of ergodicity that involves the distribution function. Here, again, we merely sketch out the problem and some general results.

Consider again the practical problem of measuring the probability distribution or density function of X_t for a fixed t. We have available one or a few realizations of the process and from them would like to make some statement about the distribution function that is associated with the process. Let us now assume that the process is strictly stationary, and let us consider the first order distribution function.

$$F(x; t) = P[X_t \le x] = F(x)$$

which will not involve t, since the process is strictly stationary.

Suppose that we have one realization of the process. How do we estimate $F(x)$ from this? One way that this can be done (see Figure 14.1) is to plot the realization, draw a horizontal line at level x, and record the lengths t_1, t_2, ... of those intervals of t values for which the process remains below x. Then the time average

$$y_T = \frac{t_1 + t_2 + \cdots t_n}{T}$$

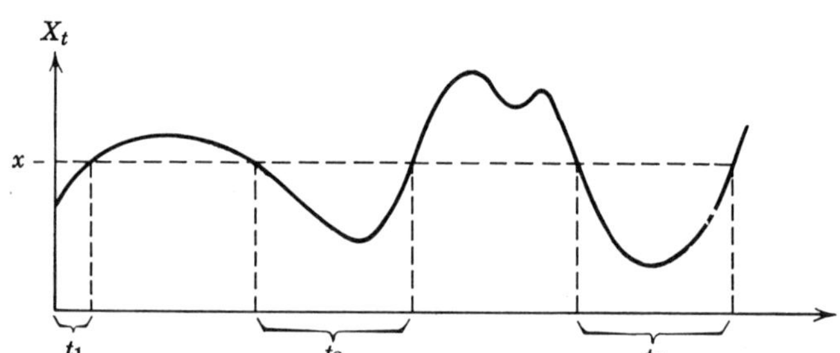

Figure 14.1 Estimating a stationary distribution function $F(x)$ from a single realization of X_t.

gives the average amount of time that the process stays below x. Here $t_1, t_2, \ldots t_n$ represent the lengths of all intervals between times 0 and T in which $X_t \leq x$. The question then is: How is y_T—the *time average* or *proportion* of time that the process is below x—related to $F(x)$, the *probability* that the process is below x? Notice that, in general, they are quite different concepts and, hence, we might not expect them to be related. The stochastic process is said to be *ergodic in distribution*, if

$$y_T \rightarrow F(x)$$

with probability 1 as $T \rightarrow \infty$. That is, for an ergodic process (ergodic in distribution) the time average y_T is a "good" estimate of the population probabilities.

Bibliography

There are many excellent references for further study in probability theory. This bibliography lists some of them.

1. Feller, W., *Introduction to Probability Theory and its Applications*, Vol. 1, 3rd ed., John Wiley and Sons, Inc., 1968.
2. Harris, B., *Theory of Probability*, Addison-Wesley, 1966.
3. Meyer, P., *Introductory Probability and Statistical Applications*, Addison-Wesley, 1965.
4. Parzen, E., *Modern Probability Theory and its Applications*, John Wiley and Sons, Inc., 1960.
5. Pfieffer, P., *Concepts of Probability Theory*, McGraw-Hill Book Co., 1965.

An introductory text for finite Markov chains with a discrete state space and discrete parameter is:

6. Kemeny, J. and L. Snell, *Finite Markov Chains*, D. Van Nostrand Co., Inc., 1959.

An introduction to stochastic processes including both those with discrete and those with continuous state spaces is:

7. Parzen, E. *Stochastic Processes*, Holden-Day, Inc., 1962.

More advanced works on stochastic processes are:

8. Chung, K., *Markov Chains with Stationary Transition Probabilities*, Springer-Verlag, 1960.
9. Karlin, S., *A First Course in Stochastic Processes*, Academic Press, 1966.

Some texts with primary interest in applications of the theory of stochastic processes are:

10. Barlow, R., and Prochan, F., *Mathematical Theory of Reliability*, John Wiley and Sons, Inc., 1965.
11. Bharucha-Reid, A., *Elements of the Theory of Markov Processes and their Applications*, McGraw-Hill Book Co., 1960.

12. Cox, D. R. and Smith, W. L., *Queues*, Methuen, 1961.

13. Morse, P. *Queues, Inventories, and Maintenance*, John Wiley and Sons, Inc., 1958.

Index